Springer Series in Advanced Manufacturing

Series editor

Duc Truong Pham, University of Birmingham, Birmingham, UK

The **Springer Series in Advanced Manufacturing** includes advanced textbooks, research monographs, edited works and conference proceedings covering all major subjects in the field of advanced manufacturing.
The following is a non-exclusive list of subjects relevant to the series:

1. Manufacturing processes and operations (material processing; assembly; test and inspection; packaging and shipping).
2. Manufacturing product and process design (product design; product data management; product development; manufacturing system planning).
3. Enterprise management (product life cycle management; production planning and control; quality management).

Emphasis will be placed on novel material of topical interest (for example, books on nanomanufacturing) as well as new treatments of more traditional areas.

As advanced manufacturing usually involves extensive use of information and communication technology (ICT), books dealing with advanced ICT tools for advanced manufacturing are also of interest to the Series.

Springer and Professor Pham welcome book ideas from authors. Potential authors who wish to submit a book proposal should contact Anthony Doyle, Executive Editor, Springer, e-mail: anthony.doyle@springer.com.

More information about this series at http://www.springer.com/series/7113

Hanmin Shi

Metal Cutting Theory

New Perspectives and New Approaches

Springer

Hanmin Shi
Ottawa, ON
Canada

ISSN 1860-5168 ISSN 2196-1735 (electronic)
Springer Series in Advanced Manufacturing
ISBN 978-3-030-08812-5 ISBN 978-3-319-73561-0 (eBook)
https://doi.org/10.1007/978-3-319-73561-0

Printed on acid-free paper

This Springer imprint is published by Springer Nature
The registered company is Springer International Publishing AG
The registered company address is: Gewerbestrasse 11, 6330 Cham, Switzerland

To My Wife Ms. Huiying Gao

Preface

This book brings together my main achievements in metal cutting theory research and its application. It includes three parts: "Non-Euclidean Cutting Tool Geometry", "Non-Free Cutting Mechanics", and "Non-Linear Machine Tool Dynamics". Compared with the traditional "Euclidean Cutting Tool Geometry", "Free Cutting Mechanics", and "Linear Machine Tool Dynamics" established by previous researchers, the above three "Non-s" may be regarded as a kind of innovation. I tried to make some new attempts in the field of metal cutting theory and its application so as to adapt the rapid development of metal cutting technology.

During graduate study period, in early 60s of the twentieth century, I participated in a research project of the basic metal cutting theory. At that time, I already had some initial ideas about expressing space angle by means of plane vectors and about chip ejection interference. Some experimental and theoretical works were done. These research works belonged to "Non-Euclidean cutting tool geometry" and "Non-Free Cutting Mechanics", respectively. The results have appeared in my graduate thesis [10] and the Journal of Huazhong Institute of Technology (HIT) [11]. Restricted to the objective conditions and the subjective level of understanding, the research failed to further.

After graduation in 1966, I began to work in HIT as a teacher. Because of the prevailing circumstances, doing research work was impossible. Nevertheless, I still conducted a great deal of mathematical deduction and delved into a number of theoretical books. In addition, in teaching "students of workers, peasants and soldiers" of the day, I taught them cutting tool geometry by means of plane vector representation of space angles, so as to turn a difficult problem into an easier one.

In the late 70s of the twentieth century, the scientific research of China began to recover. I was, therefore, able to continue my research on the plane expression of the space angles on cutting tools. I was surprised to find that the problem studied actually belongs to the mapping from 3-D Euclidean space onto a 2-D Riemann (non-Euclidian) plane. With a broadened scope of mind, in 80s and the early 90s of the twentieth century, a series of scientific papers were published in journals at home and abroad [1–9]. The research of "Non-Euclidean cutting tool geometry" was mainly completed during this period.

In the late 70s of the twentieth century, China began to open to the outside world. In 1980, I had the honor of visiting the Department of Mechanical Engineering, the University of Birmingham, United Kingdom for 2 years. Birmingham University is a historic, internationally renowned university. The head of Department of Mechanical Engineering at that time was Prof. S. A. Tobias, and the Group of Machine Tool Dynamics under his leadership was world famous for their achievements in the area. I had the opportunity of joining the group and entered the research of machine tool dynamics. In view of the factor that linear theory of machine tool dynamics had already been mature at the time, I chose such a challenging subject as “Non-linear Theory for machine tool chatter”.

At first, progress did not go smoothly. On one hand, due to the Sino-British science, technology and cultural exchanges at the time were still in the initial stage, the British side personnel to Chinese visiting scholars in the knowledge and ability took a wait-and-see attitude, and even a skeptical attitude. On the other hand, the main nonlinear factor in the chatter was supposed by the previous researchers of the group to be the nonlinear stiffness of machine tool structure. I discovered that experimental data did not support this. The structural nonlinearity is far too small to have any significant effect. Research work was mired in stalemate for a time.

Later, with large amounts of data, charts, I persuaded the British personnel to adjust the research program, to make the research work on the right track. And the research work quickly made substantive progress. In 1982 before I went home, Professor Tobias said, “I greatly enjoyed having you in the Department and will miss you. You have done excellent work which will be appreciated all over the world.”

Around 1984, Professor Tobias was planning to invite me to visit his department again, and planned to arrange me working in the machine tool dynamics laboratory of his department, continuing to research on the theory of machine tool chatter. Just then, I received a letter from a friend in the department, telling me that Professor Tobias had passed away due to heart disease. Later, I heard from a staff of the department that after a heart attack his doctor told him that he must absolutely repose in bed to have a good rest. He, however could not lay down his work and still invited his colleagues to his home to discuss the departmental work. This resulted in deterioration of his heart disease and led to his sudden death. His death made me lose a mentor, and also lose the chance to return to their machine tool dynamics laboratory to work.

After coming home, funded by the National Natural Science Fund and other funds, I continued the research work, focused on on-line monitoring and control for machine tool chatter. This research resulted in a series of scientific papers [32–56] published in journals that had impacts at home and abroad.

The achievements of research on “Nonlinear theory for machine tool chatter” won me the National Natural Science Award of China in 1987.

In 1987 and 1990, I was invited to Birmingham University in United Kingdom and Toyohashi University of technology in Japan respectively for lectures and academic exchanges on nonlinear chatter of machine tools and its on-line monitoring techniques.

The research on “Non-Free Cutting Mechanics” stopped in 1966, and then there was much hurry, and the project never recovered. The months and years passed by like a stream, 26 years just in an instant. Time is the autumn of 1992, I was invited by Professor Sam Wu, Department of Mechanical Engineering, The University of Michigan, U. S., to be a visiting professor and researcher (principal investigator), responsible for guiding the research work in metal cutting. At that time, the research team was undertaking a research project of National Natural Science Fund Committee of U.S. “On Development of a Science Base for Drills and Drill Grinding Processes”. They needed to build a cutting force model for complex drill bit (multifacet drill) and the chip interference of different drill edge segments must be taken into account. Thus, the research work on “Non-Free Cutting Mechanics” was in a position to resume. I, guiding a Ph.D. student of the department, successfully completed the task. Meanwhile, a lot of research work on the basic theory of “Non-Free Cutting Mechanics” was also done and good progresses were made [12–14].

At the same time, I also directed another Ph.D. student of the team to research on-line chatter prevention of numerical control machine tools. And made some new achievements about the application of nonlinear chatter theory to chatter prevention [57–58].

Returning from United State, funded by the National Science Foundation of China, I continued the “Non-Free Cutting Mechanics” research work, and published a series of scientific papers on domestic and international authoritative journal [15–31]. Under the guidance of non-free cutting theory established, in cooperation with the manufacturers, a new type of free cutting carbide chips was developed. The undertaken National Natural Science Fund project “Free cutting method and the design theory and method of free cutting tools” (Grant No: 59675058) completed in 1999, and among all the 120 projects in the mechanical discipline of the year, the above-mentioned project was rated as one of the nine excellent closed projects.

I retired in 2007, and then all the professional activities (including scientific research and teaching works) came to the end. I immigrated abroad and my main interest turned to the development of basic science, such as mathematics, physics, astronomy, geography, and biology. I wrote and published some popular science books. However, I still care about the latest development in the field of metal cutting technology and theory.

I miss all the colleagues, Ph.D. students and Master Degree students, who joined my team in various periods of times and were with me through thick and thin, tilled in this area. They are Yongjie Chen, Huashu Zhang, Liangshan Xiong, Zhijian Mei, Yueyun Xie, Jingyan Liu, Tiankuang Lu, Song Chang, Zhenglin Wang, Haiyang Zhang, Ning Fang, Xibin Wang, Tao Lu, Xing Yan, etc. Here, I would like to make my deep appreciation and say “take good care of you” to each of them, no matter where they are.

I was grateful to Professor Horsh, Tetsutaro (Japan). He and I had deepgoing discussions and academic exchange for machine tool chatter theory and vibration control and suppress. I reaped no little benefit from him.

Many thanks to Prof. Shuzi Yang, for his support, help, and cooperation in my research work of on-line chatter monitor and in real-time control.

I remember late Prof. S. A. Tobias (UK) and Professor Sum Wu (US), they fully understood and supported my ideas, thoughts, and research results; although they looked so incredible.

Finally, I should express my deep appreciation for the support given by the National Natural Science Fund of China for many years. Without the support, such kind of research projects focusing on the basic theory was impossible.

In quite a long time span, although I have published a number of scientific papers related to the contents of this book, this book is not a simple collection of scientific papers published by me, but an outcome of new rethink, new exploration, and a new description for the achievements attained in the past, with today's understanding profundity and level.

I collected results in metal cutting area and published them in this book as an academic viewpoint and technical approaches, in order to discuss with peer experts or scholars from home and abroad. Commands and criticisms are welcome.

Ottawa, Canada Hanmin Shi

Permissions and Acknowledgements

Grateful acknowledgement is hereby made to the following individuals and publishers for giving permission to use the material that they have provided or to quote from sources for which they hold the rights. Every effort has been made to locate the copyright owners of material reproduced in this book.

Thanks to华中工学院学报、华中理工大学学报、华中科技大学学报 (Journal of Huazhong University of Science and Technology) for Figs. 2.3, 2.6, 2.7, 3.7, 3.10–3.13, 5.3, 9.1–9.6, 9.9, 13.2, 13.5, 14.15–14.24, 15.9–15.12, 15.14,15.15 and the relevant text or equation quotations from the sources.

Thanks to 华中科技大学出版社 (Huazhong University of Science and Technology Press) for all the figures in Chap. 11.

Thanks to International Journal of Machine Tool Design and Research for Figs. 3.8, 3.9, 5.4, 13.4, 13.7, 13.8, 13.13 and 14.1–14.13.

Thanks to Journal of Engineering for Industry, Transactions of The ASME for Figs. 4.1–4.9, 2.10, and relevant text.

Thanks to International Journal of Machine Tool and Manufacturing for Figs. 6.6–6.8, 7.1–7.4, and the relevant text or equation quotations from the sources.

Thanks to Science in China (Series E) for Fig. 7.6.

Thanks to 中国机械工程 (Chinese Mechanical Engineering) for Figs. 9.8, 9.13 and 10.1.

Thanks to 应用力学学报 (Journal of Applied Mechanics) for Figs. 13.6, 13.9, 13.10, 13.14, 14.14 and 14.6.

Finally, I would like to express my heartfelt thanks to Dr. James Janelin Wang, Dr. Xing Yan, and Dr. Haiyan Henry Zhang for their permissions of my quotations of some materials from their Ph.D. dissertations, which make part of the contents of Chaps. 5, 7, 9, 10, and 15 of this book.

About the Book

This book summarizes author's main achievements in his lifetime in the field of metal cutting theory and its applications, including "Non-Euclidian Cutting Tool Geometry", "Non-Free Cutting Mechanics", and "Non-Linear Machine Tool Dynamics". All the achievements and their practical significance have been theoretically proved and experimentally verified. Compared with traditional "Euclidean Cutting Tool Geometry", "Free Cutting Mechanics", and "Linear Machine Tool Dynamics", the above-mentioned contents can be said to be new perspectives and new approaches.

The main benefits of reading the book are as follows:

To know how to turn a difficult 3-D problem of cutting tool geometry into a 2-D one, and make analyzing and designing complex-shaped cutting tools more intuitive and easier;
To know how to realize free cutting by means of dredging chip ejections from different parts of cutting edge so as to reduce chip ejection interference, tool wear, and power consuming;
To know there are some chatter free area of cutting parameters with high productivity. Therefore, chatter prevention is not necessarily at the cost of sacrificing productivity.

Contents

Symbols and Nomenclature

a	In-phase component of operative receptance (mm/N)
a	Unit vector in tool face and perpendicular to the cutting edge (–)
a, b, c ...	Space strait lines, images of space strait lines (–)
$\boldsymbol{a}, \boldsymbol{b}, \boldsymbol{c}$...	Column vectors of homogeneous coordinates of line images (–)
a_1, a_2, a_3	Homogeneous coordinates of image line a (–)
a_c	Thickness of cut (mm)
a_{ch}	Chip thickness (mm)
a_{cr}	Critical amplitude when the cutting edge pass begins to leave the workpiece material, (mm)
a_o	Amplitude of stabilized chatter vibration (mm); orthogonal projection of space line a on the projective plane (–)
a_p	Depth of cut (mm)
a_x, a_y	Cartesian coordinates of the intersection point of space line a and projective plane (–)
a_{pcr}	Critical depth of cut when chatter starts (–)
a_w	Width of cut (mm)
a_{wcr}	Critical width of cut when chatter starts (–)
A	Cutting power (kw)
A, B, C ...	Space planes, image of space planes, module of obliquity vector of planes (–)
$\boldsymbol{A}, \boldsymbol{B}, \boldsymbol{C}$ ...	Column vectors of homogeneous coordinates of planes images (–)
$\mathbf{A} = \boldsymbol{OA}$	Obliquity vector of plane A (–)
A_1, A_2, A_3	Homogeneous coordinates of image point A (–)
A_x, A_y	Cartesian coordinates of image A (–)
A_γ	Tool face, image of tool face (–)
A_α	Major flank, image of major flank (–)
$A_\alpha{}'$	Minor flank, image of minor flank (–)
A_+	Power made by cutting force to vibration system in one cycle (–)
A_-	Energy consumed by positive damping of machine tool structure in one cycle (–)

$\overline{AB}$	Euclidean distance between image points A and B (–)
A^*, A^{**}	Images of plane A after rotating once, twice (–)
b	Out-of-phase component of operative receptance (mm/N)
$\mathbf{b}$	Unit vector in the direction of cutting edge (tool face is located at the left side of the vector) (–)
c	Penetration rate factor, hyperbolic real half-axis (–)
c_1, c_2	Nonlinearity coefficients of cutting force (1/mm, 1/mm^2)
dr	Penetration rate (mm/sec)
D	Diameter of twist drill (mm)
$\mathbf{e}$	Unit vector perpendicular to shear plane (–)
f	Feed rate (mm/r); striate line in feed direction; frequency of vibration, Hz; perturbation frequency of spindle rotation (Hz)
$\mathbf{f}$	Unit vector in feed direction (–)
f_o	Natural frequency of machine tool structure (Hz)
f_N	Nyquist frequency (Hz)
f_z	Feed rate per tooth (mm/z)
$\mathbf{f'}$	Unit vector in direction of projection of feed on reference plane (–)
F	Resultant cutting force (N)
F_c	Cutting force (component of resultant cutting force in the direction of cutting speed vector V) (N)
F_f	Feed force (component of resultant cutting force in the direction of feeding motion f) (N)
F_o	Average (static) cutting force (N)
F_p	Back force (component of resultant cutting force in the direction of depth of cut) (N)
g	Generatrix of tool face and flank (–)
$\mathbf{g}$	Unit vector normal to tool face (–)
G_{xy}	Cross-spectral density of signal x and y (–)
h	Normal of tool flank (–)
$\mathbf{h}$	Unit vector normal to tool flank (–)
h'	Normal of tool minor flank (–)
$\mathbf{i}$	Unit vector in x axis direction (–)
$\mathbf{j}$	Unit vector in y axis direction (–)
$\mathbf{k}$	Unit vector in z axis direction (–)
k	Static stiffness of vibration system (–)
K	Coefficient showing the influence of chip ejection vector on cutting force (–)
k_s	Cutting force coefficient showing the influence of unit change of thickness of cut on cutting force, under unit depth (or width) of cut (N/mm^2)
l	Length of cutting edge (–)
$\boldsymbol{l} = \mathbf{b}l$	Vector of cutting edge (–)
m	Mass of a model (–)
N	Machine tool spindle revolutions (r/min)

O	Projective plane, coordinate origin (–)
$\mathbf{o}$	Unit normal vector of orthogonal plane (–)
P	Shear force (in shear plane) (N)
P_f	Assumed working plane, image of assumed working plane (–)
P_{fe}	Working plane, image of working plane (–)
P_p	Tool back plane, image of tool back plane (–)
P_{pe}	Working back plane, image of working back plane (–)
P_n	Cutting edge normal plane, image of cutting edge normal plane (–)
P_o	Tool orthogonal plane, image of tool orthogonal plane (–)
P_{oe}	Working orthogonal plane, image of working orthogonal plane (–)
P_r	Tool reference plane, image of tool reference plane (–)
P_{re}	Working reference plane, image of working reference plane (–)
P_s	Tool cutting edge plane, image of tool cutting edge plane (–)
P_{se}	Working cutting edge plane, image of working cutting edge plane (–)
P_{sh}	Shear plane (–)
P_n'	Minor cutting edge normal plane, image of minor cutting edge normal plane (–)
P_o'	Tool orthogonal plane for minor cutting edge, image of tool orthogonal plane for minor cutting edge (–)
P_{oe}'	Working orthogonal plane for minor cutting edge, image of working orthogonal plane for minor cutting edge (–)
P_s'	Tool cutting edge plane for minor cutting edge, image of tool cutting edge plane for minor cutting edge (–)
P_{se}'	Working cutting edge plane for minor cutting edge, image of working cutting edge plane for minor cutting edge (–)
$\mathbf{p_a}$	Obliquity vector of line a (–)
p_{ax}, p_{ay}	Cartesian coordinates of obliquity vectors (–)
r	Intersection line of tool orthogonal plane P_o with tool reference plane P_r, cutting ratio, corner radius (mm)
$\mathbf{r}$	Unit vector in direction of r (–)
r'	Intersection line of tool orthogonal plane P_o' for minor cutting edge with tool reference plane P_r (–)
r_e	Intersection line of working orthogonal plane P_{oe} with working reference plane P_{re} (–)
r_e'	Intersection line of working orthogonal plane P_{oe}' for minor cutting edge with working reference plane P_{re}' (–)
$\mathbf{r_e}$	Unit vector in direction of r_e (–)
r_ε	Radius of rounded tool corner (–)
R	Radius of twist drill (mm)
R_n	Force normal to tool face (N)
R_t	Friction force in contact area of tool face and chips (N)
R_x	Radius of twist drill at point x (mm)
$2R_o$	Web thickness of drill (mm)
$\boldsymbol{R} = (R_{ij})$	Receptance matrix of machine tool structure (mm/N)

R_{ij}	Direct and cross receptances of machine tool structure exciting in j direction and measure vibration in i direction (i, j = 1, 2, 3) (mm/N)
R_{XF}	Operative receptance (exciting in cutting force direction and measure vibration in nomal direction to the cutting surface) (mm/N)
s	Major cutting edge, image of major cutting edge (–)
$\mathbf{s}$	Unit vector in direction of s (–)
s'	Minor cutting edge, image of minor cutting edge (–)
s_o	Thickness of cut per tooth of milling cutter (mm)
$s(t)$	Instantaneous depth of cut (mm)
S_x	Power spectral density of signal x (–)
S_{AB}	Non-Euclidian distance between image points A and B (i.e., angle between space planes A and B) (–)
t	Time (sec), intersection line of tool cutting edge plane P_s with tool reference plane P_r (–)
$\mathbf{t}$	Unit vector in direction t (–)
T	Torque (Nm), surface energy of chip material, or force required for tearing chip of unit depth (N/mm)
$\boldsymbol{T}$	Chip's translation speed vector (–)
t_e	Intersection line of working cutting edge plane P_{se} with working reference plane P_{re} (–)
t'	Intersection line of tool cutting edge plane for minor cutting edge P_s' with reference plane P_r (–)
t_e'	Intersection line of working cutting edge plane for minor cutting edge P_{se}' with working reference plane for minor cutting edge P_{re} (–)
$\mathbf{u}$	Unit vector in chip ejection direction (–)
$\boldsymbol{U}$	Chip ejection vector (m/min)
U	Chip ejection speed (m/min)
$\boldsymbol{U_o}$	Natural chip ejection vector (m/min)
$\boldsymbol{V}$	Cutting speed vector (m/min)
V	Cutting speed (m/min)
$\boldsymbol{V_e}$	Resultant cutting speed vector (m/min)
V_e	Resultant cutting speed (m/min)
$\boldsymbol{V_f}$	Feed speed vector (m/min)
V_f	Feed speed (m/min)
$\boldsymbol{V_f'}$	Component of feed speed vector in reference plane P_r (m/min)
w	Width of cut (mm)
w_{cr}	Critical width of cut (mm)
$\boldsymbol{W}$	Shear velocity vector (m/min)
W	Shear speed (m/min)
W_p	The work done by the cutting force in a vibrating cycle (–)
W_d	The vibration energy dissipated by the internal damping of the machine tool structure in a vibration cycle (–)
$X(\gamma)$	Moving operator while rotating around axis x over an angle γ (degree)
x	x axis of Cartesian coordinate, intersection line of assumed working plane P_f with tool reference plane P_r (–)

x_e	Intersection line of working plane P_{fe} with working reference plane P_{re} (–)
$x_o(t)$	Wave cut of nth cutting tooth in workpiece surface (mm)
$x_i(t)$	Wave cut of $(n\text{-}i)$th cutting tooth in workpiece surface, $i = 1, 2, \ldots$ (mm)
$Y(\gamma)$	Moving operator while rotating about axis y over an angle γ (degree)
y	y axis of Cartesian coordinate, intersection line of tool back plane P_p with tool reference plane P_r (–)
y_e	Intersection line of working back plane P_{pe} with working reference plane P_{re} (–)
$y_o(t)$	Wave cut on cutting surface before wave cut of n-th cutting tooth (mm)
z	z axis of Cartesian coordinate, number of teeth of a milling cutter (–)
z_c	Number of cutting edges engaged in cutting (–)
$Z(\gamma)$	Moving operator while rotating about axis z over an angle γ (degree)
α_f	Tool side clearance (degree), drill tangential clearance angle (degree)
α_f'	Tool side clearance for minor cutting edge (degree)
α_o	Tool orthogonal clearance (degree)
α_p	Tool back clearance (degree)
α_o'	Tool orthogonal clearance for minor cutting edge (degree)
α_p'	Tool back clearance for minor cutting edge (degree)
β	Friction angle between chip and tool face (degree), phase difference between wave cuts in successive passes (degree)
γ_f	Tool side rake (degree)
γ_n	Cutting edge normal rake (degree)
γ_o	Tool orthogonal rake (degree)
γ_{oe}	Working orthogonal rake (degree)
γ_p	Tool back rake (degree)
γ_n'	Minor cutting edge normal rake (degree)
γ_o'	Tool orthogonal rake for minor cutting edge (degree)
γ_{fe}	Working side rake (degree)
γ_{ne}	Working cutting edge normal rake (degree)
γ_{oe}	Working orthogonal rake (degree)
γ_{pe}	Working back rake (degree)
γ_{ne}'	Working minor cutting edge normal rake (degree)
γ_{oe}'	Working orthogonal rake for minor cutting edge (degree)
γ_ψ	Tool orthogonal rake in direction of chip ejection (degree)
γ_g	Maximum rake (in the direction of obliquity vector of rake face (degree)
Δb	Length of elementary cutting edge (mm)
ε	Angle between friction marks on tool face and the generatrix of cylindrical tool face (degree)
ε_r	Tool included angle (degree)
ζ	Relative variation of chip ejection vector $\boldsymbol{U}$ in the direction of cutting edge (–)

η	Chip ejection angle (in reference plane), i.e., angle between the projections of chip ejection vector $\boldsymbol{U}$ and main cutting edge on reference plane (degree)
Θ	Moving operator (3×3 matrix) (–)
θ	Tool edge angle deviation angle of real chip ejection vector from natural chip ejection vector (degree)
θ_{ab}	Non-Euclidean angle between image lines a and b, i.e., the angle between space lines a and b (degree)
θ_{AQB}	Non-Euclidean angle decided by three image points A, Q and B, with point Q as the vertex of the angle, i.e., the angle between space planes A and B measured in plane Q (degree)
κ_r	Tool cutting edge angle (degree)
κ_r'	Tool minor cutting edge angle (degree)
κ_{re}	Working cutting edge angle (degree)
κ_{re}'	Working minor cutting edge angle (degree)
λ	Stiffness of machine tool structure (N/mm)
λ_s	Tool cutting edge inclination (degree)
λ_s'	Tool minor cutting edge inclination (degree)
λ_{se}	Working cutting edge inclination (degree)
λ_{se}'	Working minor cutting edge inclination (degree)
μ_1	Coefficient indicate the strength of influence of ζ on cutting power (–)
μ_2	Coefficient indicate the strength of influence of ξ on cutting power (–)
ξ	Relative variation of chip ejection vector $\boldsymbol{U}$ in the direction perpendicular to cutting edge, damping ratio of a vibration system (–)
λ_{TX}	Cutting edge inclination of twist drill in end face (degree)
τ	Stress at shear plane (N/mm^2)
φ	Shear angle (degree)
φ_x (κ_r)	Cutting edge angle at x point on main cutting edge of a twist drill (degree)
Φ	Half-point angle of a twist drill (degree), coefficient of non-free cutting (–)
ψ_λ	Chip ejection angle, i.e., angle between chip ejection direction and the normal to the cutting edge (degree)
$\psi_{\lambda o}$	Natural chip ejection angle (degree)
ω	Helical angle (degree), angular frequency of a vibration system (rad/sec)
ω_o	Natural angular frequency (rad/sec)
$\boldsymbol{\Omega}$	Rotation component of chip ejection motion vector of angular speed of spindle revolutions (sec^{-1})
Ω	Angular speed of spindle revolutions (sec^{-1})
$\angle AQB$	Euclidean angle decided by three points A, Q and B, with Q as vertex (degree)
$\angle ab$	Euclidean angle between lines a and b (degree)

List of Figures

Chapter 1
Introduction

Cutting processing is an operation of removing a layer of allowance from a work piece surface using cutting tools or grinding wheels to reach the required size accuracy, shape accuracy and surface quality of the work piece. In 21st century, cutting will still be the most primary processing method for precision mechanical parts in volume production.

In a machining operation, cutting tool is directly used for processing work piece; the machine tool is used for realizing the exact required motion of the cutting tool relative to the work piece. Throughout the manufacturing cost, although cutting tools' cost is very small, only a few percent, however, because tool-design or selection strongly influences the cutting efficiency, shutdown time for tool changing and the machine tool utilization, cutting tools have a substantial impact on production cost. Meanwhile, proper tool-design and the correct tool choice also have a direct impact on the quality of the product. Unqualified tools will make complicated and expensive machine tools or machining system become useless, totally ineffective.

Cutting processing dates back to prehistoric times. In the "Stone Age", rough stone tools left over like "choppers" or "scrapers" and other cutting tools reflect the remains that the ancestors engaged in original "cutting processing" labor, such as cutting sticks, peeling animal skins. In "Neolithic" polished and elaborate "knives", "axes" reflect the progress of cutting technology. According to historical records, in the Shang dynasty of China, a variety of bronze cutting tools, such as knives and drills etc. were adopted; In the Spring and autumn period of China, the 8th century BC, iron tools such as saws and chisels were used; in 1668, horse traction Miller and pedal-powered grinders were used.

Modern metal cutting technology was developed after the industrial revolution and has a history of more than 200 years. In the 60s of the 18th century, the British James Watt improved the original steam engine, in 1775, the British J. Wilkinson developed the boring machine for processing steam engine's cylinder and in 1818 the American Eli Whitney invented a milling machine. In 1865, before and after the Paris International Exposition, various types of lathes, boring machines, slotting

H. Shi, *Metal Cutting Theory*, Springer Series in Advanced Manufacturing,
https://doi.org/10.1007/978-3-319-73561-0_1

machines, gear and threading machines appeared in succession. In recent years, materials processed by cutting or grinding, have been expanded from metal to non-metal materials, such as ceramics, silicon, precious stones and glass. But the most important material machined by cutting is still metal. Today, cutting technology is quite mature. However, the variety of machinery products has increased, complexity and machining precision of parts are constantly improving, and new engineering materials have been put into practical applications, as well as the automated production being rapidly developing, all these have raised new and higher requirements on cutting technology. Facing the challenges of new technologies, cutting technology is still in development, and continuously innovates new potential, showing a good ability to adapt.

Adapting to the development of cutting technology, the systematic research on the basic theory of metal cutting also has a history of more than 100 years. In 1851, the French Coequilhat researched drilling energy consumption; in 1864, the French Joessel studied the effects of tool geometry on the cutting force; 1870–1877 Russian И. А. Тиме studied on chip formation and types; in 1907, American F. W. Taylor studied the relationship between tool life and cutting speed. They were the early researches on the theory of metal cutting. In the following more than 100 years, countless researchers continued to conduct a lot of researches in this area, which guided and facilitated the development of cutting technology and machining production.

After the end of World War II, mechanical processing, including machine tools, precision machining technology, high speed cutting technology and automation technology has developed by leaps and bounds. These developments require new development and breakthroughs in the research of basic theories for metal cutting.

The research work of the author in the area of elementary theory for cutting processing focused on three aspects: cutting tool geometry, mechanics of cutting process and chatter of machine tools (cutting dynamics). In these three aspects the traditional research basically confined to Euclidean geometry or general mathematical tools, free cutting mechanics and linear vibration theory; from modern machining practice such research works already have a sizeable gap. Based on previous research work the author attempts to make somewhat advance and hopes to give today's machining production more practical guidance and facilitating roles.

The follows are a brief introduction of the author's basic idea, the main achievements reflected in the book and their values for application.

1.1 Non-Euclidean Cutting Tool Geometry

Cutting tool geometry has a decisive influence on its cutting performance. Improper tool angles will worsen their performance, or even completely lose the ability to cut. Tool angles involve the angles between a series of space inclined planes (such as coordinate planes, tool face and tool flank, etc.) and straight lines (such as cutting edges). Such angle relations are not easy to be clearly expressed in a mechanical drawing which is based on "orthographic projection". It is also inconvenient, troublesome, lack of intuition and visibility, and prone to errors to deal with such a problem by conventional mathematics tools, such as solid geometry, space analytic geometry or differential geometry. In fact, the tool angles description, design, calculation, analysis and measuring are always difficult points in machinery manufacturing science and technology, becoming a very abstruse knowledge and skills, which only a few experts can really master. The complicated trend in modern cutting tool geometry and in machining movement also deepens this contradiction.

The main result of the author's work was that he found a method of mapping and isomorphism principle which can project inclined planes and straight lines in three-dimensional Euclidean space onto a two-dimensional elliptic (also called, Riemannian) plane, that is, projective plane, as image points and image lines in the plane, respectively. Then, based on the isomorphism of angle relations of planes and straight lines in three-dimensional space with the metric (distance or angles) relations of their images in Riemann plane (projective plane), the angles of space planes and lines can be analyzed and calculated on the projection plane.

This approach has the following significance:

First, it explains profound theories in simple language. Although the use of the more difficult mathematical tools, but the problem of three dimensions is thus reduced to a two-dimensional problem. And, under this mapping, some space angle relations obtained through extremely lengthy deduction in the past become clear geometric fact which may be seen through almost at a glance. For example, the reason why the "Qun Drill"[1] performs better than standardly ground drill becomes clear on the projection plane (see Sect. 2.2.2).

Second, this approach can visually and synthetically display the effects of tool rotation in the space on their geometric angles. Consequently it provides a mathematical model, theoretical basis and an operating plane, i.e., the projective plane, for the cutter working angle and dynamic angle analysis and calculation, for angle online monitoring, and for optimum cutter design.

Based on the above method of mapping and isomorphism, the relationships of the cutting edge shape and the rake angle distribution along the edge of a standard twist drill was studied. It was discovered in the research that there is a "maximum rake edge line", along which at every point of the edge line the rake angle can reach the possible highest value of the point, so the cutting performance of the drill was obviously improved. If based on the traditional mechanical drawing or general mathematical methods, such a problem could not be solved, even the problem was difficult to explain.

Mapping space angles of cutting tools on a project plane also provides a simple and intuitive tool for the technical expression, professional education or technical training for complex shaped cutting tools. The author's teaching practice has shown that by using this method to turn a three-dimensional problem into a two-dimensional one, some space angle relations, which were difficult to imagine

[1]Qun-Drill is a kind of high-efficiency twist drill developed by Chinese workers and technicians.

and hard to express, can be clearly and intuitively expressed on the black board or on a computer projector screen. And this makes those, who are lack of space imagination training, feel easier to accept and understand angle relations in the space.

1.2 Non-Free Cutting Mechanics

Metal cutting mechanics are the theoretical base of machine tool design and cutting tool design. Correctly summing up and utilizing cutting mechanics law directly relates to the machining precision, efficiency, and cost. Due to the high development of Modern cutting tool manufacturing and grinding technology, especially the development in non-regrinding carbide inserts and solid carbide tool designing technology and hot-pressing technology, and NC grinding technology, the cutting part of cutting tool can be made into very complex shapes. The cutting processes of such kind of tools belong to very complex non-free cutting. Non-free cutting is basically characterized by conflict and interference of chips ejected from different segments of the cutting edge. Such a problem complicates chip deformation; increases machine tool energy consumption; shortens the tool life and deteriorates the quality of the machined surface. Disadvantages of cutting processes often originate from this issue. On the other hand, however, the research work regarding to cutting theory and chip deformation has largely confined to the exploration of the free cutting. The complexity of cutting tools and chip-ejection interference are ignored. The theory has divorced from the production practice. Today, we urgently need a theoretical tool as a bridge between the basic laws in free cutting and the various complex phenomena in non-free cutting, so as to guide the tool design and machining process modeling and optimization.

The traditional mechanical model of cutting tools is a "parametric model". This is because in the past among the cutting tools of the same category there was an overall similarity in geometry, and several dimensional parameters were sufficient to define their geometric characteristics. For example, for standard twist drills (under conventional grinding), regardless of their size, their tip shapes are approximately similar, so in traditional drilling torque formula, only one dimensional parameter, i.e. their diameter D, as independent variables, is enough to describe their geometric characteristics. Modern cutting tools change greatly in their shapes, even in the same category, such as twist drills, little overall similarity can be accessed. Therefore, parameter model has not been able to meet the demand. The possible approach is to build tools' "functional model" on the basis of local similarity (or differential similarity) among the elementary cutting tools to reflect the effects of tool shape on their cutting properties.

If the traditional parametric models can be built through the method of experiments and empirical formula fitting, then, in such a way to build "functional model" is clearly unrealistic. Cutting theory needs new synthesis and new sublimation.

This book reflects the author's research work on non-free cutting; the main development is as follows.

First, through experiment, observation and non-balance state thermal dynamics analysis, "minimum energy dissipation principle", which was used by predecessors for particular cases, is expanded to a universal mechanical law, a basic rule restricting cutting process, that is, on the premise of satisfying the constraint conditions set by the control parameters (i.e. cutting parameters, such as cutting speed, feed rate and depth of cut, and the tool geometrical parameters, such as rake, inclination and clearance etc.), the state parameter (i.e. the physical parameters, such as chip-ejection angle and ejection velocity, shear angle, and cutting ratio, etc.) of a cutting process always makes cutting power consumption take minimum value. This law has imposed a strong restriction on the value of status parameters, and has been in a number of specific conditions verified as correct.

Second, "Elementary Cutting Tool" concept is used as a basis for analysis of complex cutting tools, and through theoretical analysis and specially designed cutting experiments it has been proved that natural values of chip-ejection vector (including chip-ejection direction and chip-ejection velocity) conforms to the principle of minimum energy consumption, and in any way the chip-ejection vector being forced away from its natural value will inevitably lead to cutting power consumption to rise.

Third, in order to establish the functional model for modern complex shaped cutting tools, a method of "Elementary Cutting Tool nonlinear synthesis" is proposed. The key to this approach is that the problem of chip-ejection interference and coordination should be successfully handled. The author concluded that in order to maintain the chip integrity, all the Elementary Cutting Tools, each taking his own way in chip-ejection direction and ejection velocity, must coordinate with each other. The principle of coordination is also "minimum energy dissipation", i.e. the overall chip movement of cutting tool must minimize the sum of the energy consumed by all the Elementary Cutting Tools working simultaneously. Based on this, the general equation conditioning the chip-ejection motion was derived; the solutions of the equation were studied. The correctness of the solutions under specific conditions was experimentally verified. This method provides a feasible way of establishing "functional model" for arbitrarily shaped cutting tools based on theoretical or empirical Elementary Cutting Tool models. Along the way, a mechanical model for twist drills with arbitrary edge shapes was established, and the actual cutting tests proved the correctness of the model.

Fourth, from the above model the so-called "law of averages" was deduced. The law says that a certain state parameters (such as chip-ejection vector) affiliated with the whole tool is equal to the weighted average of the corresponding parameters affiliated to each Elementary Cutting Tool with the power consumption of each Elementary Cutting Tool while working individually (free cutting) as its weight. The law provides a theoretical basis and a simple tool for analyzing influence of various parameters of the modern tools on their performance.

Fifth, results of chip-ejection coordinating for non-free cutting tools ensure a minimum value of total cutting power under the given constraints, but do not guarantee that each Elementary Cutting Tool's cutting power tends to be the minimal. Therefore, the aggregate power consumption of a non-free cutting tool is often greater than the simple addition of the cutting power consumption of all the Elementary Cutting Tools contained in the cutting tool while each of them working alone (free cutting). The discovery reveals nonlinearity characteristics of cutting mechanics. This characteristic is the essence of the many issues that arise in the cutting process. This feature shows the complexity of cutting process. It indicates that a tool with complex shape cannot be seen as a simple addition of a series of independent concurrent Elementary Cutting Tools. We must study and grasp the phenomena and laws of the chip-ejection interfere and coordination of different parts of the cutting edge on the whole.

Sixth, a non-freedom degree coefficient of a cutting process is proposed to quantify chip-ejection interference intensity in the non-free cutting and the rationality of tool-design.

Seventh, the necessary and sufficient conditions for realizing free cutting are derived. The "free cutting" method is proposed. Tool design principle of dredging chip-ejection, and eliminating or reducing chip-ejection interference was studied. The corresponding software was developed. Excogitated "free cutting tool" of good performance obtained the national patent. In cooperation with manufacturers, a kind of free cutting carbide tips was designed and developed through two rounds of trial. Preliminary cutting tests showed their cutting performance (including cutting capacity, power consumption, and chip control and tool life) was really better than that of comparative tips provided by manufacturers. The results also clarify a widespread misunderstanding about free cutting, deepening the understanding of non-free cutting nature and its law.

Eighth, based on the above model, it is predicted that there are bifurcation and catastrophe in the process of cutting, and it has been shown by experiments that they do exist. The finding of bifurcation and catastrophe phenomena in cutting process has impacted people's traditional view point of cutting process: on one hand, the physical state of a cutting process (as chip-ejection direction, and cutting ratio, etc.) cannot be completely and immediately controlled by the control parameters (such as cutting parameters or tool geometry), but closely depends on the changing history of control parameter (such as cut-in process); On the other hand, continuous changes in the control parameters at some bifurcation points can lead to cutting state jumping. The findings reveal the complexity of cutting process, which has functional significance for cutting process control.

1.3 Non-Linear Machine Tool Dynamics

Machine tool chatter is a type of self-excited vibration phenomena in the cutting process, which seriously affects machining quality and limits productivity of machining. In addition, the shrill noise generated by cutting chatter pollutes environment, damages the operator's health.

Theory of machine tool chatter, which appeared in the 40s of the 20th century, studied the dynamic stability of metal machining system, chatter generating mechanism and prevention methods.

In recent years, the development of machining technology has presented some new, more urgent requirements for machine tool chatter control:

First, the development of the ultra-precision machining technology requires strict control of vibration during the cutting process to ensure accuracy and surface quality;

Secondly, ceramics and other super hard cutting tool materials have been widely used. This type of tool materials with high hardness, thermal stability and long life can perform high speed cutting, which is the advantages. However, this type of tool materials are brittle, afraid of shocks, vibrations must be strictly controlled in its cutting process;

Third, the application of super high speed cutting and heavy duty cutting technology, the cutting of difficult machining materials such as titanium processing and the development of machine parts towards lightweight and thin-walled, all these greatly increase the likelihood of inspiring chatter;

Finally, in modern automatic machining system, such as machining centers, flexible manufacturing systems or intelligent manufacturing systems due to the diversity of operations, the machining system's state more likely falls into the unstable region. Furthermore, this automated system is not under constant monitoring of operators and the processing area is often hidden, not easy to observe. Once chatter happens it likely leads to very serious consequences. Therefore, for modern machining systems chatter becomes more prominent and complex.

Traditional theory of machine tool chatter is linear theory; it succeeded in explaining and predicting the threshold of chatter, but could not describe the whole process of chatter generating, chatter development and finally, its self-stabilizing. Starting from 60s of the 20th century some researchers were attempting to develop a nonlinear theory of machine tool chatter, but did not make substantive progress. Author's work, however, focuses on the nonlinear theory of machine tool chatter, especially nonlinear regenerative chatter.

Author's work was done during his visit to England and the main results achieved were:

The first, two basic universally existing nonlinear factors constraining chatter process and a way to mathematically describe them were found;

Second, based on the above factors, a nonlinear variable delay differential equation was created to describe machine tool chatter, and its solutions were studied; the physical meaning of these solutions and their technological significant explained;

Third, the correctness of the non-linear model and its conclusions was verified with a large number of experimental facts and data.

After returning home from England the main works done in the area of machine tool chatter online monitoring and control are as follows:

First, the effective and practicable criteria for early diagnosis of chatter were found;

Second, the theory and methods for online adaptive adjusting cutting parameters to curb or eliminate chatter were studied; the first on-line monitoring and controlling test system for machine tool chatter was established in China.

The author was invited to visit the University of Michigan in The United States. During his visit he was engaged in the research of on-line chatter prevention technics for NC machine tools. Specifically, based on non-linear chatter theory established by him, through computer simulation and experiments the effects of fluctuation of machine tool spindle speed on regenerative chatter suppression was carefully studied.

The significance of the author's work on machine tool chatter, as above mentioned may be expressed in the following aspects:

First, the traditional linear theory of machine tool chatter was promoted into non-linear theory. The non-linear model established can more deeply reveal the physical mechanism and the mathematical law for machine tool chatter generation and chanter development. So the non-linear model has richer connotations than the linear one. With minimal, common assumptions, this model is able to give a satisfactory explanation and prediction to a series of important phenomena in machine tool chatter, which was unable to be explained before.

Second, the study reveals some little-known important rules with practical significance. For example: the sensitively dependence of chatter amplitude on the cutting parameters, and the relationship of stability of machining system with the intensity of disturbance, and so on. One of the most amazing discoveries was that sometimes a load increase on the machine helps to improve the system stability on the contrary. Therefore chatter prevention is not always at the expense of productivity, as asserted by the previous linear theory.

Third, the discovered two basic nonlinear factors, which constraints machine tool chatter, become elements which must be taken into account and cannot be avoided in research of the whole process of machine tool chatter. In fact, the author's research in this area has been referenced or included by literature, abstracts, and newly published books at home or abroad.

Fourth, based on the theory of nonlinear models established, the principles, the methods and the processes for adjusting the cutting parameters to suppress machine tool chatter were studied and a simple way for chatter prevention or avoiding in cutting process has been found.

Fifth, the traditional off-line modeling approach was advanced to online monitoring and control. Thus the theoretical foundation and technical guidance are provided for cutting chatter online monitoring and control in real time of modern CNC machining equipment and automatic production system.

The contents of this book include three "Non-s", that is Non-Euclidean Cutting Tool Geometry, Non-Free Cutting Mechanics and Non-Linear Machine Tool Dynamics. Throughout they have in common is "Non-Traditional". The author hopes in this area on the basis of traditional knowledge, make some new achievements and new advance in order to meet the needs of today's scientific and technological development.

Part I
Non-Euclidean Cutting Tool Geometry

Chapter 2
The Way of Expressing, Analyzing and Calculating Space Angles in a Plane

Based on mapping method and the principle of non-Euclidean geometry, this chapter presents a novel method for expression, analysis and calculation of space geometric angles. According to this method, a space inclined plane is mapped onto a projective plane as an image point, and a space line is mapped on the projective plane as an image line. The problem of calculating the angle between two space inclined planes is thus converted into the problem of calculating the non-Euclidean distance between their image points on the projective plane, and the problem of calculating the angle between two space inclined lines is converted into the problem of calculating non-Euclidean angle of their image lines. The problem of a rigid body rotating in the space is converted into the problem of moving and rotating in the projective plane of its image diagram. Compared with conventional ways, this method is simple and intuitive. It is a more effective approach for designing and calculating the cutting tool angles or angles in machine parts.

In designing, manufacturing, testing, installation or adjusting cutting tools or machine parts, a problem of analyzing and calculating the angles between space inclined planes and lines is often involved. If only the angular relationships are taking into consideration, and ignore the linear distance, all the planes and lines can be translated into an intersection point, to form a geometrically so-called "bundle", i.e., the totality of all the inclined planes and lines passing through a particular point in the space. So the problem of space angle analysis and calculation comes down to geometry of a bundle.

H. Shi, *Metal Cutting Theory*, Springer Series in Advanced Manufacturing,
https://doi.org/10.1007/978-3-319-73561-0_2

2.1 The Geometry of a Bundle and Its Images on Projective Plane

2.1.1 Mapping Method

At first glance the geometry of bundles belongs to space geometry problem, it is actually a kind of plane geometry. By means of following mapping method, we can build one to one relationship between elements of a bundle (space inclined planes and lines) and elements of a projective plane (points and planar lines).

First, taking the bundle center O (i.e. the meeting intersection of all planes, and lines in the bundle) as sphere center, make a spherical surface of unit radius (see Fig. 2.1). For any plane A, passing through O, make its normal line a, which intersects with the spherical surface at point A'. For any line a, passing through O, make its normal plane A, which intersects with the spherical surface at great circle a' (great circle is a circle on a spherical surface, the center of which coincides with the center of spherical surface). In this way all the planes and lines in the bundle are mapped onto spherical surface as points or great circles, respectively.

Second, make a plane O' contacting the spherical surface at point O', by means of gnomonic projection the point A' and great circle a' on the spherical surface can be re-mapped onto plane O' as point A'' and strait line a'' respectively. Point A'' and strait line a'' are called the images of space plane A and space line a respectively. And plane O' is called projective plane.

In this way, through twice mappings, all the inclined planes and lines in the bundle are one to one projected onto projective plane as points or strait lines. The

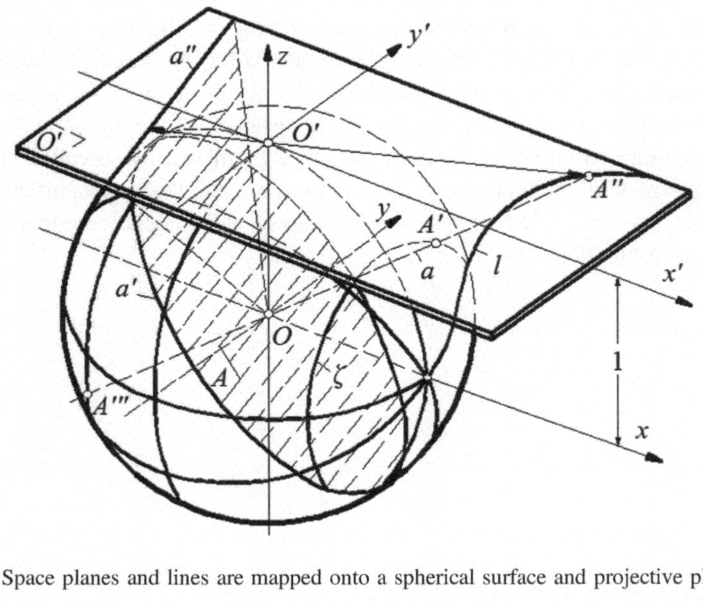

Fig. 2.1 Space planes and lines are mapped onto a spherical surface and projective plane

planes perpendicular to projective plane O' are projected to infinite points at corresponding directions, such as planes xOz and yOz, are projected to the infinite points of directions of axis y and axis x, respectively. Line perpendicular to projective plane O', i.e. axis z, is projected to the infinite line of projective plane O'.

Let's go a step further to introduce a metric system in productive plane. We define the "distance" S_{AB} between two image points A and B on projective plane as the real (in the sense of Euclidian geometry) angle between the two space planes represented by image points A and B; and the "angle" θ_{ab} between two image lines a and b on projective plane is defined as the real (in the sense of Euclidian geometry) angle between the two space lines represented by image lines a and b. It is easy to prove that after introducing above definitions the geometric property of projective plane satisfies the axiom system of a sort of non-Euclidian plane geometry: elliptic geometry (also called Riemannian geometry), and projective plane O' is thus called elliptic plane or Riemannian plane.

2.1.2 Non-Euclidean Geometry Compared with Euclidean Geometry

Engineers are aware of Euclidean geometry and use this geometry in their jobs or in their daily life. Euclidean geometry is, however, not the only possible geometry. Non-Euclidean geometries be also feasible and self-consistency. They are applied in solving various problems in engineering.

Euclidean geometry is based on the following 5 postulates:

1. A straight line segment can be drawn joining any two points.
2. Any straight line segment can be extended indefinitely in a straight line.
3. Given any straight line segment, a circle can be drawn having the segment as radius and one endpoint as center.
4. All right angles are congruent.
5. In a plane, give a line and a point not on it, at most one line parallel to the given line can be drawn through the point.

Elliptic geometry, as a kind of non-Euclidean geometry, keeps Postulates 1–4 unchanged except Postulate 5, which is changed as "there are not any parallel lines on elliptic plane".

A great deal of Euclidean geometry carries over directly to elliptic geometry. For example, the statements "two points uniquely determine a line" and "two lines uniquely determine a point" are also true in the elliptic geometry. This similarity between elliptic geometry and Euclidian geometry let us can conveniently analyze and calculate space angles on the projective plane.

There are some special things in elliptic geometry which are different from Euclidean geometry, such as "the sum of the interior angles of a triangle is greater than 180°", "the ratio of the circumference of a circle to its diameter is less than

$\pi = 3.14159$", "the segment of a straight line can be infinitely elongated, but the total lengths cannot exceed π". This differences make elliptic geometry look foreign, strange and not understandable.

To make the elliptic geometry understood, let's see Fig. 2.1, geometric graphics on projective plane O' are the projections of the geometric graphics on spherical surface. The non-Euclidean geometric properties of the projective plane can trace back to the Euclidean geometric properties of a spherical surface. So spherical surface geometry is called a model of elliptic geometry on Euclidean spherical surface. In this way, we can understand the peculiar properties of elliptic geometry through the properties of spherical surface geometry.

In the Fig. 2.1 we can see that two points A' and A''', the intersection points of a diameter a with the spherical surface are projected onto the same point A'' on the projective plane. So the two end points of any sphere diameter (called antipodes) should be considered as the same point. Bearing that in mind, let's carefully check the geometry of a spherical surface.

As mentioned above, that straight lines on the projection plane are the projections of great circles on a spherical surface, as showed in Fig. 2.2a. However, any two great circles b and d on a spherical surface will certainly intersect at two antipodes A and A' (remember, they are regard as one point). This means that there are no parallel lines in the projective plane. When two great circles intersect at antipodes of a diameter which is parallel to plane O' (Fig. 2.1), the image lines will look like parallel, they actually intersect at infinite point.

Figure 2.2b shows that the triangle made of the equator a and two longitudes b and c, which are 90° apart from each other on the spherical surface, has a sum of the interior angles equal to 270°. The projection on the projective plane of such a triangle is one of the sum of interior angles equal to 270°. This is only an example. Actually all the triangles on projective plane have a sum of the interior angles bigger than 180°. When a triangle is smaller, the difference of its sum of the interior angles from 180° will be also smaller.

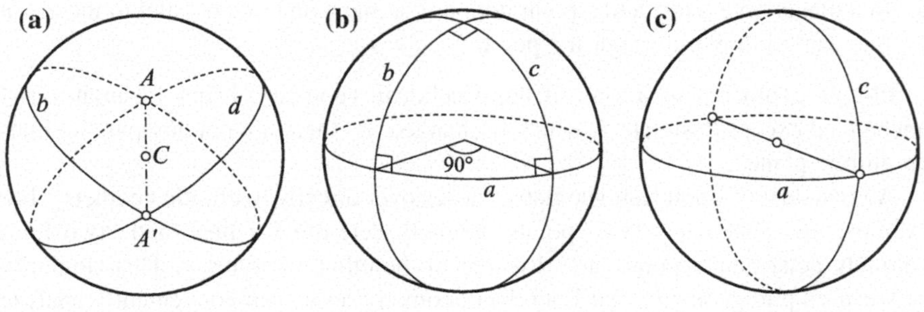

Fig. 2.2 Geometric properties of spherical surface. **a** Any two great circles on a spherical surface will certainly intersect at two antipodes. **b** The sum of the interior angles of a triangle is greater than 180°. **c** The ratio of the circumference of a circle to its diameter is less than π

Figure 2.2c shows that a circle a (the equator) has a circumference of 360°, its diameter is, however, half of the circle b, its "length" is 180°, so the ratio of the circumference of circle a to its diameter is 2, which is less than $\pi = 3.14159$.

This is not a book of geometry, but is a book of engineering, and we will not go further to elaborate on those geometric problems.

2.1.3 Isomorphism

Isomorphism is the most important relation between the two systems. If in system Π (e.g., a bundle) three elements a, b and C (e.g., lines and plane) have a relation $a \times b = C$ (e.g., lines a and b determine a plane C), in corresponding system Π' (e.g., the projective plane) we have $a' \times b' = C'$, where a', b' and C' are the corresponding elements of a, b and C in Π' (e.g., a', b' and C' are the images of a, b and C respectively). This means that if in the space lines a and b are on plane C, then on the projective plane the intersection point C' of image lines a' and b' is the image point of plane C.

The purpose of this chapter is based on isomorphic relationships between two-dimensional elliptic geometry and bundles geometry to find a shortcut, so as to effectively solve problems of expressing, analyzing and calculating space angles on cutting tools or in mechanical engineering.

2.1.4 The Images of Space Planes and Straight Lines on an Elliptic Plane

We can, in fact, remove the unit spherical surface and directly map planes and lines onto projective plane O as showed in Fig. 2.3a, b. O is bundle center, xOy plane is one of planes in the bundle selected as the base plane used to measure inclination angles of other planes and lines with respect to it. Plane $x'O'y'$ is projective plane, which is parallel to base plane xOy, and the two are one unit distance apart.

Suppose there is a plane A in the bundle (see Fig. 2.3a), draw its normal line a, which intersects with projective plane $x'O'y'$ at point A'. A' is the image of plane A. Obviously, if the orientation of plane A in the space has been determined, then the location of its image on the projective plane (or its coordinates A_x and A_y) are also determined. Conversely, The location of image A' in projective plane can determine the orientation of plane A in the space. The image of the base plane xOy is O', the origin of the projective plane.

Suppose there is a line a in the bundle (see Fig. 2.3b), draw its normal plane A, which intersects with projective plane $x'O'y'$ at line a'. Line a' is image of line a. In like manner, the orientation of line a in the space and the location of its image line a' on the

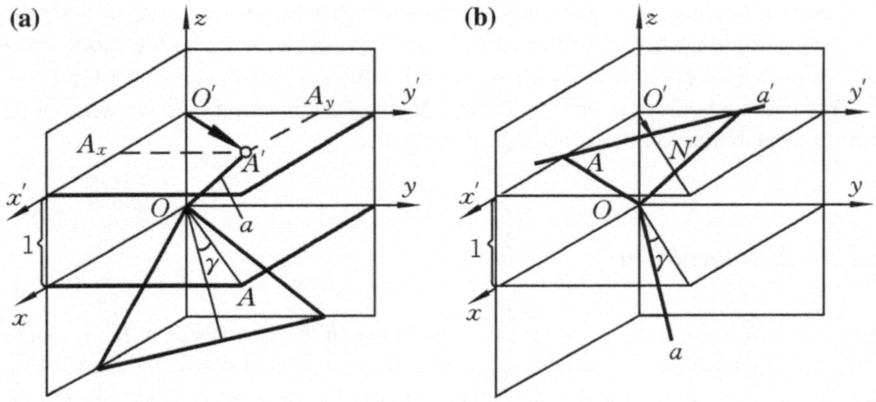

Fig. 2.3 Space inclined planes and lines are mapped into projective plane. **a** A plane's image point. **b** A line's image line

projective plane (or its equation in coordinate system $x'O'y'$) also have determinate corresponding relationship.

From now on we will concentrate our mind in the images on projective plane.

2.1.4.1 The Image Points of Plans and Their Obliquity Vectors

Suppose A is the image point of a space plane A (see Fig. 2.4a), O is origin of the projective plane, also, the image of base plane O. Vector $\boldsymbol{OA}$, or simplified as $\boldsymbol{A}$, is obliquity vector of plane A. It's easy to prove that $|\boldsymbol{A}| = A = \tan\gamma$, where γ is the inclination angle of plane A with respect to base plane O. The vector $\boldsymbol{A}$ also shows the obliquity direction of plane A with respect to base plane O.

The Cartesian coordinate $A_x = \tan\alpha_A$ and $A_y = \tan\beta_A$ of image point A have explicit geometric meaning: α_A and β_A are the angles between plane A and base O measured in planes perpendicular to O and passing through axis x and y, respectively.

The components of obliquity vector at any directions, such as $\tan\theta$ at t direction, $\tan\omega$ at l direction, etc. show the obliquity at the directions, respectively. Obviously, at direction $\boldsymbol{A}$ plane A has the biggest obliquity, and at the direction n perpendicular to direction $\boldsymbol{A}$ the obliquity of plane A is zero, this is the direction of its contour.

The end points of the obliquity vectors at various directions are all at a circle the diameter of which is obliquity vector $\boldsymbol{A}$. This circle is called obliquity circle. The rays at any directions, such as t, l, x, y, etc., cut by the circle (the chords) give the obliquities at the corresponding directions $\tan\alpha_A$, $\tan\beta_A$, $\tan\theta$ and $\tan\omega$.

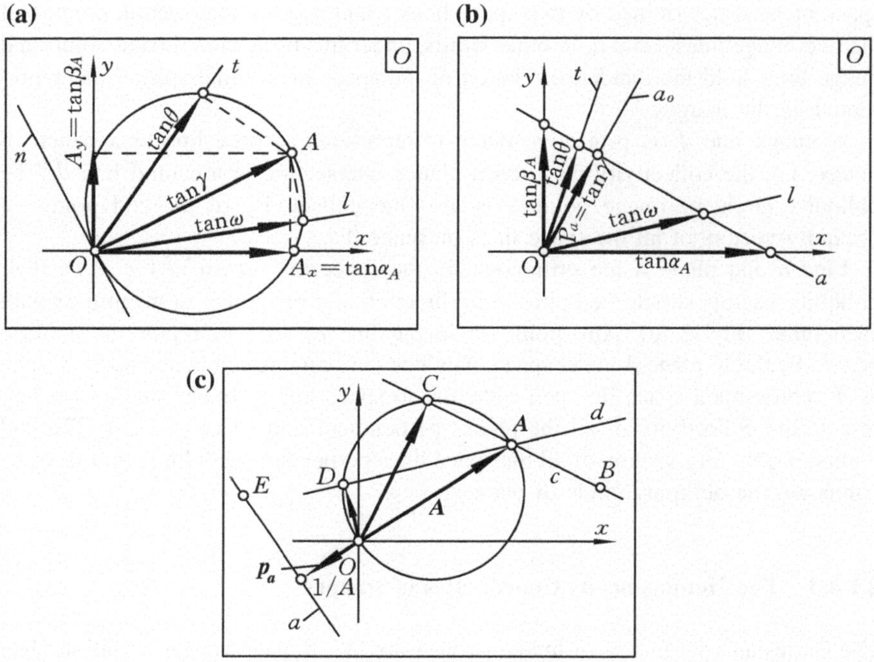

Fig. 2.4 Images of space planes and straight lines and the geometrical relations between them. **a** Image of a plane, its obliquity vector and obliquity circle. **b** Image of a line and its obliquity vector. **c** Incidence relation of images of planes and lines

2.1.4.2 The Image Lines of Straight Lines and Their Obliquity Vectors

Suppose on projective plane O line a is the image of a space line a, its normal vector $\boldsymbol{p}_a$ (see Fig. 2.4b) is the obliquity vector of the space line a with respect to base plane O, and $|\boldsymbol{p}_a| = \tan\gamma$, where γ is the angle between a and O. The orthographic projection a_o of the space line a on plane O is in the direction of $\boldsymbol{p}_a$.

The rays at any directions, such as t, l, x, y, etc., cut by the image line a give the angles $\tan\alpha_A$, $\tan\beta_A$, $\tan\theta$ and $\tan\omega$ between plane O and the orthographic projections of space line a on the planes normal to plane O and passing through corresponding rays.

2.1.4.3 The Incidence and Orthogonality of Planes and Lines

If space line c is on the space plane A, then image point A is on the image line c (Fig. 2.4c); The image line c of intersection line c of two space planes A and B is the connect line c of the two image points A and B; The image point A of the space plane A determined by two space lines c and d is the intersection point A of the two image lines c and d. In other words, under this projection, image points and image lines hold the incidence relation of the space planes and space lines represented by the images.

A image line d on projective plane O represents a space line or a pencil of planes, i.e. the collectivity of a series planes intersecting at a central line d; The obliquity circle of image point A is also the collectivity of the end points of obliquity vectors of all the space lines on plane A.

Line a and plane A are orthogonal in the space (as shown in Fig. 2.1), their obliquity vectors should be opposite in direction and reciprocal in magnitude with each other (Fig. 2.4c). Any point on image line a, e.g., E, represents a plane perpendicular to plane A in the space; Any line passing through image point A, such as d, represents a space line perpendicular to space line a. Image line a also represents the collectivity of all the planes perpendicular to space plane A; The end points of obliquity vectors of all the space lines perpendicular to line a in the space composes the obliquity circle of plane A.

2.1.4.4 The Homogeneous Coordinates of Images

The Cartesian coordinates of image point A are (A_x, A_y), as there is an infinite line and infinite points on projective plane, once image points move to infinite point, one or two of its Cartesian coordinates will become infinite and this results in inconvenience in calculation. Therefore we need to introduce homogenous coordinates as in projective geometry [1, 2, 60, 61].

In coordinate system xyz of Fig. 2.3a, the equation of plane A can be expressed as

$$A_1x + A_2y + A_3z = 0, \tag{2.1}$$

The column vector of homogeneous coordinates of image point are defined as

$$\boldsymbol{A} = \begin{pmatrix} A_1 \\ A_2 \\ A_3 \end{pmatrix}. \tag{2.2}$$

Obviously, A_1, A_2 and A_3 cannot be vanished simultaneously. When point A is not at infinite $A_3 \neq 0$, and it is easy to prove that the relation between Cartesian coordinates and the homogeneous coordinates of image point A are

$$\begin{aligned} A_x &= A_1/A_3 \\ A_y &= A_2/A_3 \end{aligned}. \tag{2.3}$$

They are mutually conversional, and we have

$$\boldsymbol{A} = \begin{pmatrix} A_1 \\ A_2 \\ A_3 \end{pmatrix} = \begin{pmatrix} A_x \\ A_y \\ 1 \end{pmatrix}. \tag{2.4}$$

However, if image point A is located at infinite, $A_3 = 0$ and A_x and A_y become infinite or infinitive 0/0, cannot be used for calculation, and only homogeneous coordinates can be used.

It should be noticed that if $(A_1, A_2, A_3)^{\mathrm{T}}$ is homogeneous coordinates of an image point, obviously, $(\rho A_1, \rho A_2, \rho A_3)^{\mathrm{T}}$ are also the homogeneous coordinates of the same image point, where ρ is any non-zero real number. This means that homogeneous coordinates can only be determined to the extent of an arbitrary constant factor.

Similarly, suppose in coordinate system xyz of Fig. 2.3b the equation of space line a can be expressed as

$$\frac{x}{a_1} = \frac{y}{a_2} = \frac{z}{a_3}, \tag{2.5}$$

It is easy to prove that the equation of image line a on projective plane O' is

$$a_1 x + a_2 y + a_3 = 0, \tag{2.6}$$

Then its homogeneous coordinates are defined as

$$\boldsymbol{a} = \begin{pmatrix} a_1 \\ a_2 \\ a_3 \end{pmatrix} = \rho \begin{pmatrix} a_1 \\ a_2 \\ a_3 \end{pmatrix}. \tag{2.7}$$

Obviously as in the case of image point, homogeneous coordinates of an image line also can only be determined to the extent of an arbitrary constant nonzero real factor ρ.

If let

$$\begin{cases} a_x = a_1/a_3, \\ a_y = a_2/a_3, \end{cases} \tag{2.8}$$

Then (a_x, a_y) is the Cartesian coordinates of intersection point A'' of space line a with projective plane O' (see Fig. 2.1). And the obliquity vector of image line a are (Fig. 2.3b)

$$\begin{cases} p_{ax} = -\frac{a_x}{a_x^2 + a_y^2}, \\ P_{ay} = -\frac{a_y}{a_x^2 + a_y^2}. \end{cases} \tag{2.9}$$

The homogeneous coordinates of plane A and line a perpendicular to each other in the space must equal

$$\boldsymbol{A} = \begin{pmatrix} A_1 \\ A_2 \\ A_3 \end{pmatrix} = \begin{pmatrix} a_1 \\ a_2 \\ a_3 \end{pmatrix} = \boldsymbol{a}, \tag{2.10}$$

Or, are difference by a constant factor

$$\boldsymbol{A} = \rho \boldsymbol{a}. \tag{2.11}$$

2.1.4.5 The Distances and Angles on Projection Plane

According to the definition previously described, the non-Euclidian "distance" S_{AB} between two image points A and B is the angle between two space planes A and B, which can be expressed as

$$S_{AB} = \tan^{-1} \frac{\sqrt{\begin{vmatrix} \boldsymbol{A}^{\mathrm{T}}\boldsymbol{A} & \boldsymbol{A}^{\mathrm{T}}\boldsymbol{B} \\ \boldsymbol{B}^{\mathrm{T}}\boldsymbol{A} & \boldsymbol{B}^{\mathrm{T}}\boldsymbol{B} \end{vmatrix}}}{\boldsymbol{A}^{\mathrm{T}}\boldsymbol{B}}, \tag{2.12}$$

where $\boldsymbol{A}$, $\boldsymbol{B}$ are the homogeneous coordinates column vector of image points A, B, and $\boldsymbol{A}^{\mathrm{T}}\boldsymbol{B}$ is the inner product,

$$\boldsymbol{A}^{\mathrm{T}}\boldsymbol{B} = (A_1, \quad A_2, \quad A_3) \begin{pmatrix} B_1 \\ B_2 \\ B_3 \end{pmatrix} = A_1B_1 + A_2B_2 + A_3B_3 \tag{2.13}$$

Similarly for $\boldsymbol{A}^{\mathrm{T}}\boldsymbol{A}$, $\boldsymbol{B}^{\mathrm{T}}\boldsymbol{A}$, $\boldsymbol{B}^{\mathrm{T}}\boldsymbol{B}$ and for hereinafter. We take the value of S_{AB} in the following interval:

$$0 \leq S_{AB} \leq \pi. \tag{2.14}$$

It must be noticed that the "distance" S_{AB} defined above is different from our familiar Euclidean distance, which is notified by $\overline{AB}$ in this book. In special case, when image A is at origin, $\boldsymbol{A} = \boldsymbol{O} = (0, 0, 1)^{\mathrm{T}}$, substituting this into Eq. (2.12) results in

$$\tan S_{AB} = \tan S_{OB} = \overline{OB} = B, \tag{2.15}$$

Non-Euclidian "distance" $\tan S_{AB}$ degrades into Euclidean distance $\overline{AB}$, which can be measured by a ruler. We are now back to the previous conclusion: the modulus

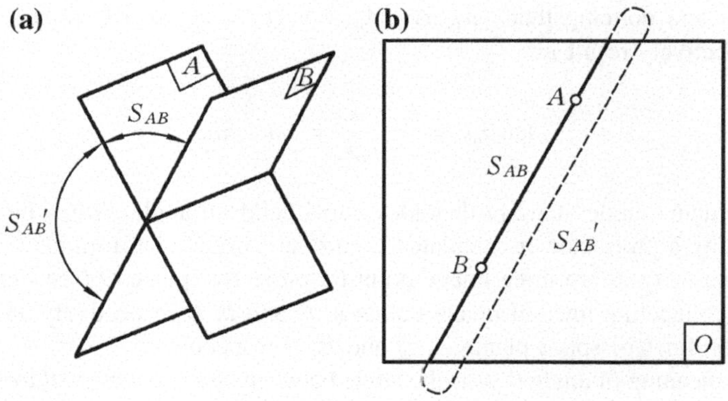

Fig. 2.5 The angle between two space planes and the distance between two image points. **a** The angle between two spatial planes. **b** The distance between two image points

of obliquity vector of image B is equal to the tangent of the angle between plane B and base plane O.

It must be noticed that every straight line on elliptic plane should be imagined to be a closed circle, the two end points at infinite meet each other, and its total length is π (see Fig. 2.5b). Any two points on a line, such as A and B, divide the line (circle) into two parts: the one is less than or equal to $\pi/2$, and the other is bigger than or equal to $\pi/2$, and the sum of two is the total length π. This is corresponding to the fact that two planes intersecting with each other make two angles (see Fig. 2.5a), the one is less than or equal to $\pi/2$, and the other is bigger than or equal to $\pi/2$, and the sum of two is the total length π. The calculating result according to Eq. (2.12) maybe any one of the two, depending on the symbol of $\boldsymbol{A}^{\mathrm{T}}\boldsymbol{B}$, and because of the property of homogenous coordinate, the symbol of $\boldsymbol{A}^{\mathrm{T}}\boldsymbol{B}$ is indeterminate.

Similarly, the "angle" θ_{ab} between two image lines a and b (i.e. the angle between two space lines a and b) can be calculated through the following formula:

$$\theta_{ab} = \tan^{-1}\frac{\sqrt{\begin{vmatrix} \boldsymbol{a}^{\mathrm{T}}\boldsymbol{a} & \boldsymbol{a}^{\mathrm{T}}\boldsymbol{b} \\ \boldsymbol{b}^{\mathrm{T}}\boldsymbol{a} & \boldsymbol{b}^{\mathrm{T}}\boldsymbol{b} \end{vmatrix}}}{\boldsymbol{a}^{\mathrm{T}}\boldsymbol{b}}, \tag{2.16}$$

where $\boldsymbol{a}$, $\boldsymbol{b}$ are the homogeneous coordinates column vector of image lines a, b, and we take the value of θ_{ab} in the following interval:

$$0 \leq \theta_{ab} \leq \pi. \tag{2.17}$$

The "angle" θ_{ab} defined above is different from our familiar Euclidean angle, which is notified by $\angle ab$ in this book. In special case when the intersection point C of image lines a and b is at the origin O, we have $a_3 = b_3 = 0$, substituting this into

Eq. (2.16) and noticing that $-a_1/a_2 = k_a$, $-b_1/b_2 = k_b$ are the slopes of image lines respectively result in

$$\tan\theta_{ab} = \frac{k_b - k_a}{1 + k_b k_a} = \tan\angle ab, \tag{2.18}$$

Non-Euclidian "angle" $\tan S_{AB}$ degrades into Euclidean angle $\angle ab$, which can be measured by a protractor or calculated according to Euclidian formula.

Suppose A, Q, B are three image point on projective plane O (see Fig. 2.6), a, b are the connecting lines of image points A, Q and B, Q respectively, namely the intersection lines of space planes A, Q and B, Q respectively.

So the meaning of angle $\theta_{AQB}$is the angle between space planes A, B measured in section Q. The angle can be calculated from the homogeneous coordinates of the three image points A, Q and B:

$$\theta_{AQB} = \theta_{ab} = \tan^{-1}\frac{\sqrt{\boldsymbol{Q}^{\mathrm{T}}\boldsymbol{Q}}\begin{vmatrix} A_1 & A_2 & A_3 \\ B_1 & B_2 & B_3 \\ Q_1 & Q_2 & Q_3 \end{vmatrix}}{\begin{vmatrix} \boldsymbol{Q}^{\mathrm{T}}\boldsymbol{Q} & \boldsymbol{Q}^{\mathrm{T}}\boldsymbol{A} \\ \boldsymbol{B}^{\mathrm{T}}\boldsymbol{Q} & \boldsymbol{B}^{\mathrm{T}}\boldsymbol{A} \end{vmatrix}}, \tag{2.19}$$

In special case when B coincides with origin O and Q becomes a section passing through x axis and perpendicular to O (i.e. plane xOz in Fig. 2.1), we have $\boldsymbol{B} = (0,0,1)^{\mathrm{T}}$ and $\boldsymbol{Q} = (0,1,0)^{\mathrm{T}}$, substituting into Eq. (2.19) results in

$$\tan\theta_{AQB} = \tan\theta_{AQO} = A_x, \tag{2.20}$$

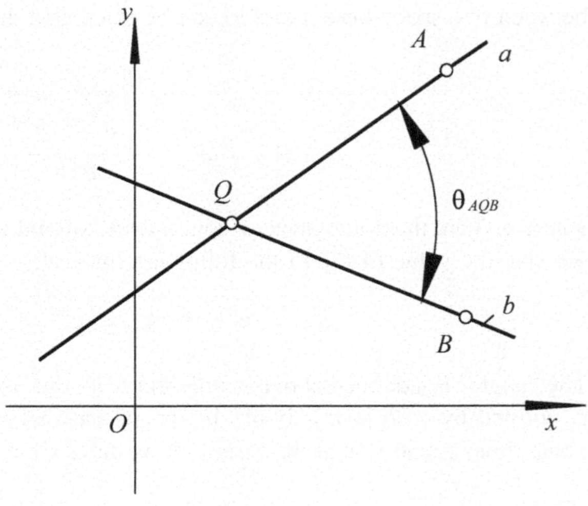

Fig. 2.6 Non-Euclidean angle defined by three image points

Therefore we are back to the conclusion: the tangent of angles between plane A and base O measured in a plane perpendicular to O and passing through axis x is A_x, the x coordinate of obliquity vector of plane A.

2.2 The Image Graphic and Its Movement on Projective Plane

In designing cutting tools and abnormal machine parts or designing their measuring tools or fixtures what is involved is not only the angular relations of a static geometrical object but also the rotations of the object in the space. While a geometrical object composed of planes and edges (i.e., lines) rotating in the 3D Euclidean space the image graphic consists of images of its planes and edges are moving on the projective plane (ellipse plane or Riemann plane). Like rigid figures moving in Euclidian plane, image figures moving in Riemann plane also keep all the non-Euclidian "distances" and "angles" unchanged.

2.2.1 *Movement of a Plane Image Point*

As shown in Fig. 2.1, while plane A is rotating in the space around axis x, its image A' on spherical surface moves along a latitude ζ, and this latitude through gnomonic projection is projected onto projective plane O' as a hyperbolic curve l (only the right branch is shown in the figure and left branch is ignored). Thus it can be seen that while a plane is rotating in the space around an axis parallel to projective plane its image is moving on the projective plane along a hyperbola. If the initial position of the image is $A_1(\tan\alpha_1, \tan\beta_1)$, after rotating around x axis by a positive angle θ (the plus-minus sign of θ is determined according to right-hand rule) as shown in Fig. 2.7a. It is easy to prove that hyperbolic curve l satisfies the following equation:

$$\frac{x^2}{c^2} - \frac{y^2}{1^2} = 1, \tag{2.21}$$

Its imaginary semi-axis is at y axis, and its length is always 1. We can see from Eq. (2.21) that when an image moving along the hyperbolic curve l its coordinates $(\tan\alpha, \tan\beta)$ satisfies the following equation:

$$\tan\alpha_1 \cos\beta_1 = \tan\alpha_2 \cos\beta_2 = c \tag{2.22}$$

That is to say that during the moving process the value of "$\tan\alpha\cos\beta$" keeps constant and identically equal to the real semi-axis c, which represents the latitude. Also we have

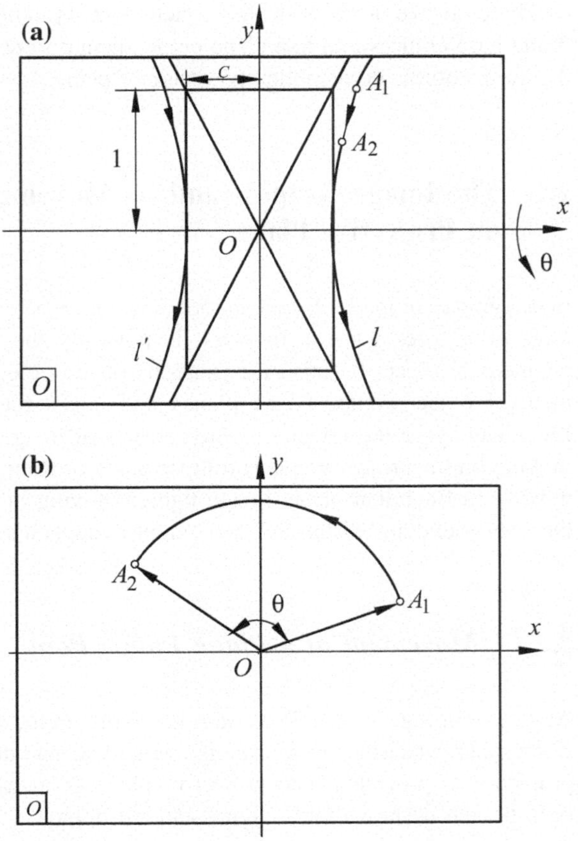

Fig. 2.7 The movement of image point on projective plane when a plane rotating about an axis in the space. **a** Rotating around x axis. **b** Rotating around z axis

$$\beta_2 = \beta_1 - \theta. \tag{2.23}$$

Based on the above two equations we can conveniently calculate the coordinates $(\tan\alpha_2, \tan\beta_2)$ of image A after rotation from the coordinates $(\tan\alpha_1, \tan\beta_1)$ of image A before rotation and the rotation angle θ.

When plane A is rotating around y axis, its image A is also moving along a hyperbola, but the real semi-axis of the hyperbola is on y axis. The initial coordinates $A_1(\tan\alpha_1, \tan\beta_1)$, final coordinates $A_2(\tan\alpha_2, \tan\beta_2)$ and the rotation θ angle have the following relations

$$\cos\alpha_1 \tan\beta_1 = \cos\alpha_2 \tan\beta_2 = c \tag{2.24}$$

$$\alpha_2 = \alpha_1 + \theta. \tag{2.25}$$

When plane A is rotating about z axis by an angle θ, its image A is moving along a circle on projective plane O by the same angle (see Fig. 2.7b), and its coordinates before and after rotating, i.e., $A_1(\tan\alpha_1, \tan\beta_1)$ and $A_2(\tan\alpha_2, \tan\beta_2)$ have the following relation:

$$\tan\alpha_2 = \tan\alpha_1 \cos\theta - \tan\beta_1 \sin\theta, \tag{2.26}$$

$$\tan\beta_2 = \tan\alpha_1 \sin\theta + \tan\beta_1 \cos\theta. \tag{2.27}$$

2.2.2 *Image Trajectory Diagram*

According to the above formulae, we can find out the moving trajectory of an image point, which is called "image trajectory diagram". See Fig. 2.8a, at the start position its image coordinates is $A_1(\tan\alpha_1, \tan\beta_1)$, after rotating about axes x, y and z in succession, its image moves along two paragraphs of parabolas and a circle arc to A_2 $(\tan\alpha_2, \tan\beta_2)$, A_3 $(\tan\alpha_3, \tan\beta_3)$ and A_4 $(\tan\alpha_4, \tan\beta_4)$ in succession. From Eqs. (2.22) and (2.23) we have the coordinates of A_2,

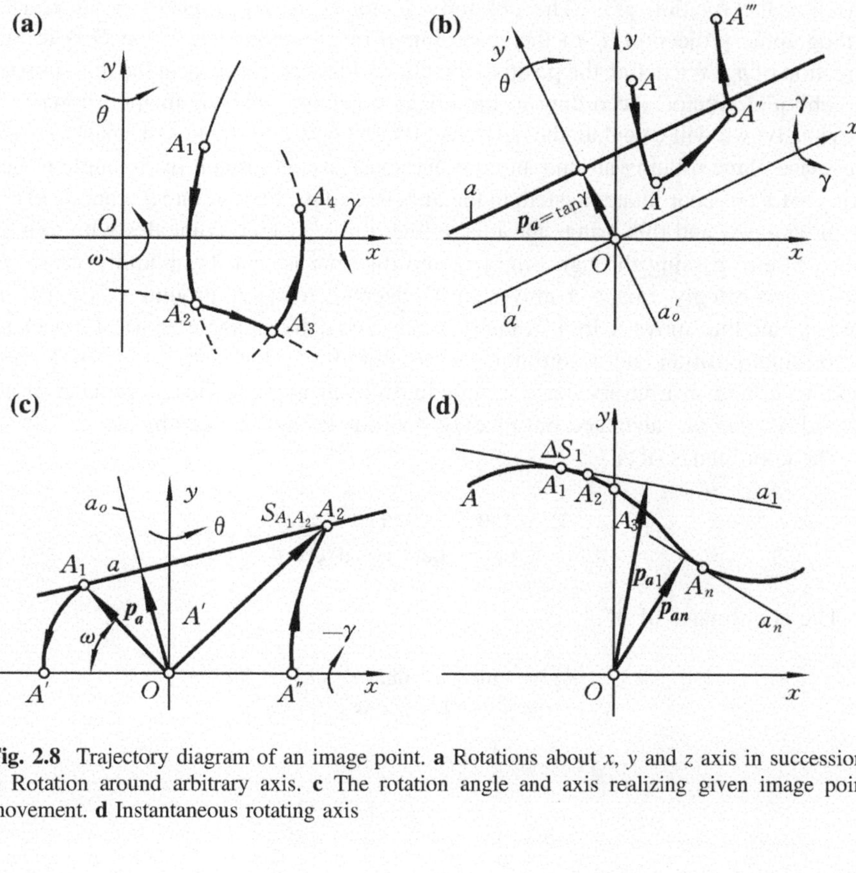

Fig. 2.8 Trajectory diagram of an image point. **a** Rotations about x, y and z axis in succession. **b** Rotation around arbitrary axis. **c** The rotation angle and axis realizing given image point movement. **d** Instantaneous rotating axis

$$A_{2y} = \tan\beta_2 = \tan(\beta_1 - \gamma),$$
$$A_{2x} = \tan\alpha_2 = \tan\alpha_1 \cos\beta_1/\cos\beta_2;$$

From Eqs. (2.24) and (2.25) we have the coordinates of A_3,

$$A_{3x} = \tan\alpha_3 = \tan(\alpha_2 + \theta),$$
$$A_{3y} = \tan\beta_3 = \cos\alpha_2 \tan\beta_2/\cos\alpha_3;$$

From Eqs. (2.25) and (2.26) we have the coordinates of A_4,

$$A_{4x} = \tan\alpha_4 = \tan\alpha_3 \cos\omega - \tan\beta_3 \sin\omega,$$
$$A_{4y} = \tan\beta_4 = \tan\alpha_3 \sin\omega + \tan\beta_3 \cos\omega.$$

2.2.3 *The Movement of Image When a Plane Rotates Around an Arbitrary Axis*

See Fig. 2.8b, A is the image of a space plane, and a is the image of a space straight line, i.e. the rotation axis. The obliquity vector of a is $\boldsymbol{p_a}$, and $p_a = \tan\gamma$. The orthographic projection a_o of the space line a on the projective plane O is in the direction of $\boldsymbol{p_a}$. We define the positive direction of image line a_o is in the direction of the obliquity vector. According to the image trajectory analysis method described previously, we will calculate the new position and new coordinates of image A after the space plane rotating around an arbitrary space straight line a by an angle θ. Set axis y' of $x'Oy'$ coordinates system in the direction of $\boldsymbol{p_a}$. First, rotate a around axis x' by an angle γ, and this brings the image line a to a', which coincides with axis y' (note, line a' passing through origin O and this means that its obliquity vector is zero). Accordingly, image A moves to A'. Second, rotate A' around axis y' by an angle θ, and thus move A' to A''. Finally, rotate axis a' by an angle $-\gamma$ to let it back to the original position and accordingly, A'' moves to A''', this is the image of A after rotating around an arbitrary space straight line a by an angle θ. The coordinates of A', A'' and A''' can be calculated out in $x'Oy'$ coordinates system step by step:

The coordinates of A',

$$A'_y = \tan\beta' = \tan(\beta - \gamma),$$
$$A'_x = \tan\alpha' = \tan\alpha \cos\beta/\cos\beta';$$

The coordinates of A'',

$$A''_x = \tan\alpha'' = \tan(\alpha' + \theta),$$
$$A''_y = \tan\beta'' = \tan\beta' \cos\alpha'/\cos\alpha'';$$

The coordinates of A''',

$$A'''_y = \tan\beta''' = \tan(\beta'' + \gamma),$$
$$A'''_x = \tan\alpha''' = \tan\alpha'' \cos\beta''/\cos\beta'''.$$

where $\tan\alpha = A_x$, $\tan\beta = A_y$ are the coordinates of image A in $x'Oy'$ system.

2.2.4 The Rotation Axis and Angle Realizing Given Image Movement

The rotation axis and rotation angles realizing the movement of the image point from A_1 to A_2 are desired. There are many different solutions to this problem. One of these solutions is to rotate the plane A about axes z, y and x in succession by angles of ω, θ and $-\gamma$ respectively, and move the image $A_1(\tan\alpha_1, \tan\beta_1)$ to $A_2(\tan\alpha_2, \tan\beta_2)$ via $A'(\tan\alpha', \tan\beta')$ and $A''(\tan\alpha'', \tan\beta'')$ as shown in Fig. 2.8c. The rotation angles can be calculated according to the geometrical relations shown in the figure,

$$\tan\omega = \left|\frac{\tan\beta_1}{\tan\alpha_1}\right|,$$

$$\tan\alpha' = -\sqrt{\tan^2\alpha_1 + \tan^2\beta_1},$$

$$\tan\alpha'' = \tan\alpha_2 \cdot \cos\beta_2,$$

$$\theta = \alpha'' - \alpha',$$

$$\gamma = \beta_2.$$

Another obvious solution is to take the intersection line a of planes A_1 and A_2 as the rotation axis (note, the image of a is the connection line of A_1 and A_2), and the rotation angle is the angle $S_{A_1A_2}$ between planes A_1 and A_2 (note, angle $S_{A_1A_2}$ is the distance between images A_1 and A_2 as shown in Fig. 2.8c).

2.2.5 Instantaneous Rotation Axis

The movement of image point A along an arbitrary curve (Fig. 2.8d) may be regarded as a series of continuous incremental rotations of plane A about a series of instantaneous axes. The images of instantaneous axes at all the moments are the tangent lines $a_1,\ldots,\ a_n,\ldots$, to the curve, or, approximately, the connect lines of adjacent points on the curve, such as A_1–A_2, A_2–$A_3,\ldots$, and the rotation angles are $\Delta S_1 = S_{A_1A_2}$, $\Delta S_2 = S_{A_2A_3},\ldots$, and so on.

2.2.6 The Movement of Image Lines

In order to determine the new position of an image line a on projection plane after the space line a has rotated around an axis by an angle, we can take any two points on image line a and move them to the new position according the methods above described, and then connect the two new image points so as to get the new image line. We also can first determine the image point of the normal plane of the space line a (see Fig. 2.4c), and move it to a new position, and in reverse to determine the new position of the image line a.

It has been proved that two space planes or two space lines after rotating by the same angle and around the same axis, the "distance" calculated according Eq. (2.12) or "angle" calculated according Eq. (2.16) would keep unchanged. This conclusion is, however, does not consistent with intuition. Let's see Fig. 2.9, the image graphic of a space trihedral angle ΔABC rotates about axis x by an angle of $-40°$ and moves to $\Delta A'B'C'$, then rotates about axis y by an angle of 40° and moves to $\Delta A''B''C''$. Superficially, these three triangles have different shapes and sizes, but from the view point of non-Euclidian geometry, it must be admitted that these are three congruent triangles, because their corresponding sides and angles all kept unchanged during moving process.

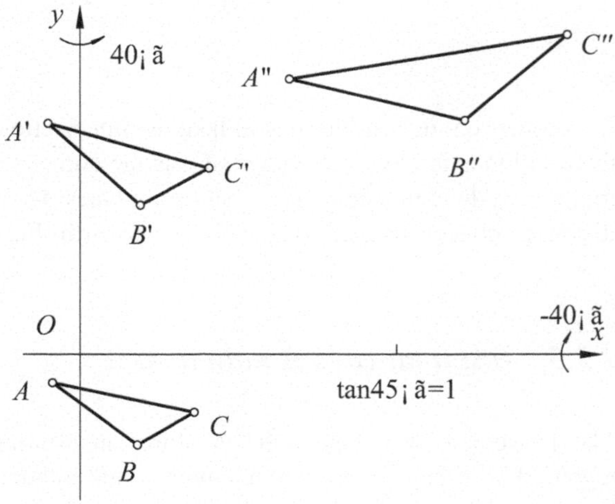

Fig. 2.9 The movement of a triangle on projective plane, congruent triangles

2.3 The Operators on Projective Plane

Operators are used to calculate unknown homogeneous coordinates from already known ones. According to different geometric relations we have different operators.

2.3.1 *Orthogonality Operator* $(\perp\cdot)$

Orthogonality operators are used to calculate the homogeneous coordinates of the normal line of a known plane or the homogeneous coordinates of the normal plane of a known line. From Fig. 2.10 and Eq. 2.10 we have

$$\boldsymbol{a} = (a_1, a_2, a_3)^{\mathrm{T}} = \perp\boldsymbol{A} = \boldsymbol{A} = (A_1, A_2, A_3)^{\mathrm{T}}, \tag{2.28}$$

$$\boldsymbol{A} = (A_1, A_2, A_3)^{\mathrm{T}} = \perp\boldsymbol{a} = \boldsymbol{a} = (a_1, a_2, a_3)^{\mathrm{T}}. \tag{2.29}$$

That is to say a plane and a line perpendicular to each other should have the same homogeneous coordinates.

2.3.2 *Composite Operators* $(\cdot \times \cdot)$

Composite operators are used to calculate the homogeneous coordinates of the intersection point of two lines or the connection line of two points as shown in Fig. 2.10,

$$\boldsymbol{A} = (A_1, A_2, A_3)^{\mathrm{T}} = \boldsymbol{b} \times \boldsymbol{d} = \left(\begin{vmatrix} b_2 & b_3 \\ d_2 & d_3 \end{vmatrix}, \begin{vmatrix} b_3 & b_1 \\ d_3 & d_1 \end{vmatrix}, \begin{vmatrix} b_1 & b_2 \\ d_1 & d_2 \end{vmatrix} \right)^{\mathrm{T}} \tag{2.30}$$

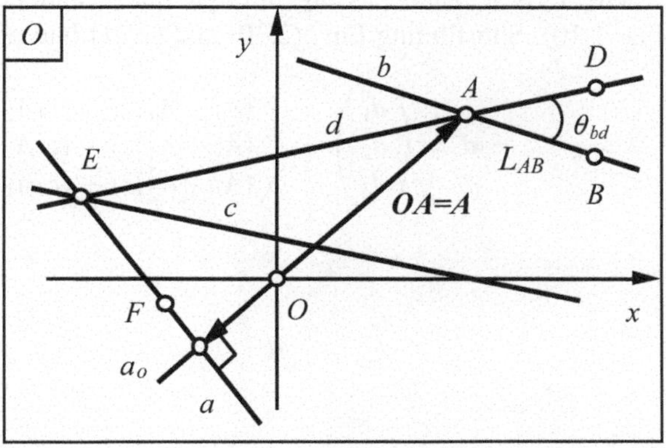

Fig. 2.10 The operators on projective plane and its geometrical meaning

$$\boldsymbol{a} = (a_1, a_2, a_3)^{\mathrm{T}} = \boldsymbol{E} \times \boldsymbol{F} = \left(\begin{vmatrix} E_2 & E_3 \\ F_2 & F_3 \end{vmatrix}, \begin{vmatrix} E_3 & E_1 \\ F_3 & F_1 \end{vmatrix}, \begin{vmatrix} E_1 & E_2 \\ F_1 & F_2 \end{vmatrix} \right)^{\mathrm{T}} \tag{2.31}$$

Combining the above two operators we have

$$\boldsymbol{E} = \boldsymbol{a} \times \boldsymbol{c} = \perp(\boldsymbol{A}) \times \boldsymbol{c} = \boldsymbol{A} \times \boldsymbol{c} \tag{2.32}$$

Expressing a space plane E which is passing through line c and perpendicular to plane A, and

$$\boldsymbol{b} = \boldsymbol{A} \times \boldsymbol{B} = \perp(\boldsymbol{a}) \times \boldsymbol{B} = \boldsymbol{a} \times \boldsymbol{B} \tag{2.33}$$

Expressing a space plane b which is in plane B and perpendicular to line a. Substituting Eq. (2.28) into Eq. (2.30) results in

$$\boldsymbol{a} = \boldsymbol{b} \times \boldsymbol{d}, \tag{2.34}$$

Which represents a space line $\boldsymbol{a}$ perpendicular to both $\boldsymbol{b}$ and $\boldsymbol{d}$. Substituting Eq. (2.29) into Eq. (2.31) results in

$$\boldsymbol{A} = \boldsymbol{E} \times \boldsymbol{F}, \tag{2.35}$$

Which represents a space plane $\boldsymbol{A}$ perpendicular to both $\boldsymbol{E}$ and $\boldsymbol{F}$.

2.3.3 *Projection Operator* $(\cdot \rightarrow \cdot)$

Projection operator is used to calculate the homogeneous coordinates of the orthographic projection $\boldsymbol{d}$ of line $\boldsymbol{c}$ onto plane $\boldsymbol{A}$.

$$\boldsymbol{d} = \boldsymbol{c} \rightarrow \boldsymbol{A} = \boldsymbol{E} \times \boldsymbol{A} = (\boldsymbol{a} \times \boldsymbol{c}) \times \boldsymbol{A} = (\perp \boldsymbol{A}) \times \boldsymbol{c} \times \boldsymbol{A} = (\boldsymbol{A} \times \boldsymbol{c}) \times \boldsymbol{A}. \tag{2.36}$$

where, E is a plane passing through line c and perpendicular to plane A (see Fig. 2.10). Substituting Eqs. (2.30) and (2.31) into above equation results in

$$\boldsymbol{d} = \begin{pmatrix} d_1 \\ d_2 \\ d_3 \end{pmatrix} = \begin{pmatrix} (A_2^2 + A_3^2)c_1 - (c_1 A_2 + c_3 A_3) A_1 \\ (A_1^2 + A_3^2)c_2 - (c_1 A_1 + c_3 A_3) A_2 \\ (A_1^2 + A_2^2)c_3 - (c_1 A_1 + c_2 A_2) A_3 \end{pmatrix}, \tag{2.37}$$

2.3.4 Moving Operator and Its Matrix Representation (Θ)

Moving operators are used to calculate the homogeneous coordinates of an image point after the space plane rotating around an axis by an angle. If the original position of the image point is $\boldsymbol{A} = (A_1, A_2, A_3)^{\mathrm{T}}$ and after rotating it moves to $\boldsymbol{A}' = (A_1', A_2', A_3')^{\mathrm{T}}$, then we have

$$\boldsymbol{A}' = \boldsymbol{\Theta}(\boldsymbol{A}). \tag{2.38}$$

Moving operator Θ may be expressed as 3 × 3 matrix. When space plane **A** about axis x rotates by an angle γ, moving operator is

$$\boldsymbol{\Theta} = \boldsymbol{X}(\gamma) = \begin{pmatrix} 1 & 0 & 0 \\ 0 & \cos\gamma & -\sin\gamma \\ 0 & \sin\gamma & \cos\gamma \end{pmatrix} \tag{2.39}$$

When space plane **A** around axis y rotates by an angle γ, moving operator is

$$\boldsymbol{\Theta} = \boldsymbol{Y}(\gamma) = \begin{pmatrix} \cos\gamma & 0 & \sin\gamma \\ 0 & 1 & 0 \\ -\sin\gamma & 0 & \cos\gamma \end{pmatrix}; \tag{2.40}$$

When space plane **A** about axis z rotates by an angle γ, moving operator is

$$\boldsymbol{\Theta} = \boldsymbol{Z}(\gamma) = \begin{pmatrix} \cos\gamma & -\sin\gamma & 0 \\ \sin\gamma & \cos\gamma & 0 \\ 0 & 0 & 1 \end{pmatrix}. \tag{2.41}$$

Knowing the above three basic operators we can deduce the moving operator when space plane **A** rotates around arbitrary axis. From Fig. 2.8b we have

$$\Theta = \mathbf{X}(-\gamma)\,\mathbf{Y}(\theta)\,\mathbf{X}(\gamma), \tag{2.42}$$

And substituting (2.37) and (2.38) into (2.40), we have

$$\Theta = \begin{pmatrix} \cos\theta & \sin\theta\sin\gamma & \sin\theta\cos\gamma \\ -\sin\theta\sin\gamma & \cos^2\gamma + \sin^2\gamma\cos\theta & -\sin\gamma\cos\gamma + \sin\gamma\cos\gamma\cos\theta \\ -\sin\theta\cos\gamma & -\cos\gamma\sin\gamma + \cos\gamma\sin\gamma\cos\theta & \sin^2\gamma + \cos\theta\cos^2\gamma \end{pmatrix}. \tag{2.43}$$

The moving operators above described can be also imposed on the homogeneous coordinates of image lings.

2.3.5 *The Invariance of Incidence Relationship Between Image Elements*

Let's see Fig. 2.10, image point A is on image line b, for example, this relation is called "incidence", which is denoted by $\boldsymbol{A} \in \boldsymbol{b}$. This relation will unchanged if both $\boldsymbol{A}$ and $\boldsymbol{b}$ are imposed by same moving operator. This fact can be mathematically described as follows: If $\boldsymbol{A} \in \boldsymbol{b}$, and $\boldsymbol{A}' = \boldsymbol{\Theta}(\boldsymbol{A})$, $\boldsymbol{b}' = \boldsymbol{\Theta}(\boldsymbol{b})$ then $\boldsymbol{A}' \in \boldsymbol{b}'$, or, $\boldsymbol{\Theta}(\boldsymbol{A} \in \boldsymbol{b}) = \boldsymbol{\Theta}(\boldsymbol{A}) \in \boldsymbol{\Theta}(\boldsymbol{b}) = \boldsymbol{A}' \in \boldsymbol{b}'$, this means that symbols "Θ" and "$\in$" can change their imposing order and operator "Θ" for operator "$\in$" satisfies distributive law.

Composite operators are, actually, showing some incidence relations, for example, $\boldsymbol{A} = \boldsymbol{b} \times \boldsymbol{d}$ expresses $(\boldsymbol{A} \in \boldsymbol{b}) \cap (\boldsymbol{A} \in \boldsymbol{d})$ and $\boldsymbol{b} = \boldsymbol{A} \times \boldsymbol{B}$ expresses $(\boldsymbol{A} \in \boldsymbol{b}) \cap (\boldsymbol{B} \in \boldsymbol{b})$, so we have

$$\boldsymbol{\Theta}(\boldsymbol{A}) = \boldsymbol{\Theta}(\boldsymbol{b} \times \boldsymbol{d}) = \boldsymbol{\Theta}(\boldsymbol{b}) \times \boldsymbol{\Theta}(\boldsymbol{d}) = \boldsymbol{b}' \times \boldsymbol{d}' = \boldsymbol{A}', \tag{2.44}$$

$$\boldsymbol{\Theta}(\boldsymbol{b}) = \boldsymbol{\Theta}(\boldsymbol{A} \times \boldsymbol{B}) = \boldsymbol{\Theta}(\boldsymbol{A}) \times \boldsymbol{\Theta}(\boldsymbol{B}) = \boldsymbol{A}' \times \boldsymbol{B}' = \boldsymbol{b}'. \tag{2.45}$$

This means that symbols "Θ" and "$\times$" can change their imposing order and operator "Θ" for operator "$\times$" satisfies distributive law.

2.4 Gnomonic Projection and Hyperbolic Coordinate Grid

As shown in Fig. 2.11a, by means of gnomonic projection the longitude lines and the latitude lines on the spherical surface are projected onto projective plane becoming a bunch of parallel lines and a bunch of hyperbolic curves respectively. They form hyperbolic coordinate grid on projective plane (Fig. 2.11b), using this grid we can conveniently and promptly solve the problems of analyzing and calculating space angles.

One of the main applications of hyperbolic coordinate grid is to measure the distance S_{AB} between two image points A and B (i.e. the angle between two space planes A and B, see Fig. 2.12). To do this, we should print the grid on a transparent sheet at the same scale as image graphic drawing on the projective plane (i.e. $\tan 45° = 1$ must equal for the both). The origins of the grid and the projective plane must coincide with each other, and the axis y of the grid must be perpendicular to the connecting line of image points A and B. The two hyperbolas passing through

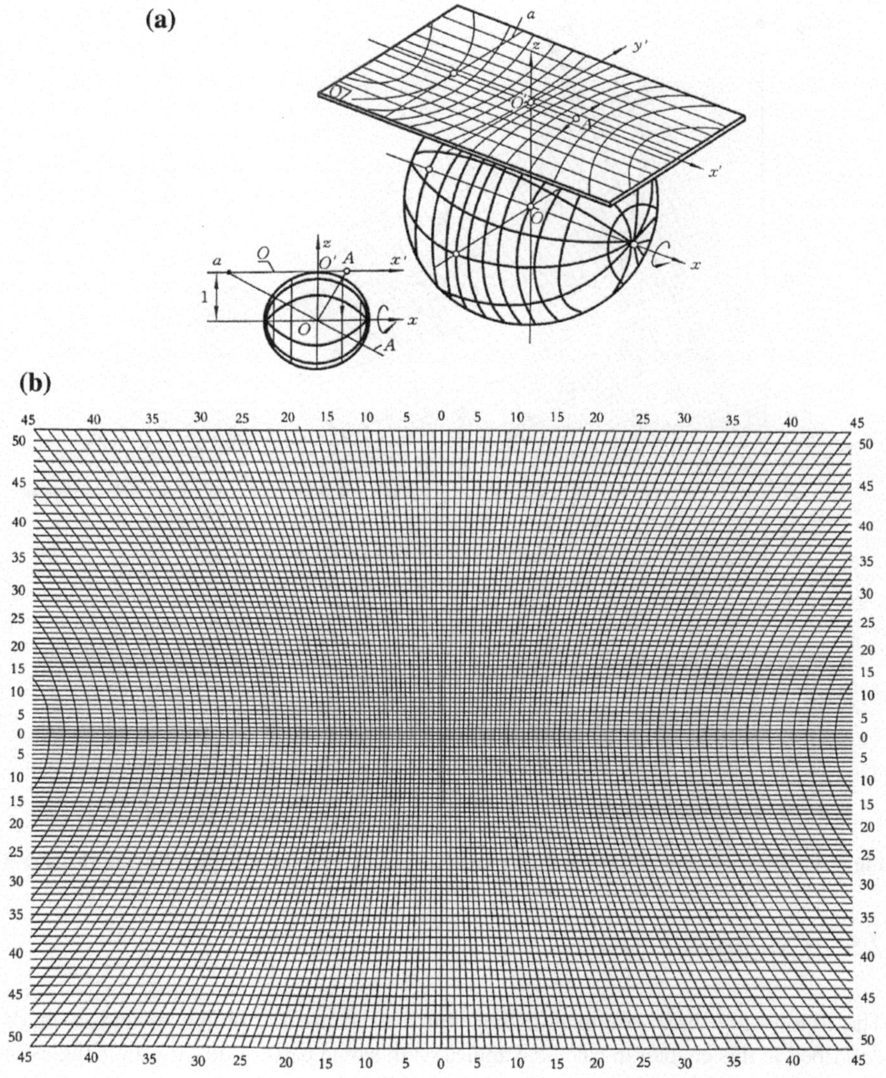

Fig. 2.11 The hyperbolic coordinate grid. **a** The generation of hyperbolic coordinate grid. **b** The hyperbolic coordinate grid generated

the image points A and B respectively are h_1 and h_2. The difference of latitude of the hyperbolas $(c_1 - c_2)$ is the distance $\tan^{-1}(S_{AB})$, or the angle between two space planes A and B. As an example, in Fig. 2.12, $c_1 = 26°$ and $c_2 = -4°$, so distance $\tan^{-1}(S_{AB}) = (c_1 - c_2) = 30°$.

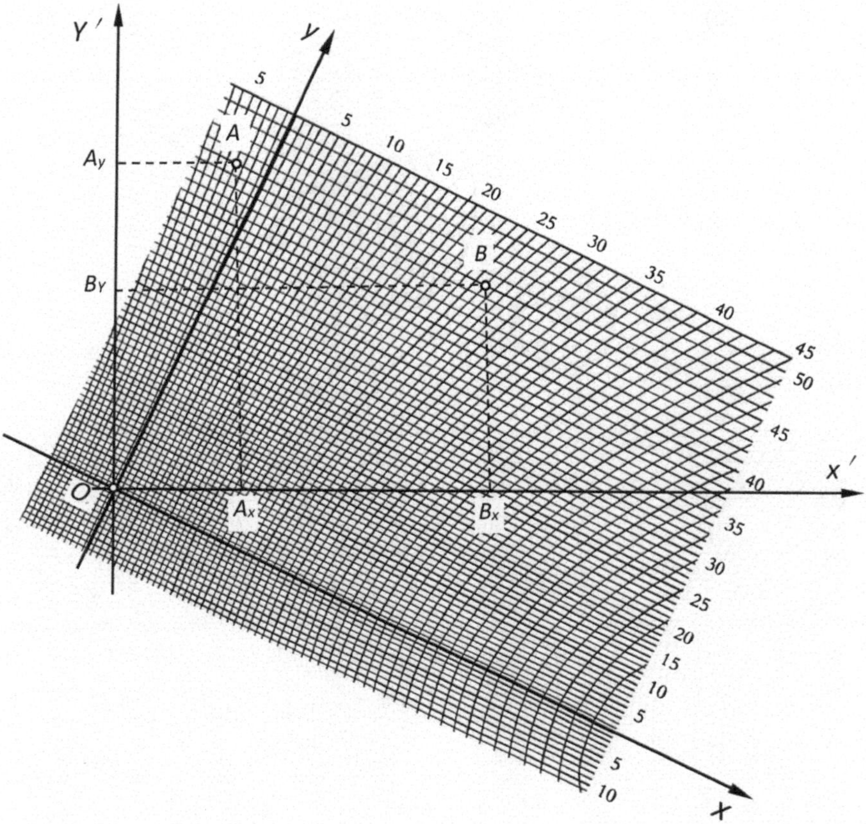

Fig. 2.12 Measurement of "distance" on projective plane by means of hyperbolic coordinate grid

2.5 Examples

The follows are three examples telling you how to use the principles and methods described in this chapter to calculate angles on machine parts or the adjusting angles of fixtures in machining processes. The application of the principles and methods in the angle analysis and calculating for cutting tools will be introduced in next chapter.

Example 2.1 There is a strengthening rib at angle steel as shown in Fig. 2.13, determine the space angle ω and the angle θ between the rib surface P and the angle steel surfaces H surfaces of the angle steel.

Solution 1: The problem can be solved through homogeneous coordinates and Eq. (2.16). Set xyz coordinate system as shown in Fig. 2.13, and take xOy plane as the base plane. The obliquity vectors of line a and line b are $\boldsymbol{p}_a$ and $\boldsymbol{p}_b$ respectively, the images of line a and b are the lines passing through the end points and perpendicular to $\boldsymbol{p}_a$ and $\boldsymbol{p}_b$ respectively. The equation of image line a is $x = \tan 54°$, or,

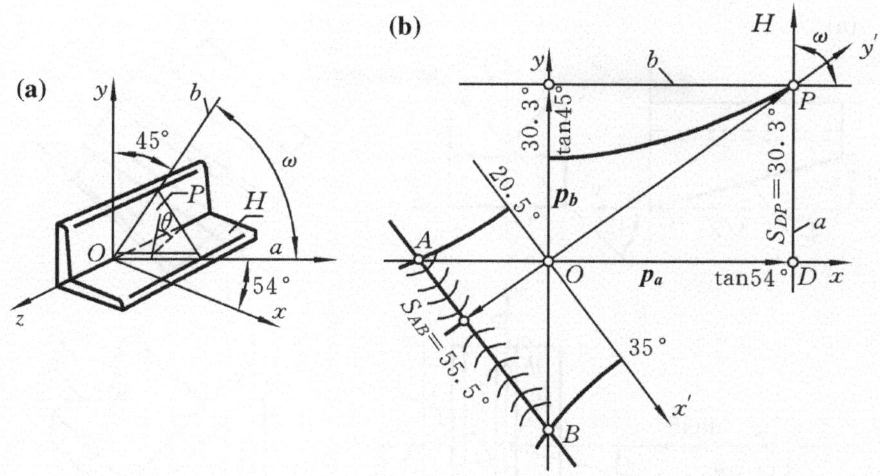

Fig. 2.13 Finding spatial angles on angle steel by calculating and measuring. **a** Angle steel and strengthening rib. **b** Vector diagram on projective plane

$1.x + 0.y - \tan 54° = 0$, so its homogeneous coordinates is $\boldsymbol{a} = (1, 0, -\tan 54°)^{\mathrm{T}}$. Similarly, the equation of image line b is $x = \tan 54°$, or, $1.x + 0.y - \tan 54° = 0$, so its homogeneous coordinates are $\boldsymbol{b} = (0, 1, -\tan 45°)^{\mathrm{T}}$. Substituting $\boldsymbol{a}$ and $\boldsymbol{b}$ into Eq. (2.16) results in $\tan\omega = \tan 55.10°$, or $\omega = 55.10°$ (the angle between line a and b on projective plane looks like a right angle from the view point of Euclidean geometry, but remember, here is non-Euclidean geometry).

The intersection point of image lines a and b is the image point of rib surface P, from Eq. (2.4) we know the homogeneous coordinates of rib surface are $\boldsymbol{P} = (\tan 54°, -\tan 45°, 1)^{\mathrm{T}}$, and the image of surfaces H surfaces of the angle steel is at the infinite point in direction of axis y, so its homogeneous coordinates are $\boldsymbol{H} = (0, 1, 0)^{\mathrm{T}}$, Substituting $\boldsymbol{P}$ and $\boldsymbol{H}$ into Eq. (2.16) results in $\theta = 59.55°$.

Solution 2: The problem can be also solved by measuring with hyperbolic coordinate grid. To do this, we should first find the image points of the normal planes A and B of image lines a and b (see Fig. 2.4c and the text in Sect. 2.1.4.3), and measure the distance between A and B with hyperbolic coordinate grid (see Figs. 2.12 and 2.13), the result is $\omega = \theta_{ab} = S_{AB} = 55.5°$.

Be aware that $\theta = S_{PH} = S_{DH} - S_{DP}$, where S_{DH} is the distance between points D and H, point H is at the infinite point of line a, so S_{DH} is the length of half line, as shown in Fig. 2.5 and described in Sect. 2.1.4.5, the length of whole line is π, so $S_{DH} = \pi/2 = 90°$, and $\theta = S_{PH} = 90° - S_{DP}$. Measured on Fig. 2.13b by hyperbolic coordinate grid that $S_{DP} = 30.3°$, so $\theta = S_{PH} = 90° - S_{DP} = 90° - 30.3 = 59.7°$.

Example 2.2 There is a V shape block, as shown in Fig. 2.14a, the working plane Q to be ground and during grinding process bottom plane F is locating surface and lateral plane P is guiding plane. The clamping way is shown in

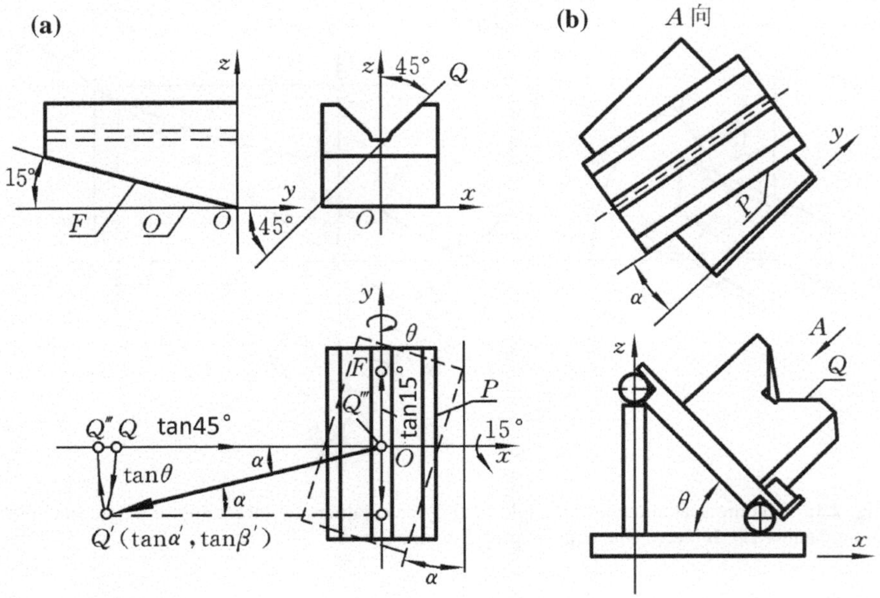

Fig. 2.14 The angle adjustment while grinding working surface of a V shape block. **a** The working drawing of a V shape block. **b** The adjustment of fixture

Fig. 2.14b. It is required to find the guiding angle α on plane O and the adjusting angle θ of the sine bar.

Solution: The initial position of the V shape block is shown in Fig. 2.14a, take xOy plane (top view) as the base plane and draw the images of Q and F on the projective plane O. The task is to move Q to the origin O, (i.e. to rotate plane Q to the horizontal position in the space). Such a problem can be solved by means of the method of image trajectory method as described in Sect. 2.2.2 and Fig. 2.8.

First, move F to O, this means to rotate plane F around axis x by 15° in the space. Because of this rotation Q moves to Q'; Second, move Q' back to axis x, becoming Q'', this means to rotate plane Q around axis z by an angle α, this is the guiding angle α; Third, move Q'' to O, becoming Q''', this means to rotate plane Q around axis y by an angle θ, this is the adjusting angle of the sine bar. From the image trajectory shown in Fig. 2.14a we immediately have,

$$\tan\beta' = -\tan 15^\circ$$

$$\tan\alpha' = -\tan 45^\circ/\tan 15^\circ$$

$$\tan\alpha = \tan\beta'/\tan\alpha' = \tan 14^\circ 3' 38'',$$

$$\tan\theta = \sqrt{\tan^2\alpha' + \tan^2\beta'} = \tan 46^\circ 55' 13''$$

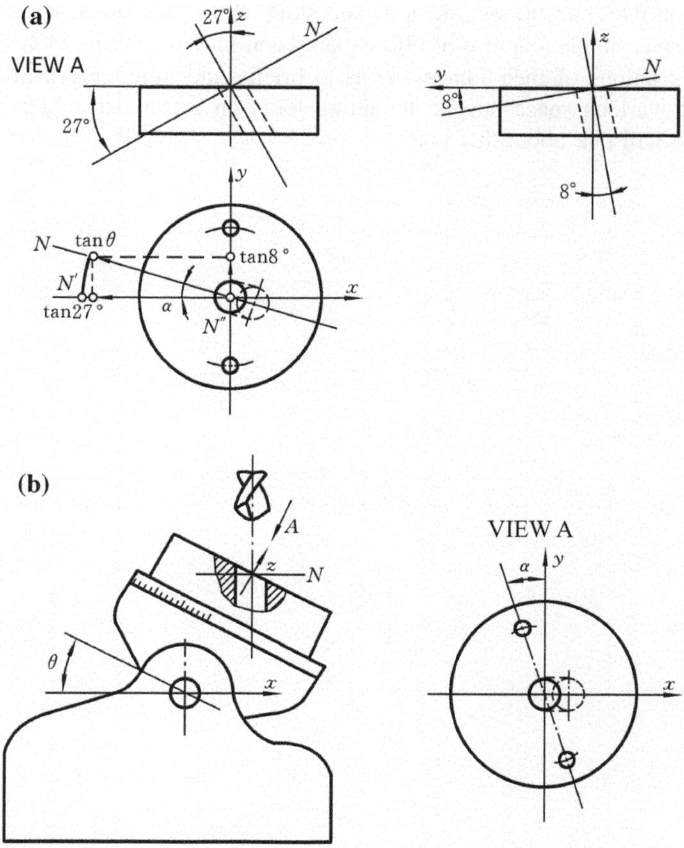

Fig. 2.15 Calculation of adjustment angles of the fixture while drilling an inclined hole. **a** The work piece and the image trajectory diagram. **b** The fixture and its adjustment

Example 2.3 An inclined hole of a part is drilling in a universal fixture as shown in Figs. 2.14a, b. It is required to calculate adjustment angles α and θ of the fixture.

Solution: Take xOy plane as the base plane and draw the image $N(-\tan 27°, \tan 8°)$ of the normal plane to the hole axis on the top view in Fig. 2.15a. What we should do is to rotate N about axes z and y in succession, so as to bring N to the origin (i.e. to move plane N to horizontal position). The adjustment angles α and θ of the fixture can be calculated from the plane vector diagram (Fig. 2.15b),

$$\tan \alpha = \frac{\tan 8°}{\tan 27°} = \tan 15°25'13'',$$

$$\tan \theta = \sqrt{\tan^2 27° + \tan^2 8°} = \tan 27°51'32''$$

Based on the conceptions, methods and skills above described, we are able to transform space angle relations of tilting planes and lines on cutting tools into plane geometry relations of their images so as to briefly and intuitively analyzing and calculating various space angles of cutting tools on a projective plane. In next chapter we will talk about this issue.

Chapter 3
The Way of Analyzing and Calculating Cutting Tool Angles in a Projective Plane

In this chapter, the space planes and lines of various cutting tools are mapped onto a plane, so-called projective plane, and then the angles of the cutting tools are analyzed and calculated through the images on the plane by means of the method introduced in last chapter. A series of examples presented in this chapter has shown the validity and visualization of such planar image method. This method explains the profound in straightforward terms, and turns difficult to easy. It can visually synthesize the relationship between the various geometric angles and it's straightforward to analyze the influence of the rotation of the cutting tool in space on its angles. This method has provided a mathematical model and a theoretical basis for the analysis and calculation of tool working angles and dynamic angles and for their on-line monitoring and optimizing design.

3.1 Synthesis of Single-Point Cutting Tool Geometry Angles

Single-point cutting tools are not only widely used in turning, planning, but also the elements which compose other complex shaped cutting tools, such as milling cutters, hole machining cutters, broach and gear cutters. In this section, the rake face, the flank face, the tool minor flank, the major cutting edge, the **minor cutting edge** and the related coordinate planes of a single-point cutting tool are mapped onto a projective plane, and then various geometric angles of the cutter are analyzed and synthesized through the images in the projective plane.

H. Shi, *Metal Cutting Theory*, Springer Series in Advanced Manufacturing,
https://doi.org/10.1007/978-3-319-73561-0_3

3.1.1 Planar Vector Diagram for Calculating Geometric Angles of Cutting Tools

Using planar obliquity vector to represent the geometric angle of a cutting tool, the relationship between various angles in various sections becomes very intuitive, so that the derivation of the angle conversion formula becomes very simple and rapid.

3.1.1.1 The Angles Related to Rake Face

See Fig. 3.1, the reference plane P_r is taken as the base plane, its image on the projective plane is the coordinate origin O. According to the definition [1, 62–64], tool orthogonal rake $\tan\gamma_o$ and tool cutting edge inclination $\tan\lambda_s$ determine the obliquities of rake face A_γ in the directions of tool orthogonal plane r and the direction of tool cutting edge plane t, respectively. And the above two, as coordinates, determine the position of the image of rake face A_γ , and the obliquity vector of image point A_γ is:

$$\boldsymbol{A}_\gamma = \boldsymbol{OA}_\gamma = \tan\gamma_o\mathbf{r} + \tan\lambda_s\mathbf{t} \tag{3.1}$$

where $\mathbf{r}$ and $\mathbf{t}$ are the unit vectors in the directions r and t. The maximum rake (i.e. the biggest obliquity of rake surface) is

$$|\boldsymbol{A}_\gamma| = A_\gamma = \tan\gamma_g = \sqrt{\tan^2\lambda_s + \tan^2\gamma_o}. \tag{3.2}$$

The projection of obliquity vector $\boldsymbol{A}_\gamma$ at any direction gives the rake in the section of that direction and perpendicular to the reference plane P_r. Based on this, we can immediately wright the formulas of various rake angle in terms of γ_o and λ_s. For example, tool orthogonal rake for minor cutting edge $\tan\gamma_o'$ is the projection of $\boldsymbol{A}_\gamma$ at the direction $\mathbf{r}'$ of tool orthogonal plane for minor cutting edge, so

$$\tan\gamma_o' = \boldsymbol{A}_\gamma\cdot\mathbf{r}' = (\tan\gamma_o\mathbf{r} + \tan\lambda_s\mathbf{t})\cdot\mathbf{r}' = \tan\gamma_o(\mathbf{r}\cdot\mathbf{r}') + \tan\lambda_s(\mathbf{t}\cdot\mathbf{r}'),$$

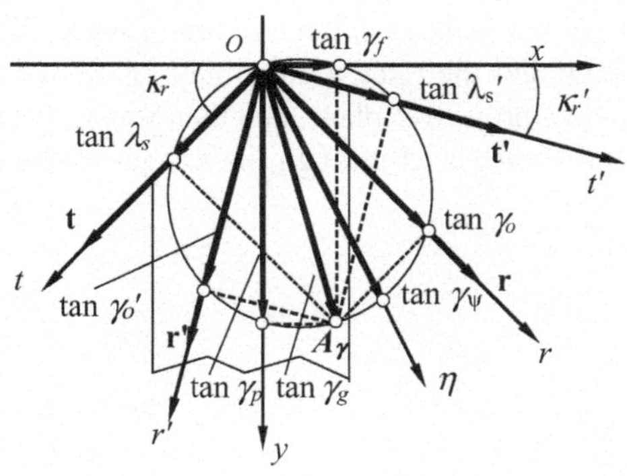

Fig. 3.1 The angles related to rake face

where, $\boldsymbol{A}_\gamma \cdot \mathbf{r}'$ means the inner product, or, the projection of $\boldsymbol{A}_\gamma$ on the direction of $\mathbf{r}'$, and the same below. So, we have

$$\tan \gamma_o{}' = \tan \gamma_o \cos(\kappa_r + \kappa_r{}') + \tan \lambda_s \sin(\kappa_r + \kappa_r{}'), \tag{3.3}$$

where κ_r and $\kappa_r{}'$ are the tool cutting edge angle and the tool minor cutting edge angle, respectively (see Fig. 3.1).

Tool minor cutting edge inclination is the projection of $\boldsymbol{A}_\gamma$ at the direction $\mathbf{t}'$ of tool cutting edge plane for minor cutting edge,

$$\tan \lambda_s{}' = \boldsymbol{A}_\gamma \cdot \mathbf{t}' = (\tan \gamma_o \mathbf{r} + \tan \lambda_s \mathbf{t}) \cdot \mathbf{t}' = \tan \gamma_o (\mathbf{r} \cdot \mathbf{t}') + \tan \lambda_s (\mathbf{t} \cdot \mathbf{t}'),$$

or,

$$\tan \lambda_o{}' = \tan \gamma_o \sin(\kappa_r + \kappa_r{}') - \tan \lambda_s \cos(\kappa_r + \kappa_r{}'). \tag{3.4}$$

Generally speaking, rays at various directions cut by the obliquity circle of $\boldsymbol{A}_\gamma$ (the chords) give the rakes at various sections, such as $\tan \gamma_o$ in tool orthogonal plane, $\tan \lambda_s$ in tool cutting edge plane, $\tan \gamma_f$ in assumed working plane, $\tan \gamma_p$ in tool back plane, $\tan \gamma_o{}'$ in tool orthogonal plane for minor cutting edge, $\tan \lambda_s{}'$ in tool minor cutting edge plane and $\tan \gamma_\psi$ in chip ejection direction, etc.

Because at the direction of $\boldsymbol{A}_\gamma$ we have the maximum rake angle γ_g (this means that in the direction the rake surface has the steepest slope), so during cutting process, if $\lambda_s \neq 0$ the chip-ejection direction η always deviates from the direction r of tool orthogonal plane and slip to the side of $\boldsymbol{A}_\gamma$, and this results in $\tan \gamma_\psi > \tan \gamma_o$. This shows vividly that tool cutting edge inclination λ_s, no matter positive or negative, pays a role of increasing the rake γ_ψ in the direction of chip ejection and makes the cutting process more light.

3.1.1.2 The Angles Related to Flank

See Fig. 3.2, the obliquity of flank in the direction $\mathbf{t}$ of tool cutting edge inclination is also $\tan \lambda_s$ the same as for rake surface. This is because the intersection line of flank with the rake, the main cutting edge is in this direction. The obliquity of flank in the direction $\mathbf{r}$ of tool orthogonal plane is

$$\tan(90^\circ - \alpha_o) = \cot \alpha_o,$$

where α_o is tool orthogonal clearance. The above two as Cartesian coordinates determine the obliquity vector of flank

$$\boldsymbol{A}_\alpha = \tan \lambda_s \mathbf{t} + \cot \alpha_o \mathbf{r}$$

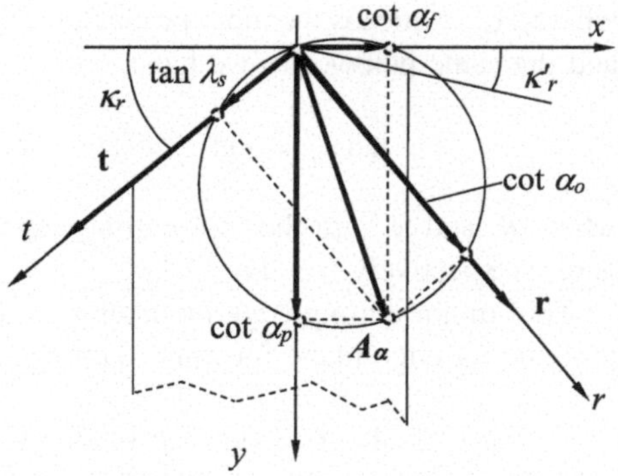

Fig. 3.2 The angles related to flank face

Also we can draw the obliquity circle of $\boldsymbol{A_\alpha}$ as shown in Fig. 3.2, and get tool side clearance α_f and tool back clearance α_p indicated by the intersection points of obliquity circle with axes x and y, respectively.

The coordinates of image points of rake and flank surfaces A_γ, A_α in coordinate systems t, r and x, y are connected by the formula of coordinate rotation,

$$\begin{pmatrix} \tan\gamma_f & \cot\alpha_f \\ \tan\gamma_p & \cot\alpha_p \end{pmatrix} = \begin{pmatrix} -\cos\kappa_r & \sin\kappa_r \\ \sin\kappa_r & \cos\kappa_r \end{pmatrix} \begin{pmatrix} \tan\lambda_s & \tan\lambda_s \\ \tan\gamma_o & \cot\alpha_o \end{pmatrix}, \tag{3.5}$$

This formula summarizes four formulae for calculating γ_f, γ_p, α_f and α_p.

The coordinates ($\tan\lambda_s'$, $\cot\alpha_o'$) of minor flank A_α' in system of tool cutting edge plane for minor cutting edge and tool orthogonal plane for minor cutting edge (t', r'), and the coordinates ($\cot\alpha_f'$, $\cot\alpha_p'$) in (x, y) system are also connected by the formula of coordinate rotation,

$$\begin{pmatrix} \cot\alpha_f' \\ \cot\alpha_p' \end{pmatrix} = \begin{pmatrix} \cos\kappa_r' & -\sin\kappa_{r1}' \\ \sin\kappa_{r1}' & \cos\kappa_{r1}' \end{pmatrix} \begin{pmatrix} \tan\lambda_s' \\ \cot\alpha_o' \end{pmatrix}, \tag{3.6}$$

where λ_s', α_o' are tool minor cutting edge inclination and tool orthogonal clearance for minor cutting edge, respectively, and α_f', α_p' are tool side clearance for minor cutting edge and tool back clearance for minor cutting edge, respectively.

3.1.1.3 Tool Geometry Matrix

In summary, the working part of a single-point cutting tool can be imaged as a triangle $\Delta A_\gamma A_\alpha A_\alpha'$ (see Fig. 3.3), the three vertexes A_γ, A_α, A_α' are the images of tool face, major flank and minor flank, respectively. Line $A_\gamma A_\alpha$ is the image of the intersection line of tool face and major flank, i.e. the major cutting edge s, which is perpendicular to the projection t of major cutting edge on reference plane, and line A_γ A_α' is the image s' of the intersection line of tool face and minor flank, i.e. the

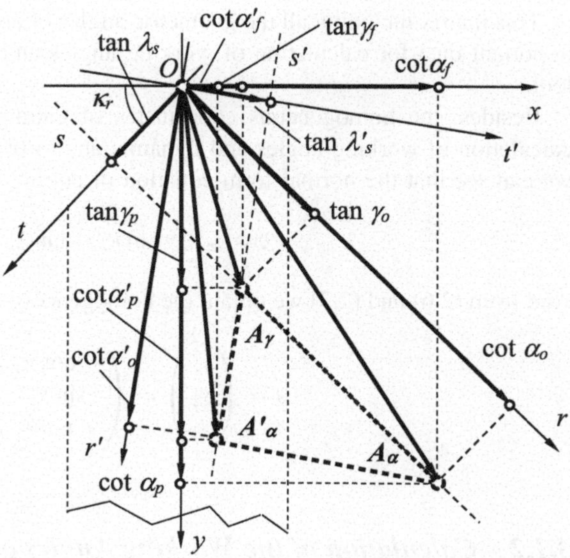

Fig. 3.3 Synthesis of the angles on a single-point cutting tool

minor cutting edge, which is perpendicular to the projection t' of minor cutting edge on reference plane. For convenience of calculation, we should use homogeneous coordinates. According to Formulae (2.1)–(2.4), we get the homogeneous coordinates of tool face, major flank and minor flank as follows,

$$\boldsymbol{A}_\gamma = \begin{pmatrix} A_{\gamma 1} \\ A_{\gamma 2} \\ A_{\gamma 3} \end{pmatrix} = \begin{pmatrix} \tan\gamma_f \\ \tan\gamma_p \\ 1 \end{pmatrix},$$
$$\boldsymbol{A}_\alpha = \begin{pmatrix} A_{\alpha 1} \\ A_{\alpha 2} \\ A_{\alpha 3} \end{pmatrix} = \begin{pmatrix} \cot\alpha_f \\ \cot a_p \\ 1 \end{pmatrix}, \tag{3.7}$$
$$\boldsymbol{A}'_\alpha = \begin{pmatrix} A'_{\alpha 1} \\ A'_{\alpha 2} \\ A'_{\alpha 3} \end{pmatrix} = \begin{pmatrix} \cot{\alpha_f}' \\ \cot{\alpha_p}' \\ 1 \end{pmatrix},$$

The homogeneous coordinates of the three points form tool geometry matrix,

$$\boldsymbol{J} = \begin{pmatrix} A_{\gamma 1} & A_{\alpha 1} & A'_{\alpha 1} \\ A_{\gamma 2} & A_{\alpha 2} & A'_{\alpha 2} \\ A_{\gamma 3} & A_{\alpha 3} & A'_{\alpha 3} \end{pmatrix} = \begin{pmatrix} \tan\gamma_f & \cot\alpha_f & \cot{\alpha_f}' \\ \tan\gamma_p & \cot a_p & \cot{\alpha_p}' \\ 1 & 1 & 1 \end{pmatrix}. \tag{3.8}$$

This matrix including all the geometric angles of a single-point cutting tool is the important data for calculation of working angles and dynamic angles of a cutting tool.

Besides, the homogeneous coordinates of cutting edge is also important for calculation of working angles and dynamic angles of a cutting tool. From Fig. 3.1 we can see that the normal form equation of cutting edge in (x, y) system is

$$-x \cos \kappa_r + y \sin \kappa_r - \tan \lambda_s = 0, \tag{3.9}$$

And from (2.6) and (2.7) we obtain the homogeneous coordinates of cutting edge is

$$s = \begin{pmatrix} s_1 \\ s_2 \\ s_3 \end{pmatrix} = \begin{pmatrix} -\cos \kappa_r \\ \sin \kappa_r \\ -\tan \lambda_s \end{pmatrix}. \tag{3.10}$$

3.1.2 *Calculation of the Working Angles or Dynamic Angles of Cutting Tools by Means of Movement Operators*

The problem of calculating working (dynamic) angles of a cutting tool from its grinding (static) angles come down to the movement of its image graphic. Such a problem may be solved by "moving operators", that is to say, to calculate working (dynamic) angles of a cutting tool from its geometry matrix multiplied by a proper moving operator.

3.1.2.1 Calculation of Working Angles in *X*, *Y* (Tool Assumed Working Plane and Tool Back Plane) System

In the coordinate system (x, y) it is convenient to calculate the variation in working angles caused by the error in cutting tool installation or feed motion. There are two typical cases: the first case is the cutting tool point installed higher than the hole central line by a distance δ, while boring a hole as shown in Fig. 3.4a. In this case, working reference plane P_{re} and working plane P_{fe} rotate by an angle $-\mu$ (working back plane P_{pe} has no change). Take static reference plane P_r as base plane, the image point of working reference plane P_{re} is at the positive direction of axis y and apart from origin by tan μ. To transfer to working reference plane P_{re}, plane P_{re}, should rotate about axis x by an angle μ, so as to move to P_{re} to the origin. Together with plane P_{re}, the tool point should also perform the same rotation and its image diagram should be also imposed by the same moving operator.

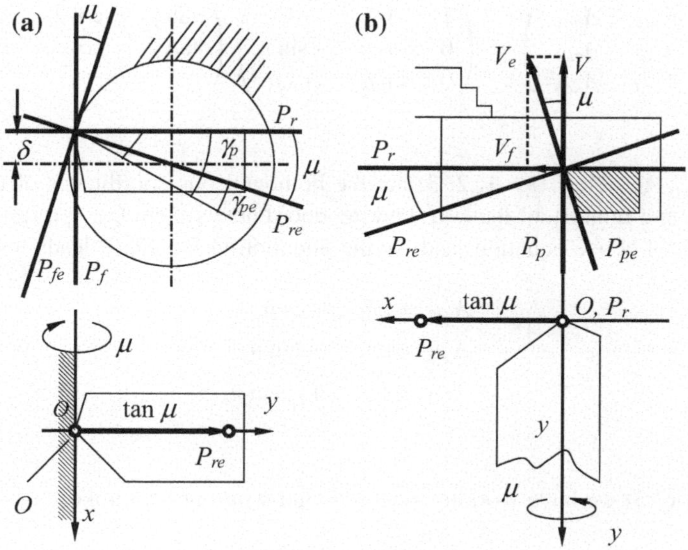

Fig. 3.4 Working angles (*x*, *y* coordinate system). **a** The cutting tool corner is higher than hole central line while boring a hole. **b** The effect of feed motion

$$\boldsymbol{\Theta} = \boldsymbol{X}(\boldsymbol{\mu}) = \begin{pmatrix} 1 & 0 & 0 \\ 0 & \cos\boldsymbol{\mu} & -\sin\boldsymbol{\mu} \\ 0 & \sin\boldsymbol{\mu} & \cos\boldsymbol{\mu} \end{pmatrix}$$

The second case is the effect of feed motion during threading, as shown in Fig. 3.4b. In this case, working reference plane P_{re} and working back plane P_{pe} rotate an angle μ (assumed working plane P_f has no change). For calculating the tool angles with respect to working reference plane P_{re}, it is necessary to rotate the tool together with planes P_{re} and P_{pe} around axis y by an angle μ, to perform this, the tool image diagram should be also imposed by the moving operator

$$\boldsymbol{\Theta} = \boldsymbol{Y}(-\boldsymbol{\mu}) = \begin{pmatrix} \cos\boldsymbol{\mu} & 0 & -\sin\boldsymbol{\mu} \\ 0 & 1 & 0 \\ \sin\boldsymbol{\mu} & 0 & \cos\boldsymbol{\mu} \end{pmatrix},$$

Now let's take the first case as an example to show how to design and calculate working angles. As described above, the tool geometry matrix left multiplied by moving operator results in working geometry matrix,

$$\begin{pmatrix} A_{\gamma e1} & A_{\alpha e1} & A_{\alpha e1}' \\ A_{\gamma e2} & A_{\alpha e2} & A_{\alpha e2}' \\ A_{\gamma e3} & A_{\alpha e3} & A_{\alpha e3}' \end{pmatrix} = \begin{pmatrix} 1 & 0 & 0 \\ 0 & \cos\mu & -\sin\mu \\ 0 & \sin\mu & \cos\mu \end{pmatrix} \begin{pmatrix} \tan\gamma_f & \cot\alpha_f & \cot\alpha_f' \\ \tan\gamma_p & \cot a_p & \cot\alpha_p' \\ 1 & 1 & 1 \end{pmatrix}, \tag{3.11}$$

where $A_{\gamma ei}$, $A_{\alpha ei}$, $A_{\alpha ei}'$ ($i = 1, 2, 3$) are the homogeneous coordinates of tool face, tool flank and minor tool flank in working coordinate system (x_e, y_e), respectively. Expansion of above equation, and paying attention to Eq. (2.7), leads to

$$\begin{aligned} A_{\gamma e1} &= \tan\gamma_{\mathrm{f}}/\cos\mu, \\ A_{\gamma e2} &= \tan\gamma_{\mathrm{p}} - \tan\mu, \\ A_{\gamma e3} &= 1 + \tan\gamma_{\mathrm{p}}\tan\mu, \\ &\ldots \end{aligned}$$

From Eq. (2.3) we have working side rake and working back rake,

$$\begin{aligned} \tan\gamma_{fe} &= A_{\gamma ex} = \frac{A_{\gamma e1}}{A_{\gamma e3}} = \frac{\tan\gamma_f}{\cos\mu\,(1 + \tan\gamma_p\tan\mu)}, \\ \tan\gamma_{pe} &= A_{\gamma ey} = \frac{A_{\gamma e2}}{A_{\gamma e3}} = \frac{\tan\gamma_p - \tan\mu}{1 + \tan\gamma_p\tan\mu} = \tan(\gamma_p - \mu), \end{aligned}$$

From second equation we get a simple relation between working back rake, tool back rake and the rotation angle μ,

$$\gamma_{pe} = \gamma_p - \mu,$$

This is easy to understand from Fig. 3.4a.

Expanding Eq. (3.11) we can also calculate $\cot\alpha_{fe}$, $\cot\alpha_{pe}$, $\cot\alpha_{fe}'$ and $\cot\alpha_{pe}'$, and determine the positions of images $A_{\gamma e}$, $A_{\alpha e}$, $A_{\alpha e}'$ in working coordinate system, and according to the geometric relations shown in Fig. 3.3 to determine all the working angles related to tool face, tool flank and minor tool flank.

The homogeneous coordinates of cutting edge in Eq. (3.10) left multiplied by moving operator results in the working angles related to the cutting edge, i.e. working cutting edge inclination λ_{se} and working cutting edge angle κ_{re},

$$\begin{pmatrix} s_{e1} \\ s_{e2} \\ s_{e3} \end{pmatrix} = \begin{pmatrix} 1 & 0 & 0 \\ 0 & \cos\mu & -\sin\mu \\ 0 & \sin\mu & \cos\mu \end{pmatrix} \begin{pmatrix} -\cos\kappa_r \\ \sin\kappa_r \\ -\tan\lambda_s \end{pmatrix}.$$

Expanding above equation results in

$$\begin{aligned} s_{e1} &= -\cos\kappa_r, \\ s_{e2} &= \sin\kappa_r\,\cos\mu + \tan\lambda_s\,\sin\mu, \\ s_{e3} &= \sin\kappa_r\,\sin\mu - \tan\lambda_s\,\cos\mu. \end{aligned} \tag{3.12}$$

Similar to Fig. 3.3, the equation of major cutting edge in working coordinate system (axes x_e, y_e, and coordinate planes P_{re}, P_{fe}, P_{pe}) must be

$$s_{e1}x_e + s_{e2}y_e + s_{e3} = 0,$$

And its normal form is

$$\frac{s_{e1}x_e + s_{e2}y_e + s_{e3}}{\pm\sqrt{s_{e1}^2 + s_{e2}^2}} = 0. \tag{3.13}$$

On the other hand, similar to Fig. 3.3 and Eq. (3.9), in working system (x_e, y_e) we also have equation

$$-x_e\,\cos\kappa_{re} + y_e\,\sin\kappa_{re} - \tan\lambda_{se} = 0.$$

Substituting Eq. (3.12) into (3.13) and comparing the resulted equation with above one, we can have working cutting edge angle

$$\tan\kappa_{re} = -\frac{s_{e2}}{s_{e1}} = \left(\tan\kappa_r + \frac{\tan\lambda_s\,\tan\mu}{\cos\kappa_r}\right)\cos\mu, \tag{3.14}$$

And working cutting edge inclination

$$\tan\lambda_{se} = \frac{s_{e3}}{\sqrt{s_{e1}^2 + s_{e2}^2}} = \frac{-\tan\lambda_s + \tan\mu\,\sin\kappa_r}{\sqrt{\left(\frac{\cos\kappa_r}{\cos\mu}\right)^2 + (\sin\kappa_r + \tan\lambda_s\,\tan\mu)^2}}. \tag{3.15}$$

3.1.2.2 Calculation of Dynamic Angles in *T*, *R* Coordinates (Tool Orthogonal Plane and Tool Cutting Edge Plane) System

In the coordinate system (*t*, *r*) it is convenient to calculate the instantaneous dynamic angles during cutting process. Take static reference plane P_r as the base, and its image *O* on projective plane is the coordinate origin.

Now let's build a working (dynamic) coordinate system on the projective plane. To do this, cutting edge inclination λ_s, which determines the orientation of the cutting tool, and the direction of its relative motion of the tool with respect to work piece should be known beforehand.

The image line of resultant cutting speed Suppose at a moment the projection of resultant cutting speed $\boldsymbol{V_e}$ on static reference plane is on the line of OA and the angle between $\boldsymbol{V_e}$ and P_r is θ (see Fig. 3.5b), then along OA draw a vector $\boldsymbol{OA}$ of the length tan θ, which is obliquity vector $\boldsymbol{p_{ve}}$ of the line on which resultant cutting speed is in (see Fig. 3.5a). Draw a line passing through the end point of $\boldsymbol{p_{ve}}$ and perpendicular the vector, which is the image line v_e of the direction of resultant cutting speed (see Fig. 3.5a).

The image of main cutting edge Suppose tool (static) cutting edge inclination is λ_s, then along the direction of projection of the major cutting edge t draw a vector $\boldsymbol{p_s}$ of the length tan λ_s, which is the obliquity vector of main cutting edge. Draw a line passing through the end point of vector $\boldsymbol{p_s}$ and perpendicular to the vector, which is the image of major cutting edge s.

The image point of working (dynamic) cutting edge plane The intersection point P_{se} of the image lines s and v_e is the image point of working (dynamic) cutting edge plane (see Fig. 3.5a).

The working (dynamic) reference plane The normal plane P_{re} of v_e is the working (dynamic) reference plane. According to Fig. 2.4c, and the description in Sect. 2.1.4.3, draw a vector $\boldsymbol{OP_{re}}$ of length $\overline{OP_{re}} = 1/(\overline{OA})$ in the opposite direction of $\boldsymbol{p_{ve}}$, we get point P_{re}, the image of working (dynamic) reference plane (see Fig. 3.5a).

The image of orthographic projection of main cutting edge the connection line l of P_{re} and P_{se} is the image of orthographic projection of main cutting edge, i.e. the image of intersection line of working (dynamic) reference plane and working (dynamic) cutting edge plane.

The image of working orthogonal plane The normal plane P_{oe} of l is the image of working orthogonal plane. It may be approved that the point P_{oe} will be certainly at the line v_e.

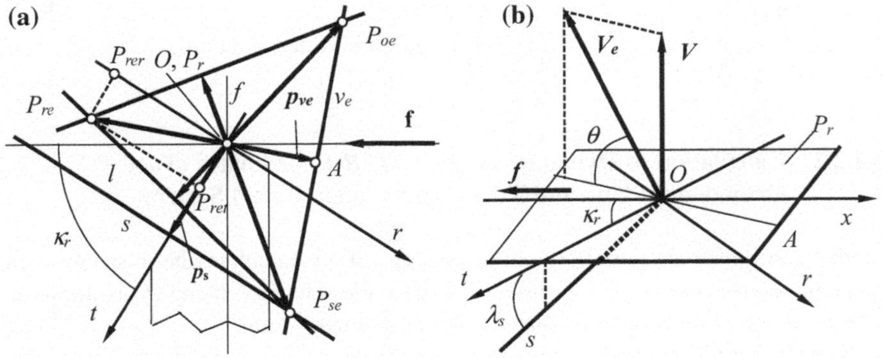

Fig. 3.5 Calculation of dynamic angles in (t, r) system. **a** The coordinate planes of tool in use system expressed in tool-in-hand system. **b** The resultant cutting speed

On the tool (static) reference plane, i.e. projection plane, three dynamic coordinate plane P_{oe}, P_{se} and P_{re} form a triangle, and the origin O is at the orthocenter of the triangle, as shown in Fig. 3.5a (Fig. 3.5a just for expressing the geometric relations of various geometric elements and doesn't need to be in strict proportion). It must noticed that the sum of interior angles of triangle $P_{oe}P_{se}P_{re}$ is 270°, which is bigger than 180° as in our familiar Euclidian geometry. We mentioned this in Sect. 2.1.2.

On static reference plane P_r (i.e. projective plane O) there is (t, r) coordinate system, where axes t, r are the intersection line of P_r with P_s and P_r with P_o respectively (Fig. 3.1). In the dynamic reference plane P_{oe} there also have coordinate system (t_e, r_e), where axes t_e, r_e are the intersection line of P_{re} with P_{se} and P_{re} with P_{oe} respectively. Our question is if the Cartesian coordinates (A_t, A_r) or homogeneous coordinates (A_1, A_2, A_3) of a space plane A in system (t, r) are known, then how to calculate its coordinates (A_{te}, A_{re}) or (A_{1e}, A_{2e}, A_{3e}) in the system (t_e, r_e). We also have the same question for image lines. Such problems may be solved through moving operators, and the way to obtain a moving operator will be briefly introduced as follows.

If the Cartesian coordinates (P_{ret}, P_{rer}) of the image of dynamic reference plane P_{re} in the tool-in-hand system (t, r) and the major cutting edge inclination $U = \tan \lambda_s$ are known, the homogeneous coordinates of dynamic coordinate planes P_{oe}, P_{se}, P_{re} can be deduced as follows on the basis of the plane geometry mentioned above and shown in Fig. 3.5a.

$$\boldsymbol{P}_{oe} = \begin{pmatrix} P_{oe1} \\ P_{oe2} \\ P_{oe3} \end{pmatrix} = \begin{pmatrix} -(1 + UP_{ret} + P_{rer}^2) \\ P_{rer}(P_{ret} - U) \\ UP^2 + P_{ret} \end{pmatrix},$$

$$\boldsymbol{P}_{se} = \begin{pmatrix} P_{se1} \\ P_{se2} \\ P_{se3} \end{pmatrix} = \begin{pmatrix} UP_{rer} \\ -(1 + UP_{ret}) \\ P_{rer} \end{pmatrix},$$

$$\boldsymbol{P}_{re} = \begin{pmatrix} P_{re1} \\ P_{re2} \\ P_{re3} \end{pmatrix} = \begin{pmatrix} P_{ret} \\ P_{rer} \\ 1 \end{pmatrix},$$

where $P^2 = P_{ret}^2 + P_{rer}^2$. After normalization of the above homogeneous coordinates and arrange them into a 3 × 3 matrix, we got the transformation matrix (or the moving operator)

$$\boldsymbol{\Theta} = \begin{pmatrix} \boldsymbol{P}_{oe1} & \boldsymbol{P}_{oe2} & \boldsymbol{P}_{oe3} \\ \boldsymbol{P}_{se1} & \boldsymbol{P}_{se2} & \boldsymbol{P}_{se3} \\ \boldsymbol{P}_{re1} & \boldsymbol{P}_{re2} & \boldsymbol{P}_{re3} \end{pmatrix}. \tag{3.16}$$

Generally speaking, because of the influence of vibration or other factors $\boldsymbol{V_e}$ is variable and so the moving operator above mentioned is also time-varying,

$$\boldsymbol{\Theta} = \boldsymbol{\Theta}(t)$$

The homogeneous coordinates of tool face in tool-in-hand system left multiplied by moving operator results in its homogeneous coordinates in tool-in-use system,

$$\boldsymbol{A}_{\gamma e} = \begin{pmatrix} A_{\gamma e1} \\ A_{\gamma e2} \\ A_{\gamma e3} \end{pmatrix} = \boldsymbol{\Theta A}_{\gamma} = \boldsymbol{\Theta} \begin{pmatrix} \tan \boldsymbol{\lambda_s} \\ \tan \boldsymbol{\gamma_o} \\ \boldsymbol{1} \end{pmatrix},$$

And the dynamic cutting edge inclination and dynamic orthogonal rake are

$$\tan \lambda_{se} = A_{\gamma et} = \frac{A_{\gamma e1}}{A_{\gamma e3}},$$

$$\tan \gamma_{oe} = A_{\gamma er} = \frac{A_{\gamma e2}}{A_{\gamma e3}}.$$

Respectively.

The homogeneous coordinates of tool flank in tool-in-hand system left multiplied by moving operator results in its homogeneous coordinates in tool-in-use system,

$$\boldsymbol{A}_{\alpha e} = \begin{pmatrix} A_{\alpha e1} \\ A_{\alpha e2} \\ A_{\alpha e3} \end{pmatrix} = \boldsymbol{\Theta A}_{\alpha} = \boldsymbol{\Theta} \begin{pmatrix} \tan \boldsymbol{\lambda_s} \\ \cot \boldsymbol{\alpha_o} \\ \boldsymbol{1} \end{pmatrix},$$

And the dynamic (working) orthogonal clearance is

$$\cot \alpha_{oe} = A_{\alpha er} = \frac{A_{\alpha e2}}{A_{\alpha e3}}$$

As shown in Fig. 3.5a, the unit vector **f** in feed direction is in static reference plane P_r, and f is the image line of line in feed direction. To calculate working (dynamic) cutting edge angle κ_{re}, the homogeneous coordinates in tool-in-hand system of the image line f must left multiplied by moving operator. The equation of image line f in tool-in-hand system is

$$t \cos \kappa_r - r \sin \kappa_r + 0 = 0,$$

where t, r are the current coordinates, so from Eq. (2.7), we have

$$f = \begin{pmatrix} f_1 \\ f_2 \\ f_3 \end{pmatrix} = \begin{pmatrix} \cos \kappa_r \\ -\sin \kappa_r \\ 0 \end{pmatrix},$$

In tool-in-use system (t_e, r_e),

$$f_e = \begin{pmatrix} f_{e1} \\ f_{e2} \\ f_{e2} \end{pmatrix} = \boldsymbol{\Theta} \begin{pmatrix} \cos \boldsymbol{\kappa_r} \\ -\sin \boldsymbol{\kappa_r} \\ \boldsymbol{0} \end{pmatrix},$$

And working (dynamic) cutting edge angle κ_{re} is

$$\tan \kappa_{re} = -\frac{f_{e2}}{f_{e1}}.$$

3.1.3 *Analyzing and Calculating Tool Angles by Using the Projective "Distance" and "Angle" Concepts*

3.1.3.1 Calculation of Normal Rake

Defined "distances" and "angles" by Formulae (2.12)–(2.19) are the invariants for coordinate transformation. So we can calculate space angles according to the homogeneous coordinates of relevant images in any coordinate system by means of the formulae. For example, normal rake γ_n is the angle between tool face A_γ and reference plane P_r and measured in cutting edge normal plane P_n. If P_r is taken as the base, its image must at the origin O (see Fig. 3.6), so normal rake γ_n is $\theta_{OP_nA_\gamma}$. Substituting the homogeneous coordinates

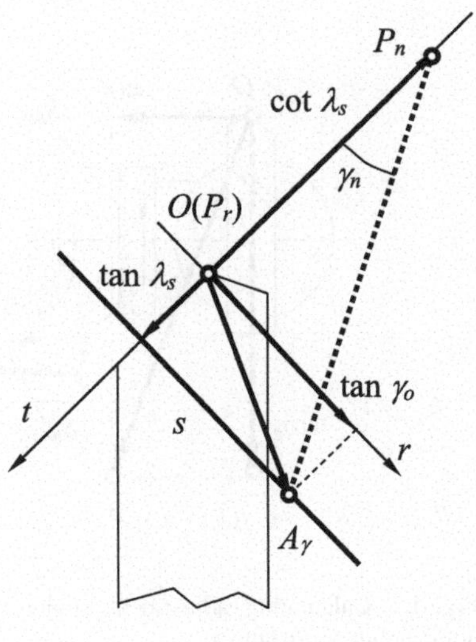

Fig. 3.6 Calculating normal rake according to homogeneous coordinates of images on projection plane

$$\boldsymbol{O} = \begin{pmatrix} 0 \\ 0 \\ 1 \end{pmatrix}, \quad \boldsymbol{P_n} = \begin{pmatrix} -\cot\lambda_s \\ 0 \\ 1 \end{pmatrix}, \quad \boldsymbol{A_\gamma} = \begin{pmatrix} \tan\lambda_s \\ \tan\gamma_o \\ 1 \end{pmatrix}$$

Into Formula (2.19) results in

$$\tan\gamma_n = \tan\theta_{OP_nA_\gamma} = \tan\gamma_o \cdot \cos\lambda_s.$$

3.1.3.2 Calculation of Adjustment Angle of a Universal Vice in Tool Sharpening

When the face of a single-point cutting tool to be ground in a universal vice as shown in Fig. 3.7, we must adjust the vice such that the tool face will be turned toward the working surface of the grinding wheel. Correspondingly, on the projective plane we must move the image point A_γ to coincide with the image of the working surface of grinding wheel. In Fig. 3.7 take the image point of working surface of grinding wheel as the origin O, and suppose the initial position of installing surface P (the undersurface of the tool) is coincide with origin O. ΔPBA_γ is a triangle formed by images of three consolidating planes, and during moving the lengths of the three sides of the triangle will keeps unchanged. One of solutions is first move ΔPBA_γ to position $\Delta P^*B^*A_\gamma^*$ along axis y, such that B will move to the position O (i.e., B^*) and the moving distance is S_{PB}. Substituting

$$\boldsymbol{B} = \begin{pmatrix} 0 \\ \tan\gamma_p \\ 1 \end{pmatrix}, \quad \boldsymbol{O} = \begin{pmatrix} 0 \\ 0 \\ 1 \end{pmatrix}$$

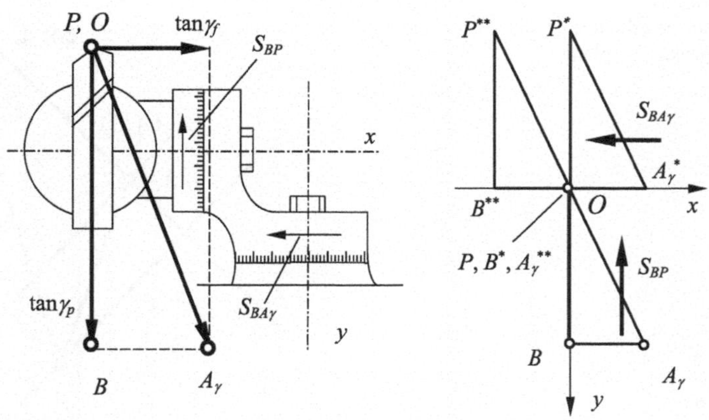

Fig. 3.7 Calculating adjustment angles of a universal vice in tool sharpening, the way of non-Euclidean distances

Into Formula (2.12), results in $S_{PB} = \gamma_p$, in the space this is equivalently about axis x rotates by angle γ_p; second, move $\Delta P^* B^* A_\gamma{}^*$ to $\Delta P^{**} B^{**} A_\gamma{}^{**}$ along axis x, so as to move $A_\gamma{}^*$ to the origin, becoming $A_\gamma{}^{**}$. Moving distance is $S_{A_\gamma^* B^*} = S_{A_\gamma B}$. Substituting

$$\boldsymbol{A}_\gamma = \begin{pmatrix} \tan \gamma_f \\ \tan \gamma_p \\ 1 \end{pmatrix}, \quad \boldsymbol{B} = \begin{pmatrix} 0 \\ \tan \gamma_p \\ 1 \end{pmatrix}$$

Into Formula (3.12) we get $S_{BA_\gamma} = \tan^{-1}(\tan \gamma_f \cos \gamma_p)$, this is the rotating angle about axis y.

3.1.4 Determining the Adjustment Angles of Universal Vice in Tool Sharpening by Means of Hyperbolic Coordinate Grid

Let's continue the example of universal vice adjustment issues, and further suppose $\gamma_o = 40°$ and $\lambda_s = -10°$. The initial position of image point A_γ in coordinate system (r, t) can be determined by γ_o and λ_s as shown in Fig. 3.8a. The problem is asking to move image point A_γ to the origin O through rotations about axes x and y. Two solutions will be given as follows and hyperbolic coordinate grid will be used.

Solution I

Let the origin O and the axis y of hyperbolic coordinate grid coincide with the origin O and the axis y of the vice, respectively, and we can find from the grid that a rotation of pane A_γ around axis x by an angle $\Phi_x = -25.2°$ will move its image point along a hyperbola to axis x and become $A_\gamma{}^*$, and then move $A_\gamma{}^*$ along axis x to the origin O, and this means to rotate $A_\gamma{}^*$ around axis y by an angle $\Phi_y = 32.8°$. The configuration and adjusting of the vice is shown in Fig. 3.8b.

Solution II

Let the origin O and the axis y of hyperbolic coordinate grid coincide with the origin O and the axis x of the vice, respectively, and we can find from the grid that a rotation of plane A_γ around axis y by an angle $\Phi_y = 35.5°$ will move its image point along a hyperbola to axis y and become $A_\gamma{}^{**}$, and then move $A_\gamma{}^{**}$ along axis y to the origin O, and this means to rotate $A_\gamma{}^{**}$ around axis x by an angle $\Phi_x = -21°$. The configuration and adjusting of the vice is shown in Fig. 3.8c.

It is noteworthy that under the two adjusting schemes the configurations of the vise, the adjusting angles Φ_x and Φ_y and their operation orders are all different.

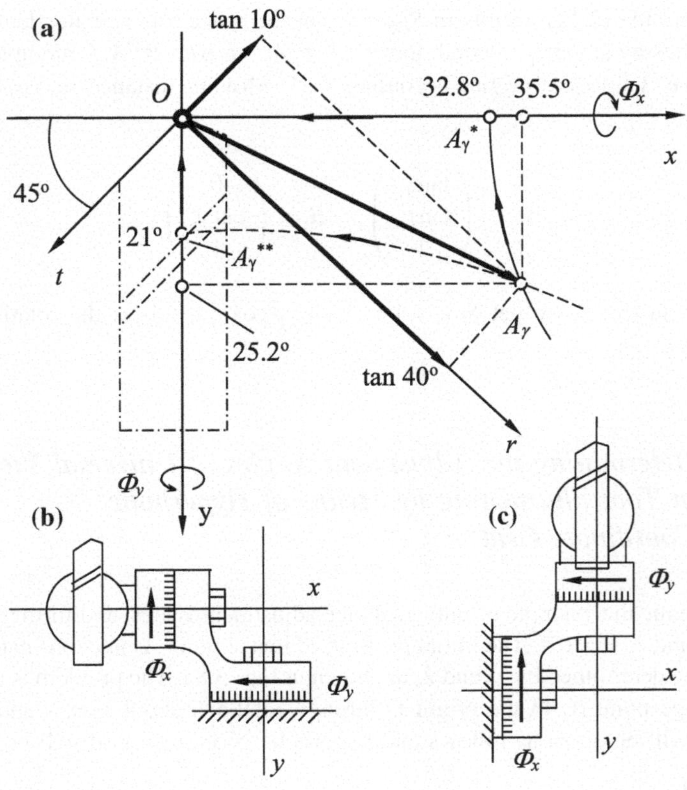

Fig. 3.8 The adjustment of universal vice, the method of hyperbolic coordinate grid. **a** Projective plane and the image trajectory diagram. **b** The configuration of the vise under the scheme I. **c** The configuration of the vise under the scheme II

3.2 Plane Vector Analysis of Geometric Angles for Other Kinds of Cutting Tools

In last section we analyzed the geometry of a single-point cutting tool on the projective plane, in this section, based on the knowledge introduced in last section we will look at the geometry of a variety of complex cutting tools by means of the methods introduced in last section.

3.2.1 Reamer's Angle Analysis

Reamers' tool orthogonal rake γ_o and tool cutting edge inclination λ_s are mainly determined according to the property of the materials being machined, and the direction of chip-ejection is also to be considered. In the design and manufacture of reamers what we normally directly designed and manufactured, however, are the other two angles: the radial rake γ_1 in end section I-I and the axial rake γ_2 in axial section II-II (i.e. the helix angle of the flute, see Fig. 3.9a). There is a problem: how to calculate γ_o and λ_s, which are directly acting in cutting process from selected γ_1 and γ_2 in design and tool manufacture, or conversely, how to choose and control γ_1 and γ_2 in design and manufacture so as to realize selected γ_o and λ_s. The method of tool face obliquity vector and its decomposition and resultant described previously can solve this problem easily and intuitively.

See Fig. 3.9, suppose radial rake γ_1 = 10°, axial rake γ_2 = 20° and tool cutting edge angle κ_r = 45° for the reamer. According to the following steps we can obtain rake γ_o and tool cutting edge inclination λ_s:

Step 1 take the reference plane at selected point *x* (i.e. the axial section *O-O* at point *x*, see Fig. 3.9a) as the base;

Step 2 draw the outline of the reamer on the projective plane, and set axes *x* and *y* in the directions of I-I and II-II, respectively (the positive direction of axis *x* is in the direction of positive radial rake obliquity; the positive direction of axis *y* is in the direction of obliquity of helix flute of the reamer). Set axes *t* and *r* in the directions of the projection of main cutting edge on the projective plane and its normal direction, respectively (the direction of axis *t* is in the direction of positive cutting edge inclination and the direction of axis *r* is in the direction of obliquity of the positive rake). The unit vectors in directions of axes *x*, *y*, *t* and *r* are denoted by **i**, **j**, **t** and **r**, respectively (they are not shown in Fig. 3.5b).

Step 3 according to designed γ_1 and γ_2 draw radial and axial obliquity vectors $\tan\gamma_1\mathbf{i}$ and $\tan\gamma_2\mathbf{j}$, respectively. Their resultant vector

$$\boldsymbol{A}_\gamma = \boldsymbol{OA}_\gamma = \tan\gamma_1 \cdot \mathbf{i} + \tan\gamma_2 \cdot \mathbf{j}$$

gives the obliquity vector $\boldsymbol{A}_\gamma$ and the image point A_γ.

Step 4 the projections of $\boldsymbol{A}_\gamma$ on axes *t* and *r* give tool cutting edge inclination λ_s and tool orthogonal rake γ_o, respectively,

$$\tan\lambda_s = \boldsymbol{A}_\gamma \cdot \mathbf{t} = \tan\gamma_1\mathbf{i}\cdot\mathbf{t} - \tan\gamma_2\mathbf{j}\cdot\mathbf{t} = -\tan\gamma_1 \sin\kappa_r + \tan\gamma_2 \cos\kappa_r,$$

$$\tan\gamma_o = \boldsymbol{A}_\gamma \cdot \mathbf{r} = \tan\gamma_1\mathbf{i}\cdot\mathbf{r} + \tan\gamma_2\mathbf{j}\cdot\mathbf{r} = \tan\gamma_1 \cos\kappa_r + \tan\gamma_2 \sin\kappa_r.$$

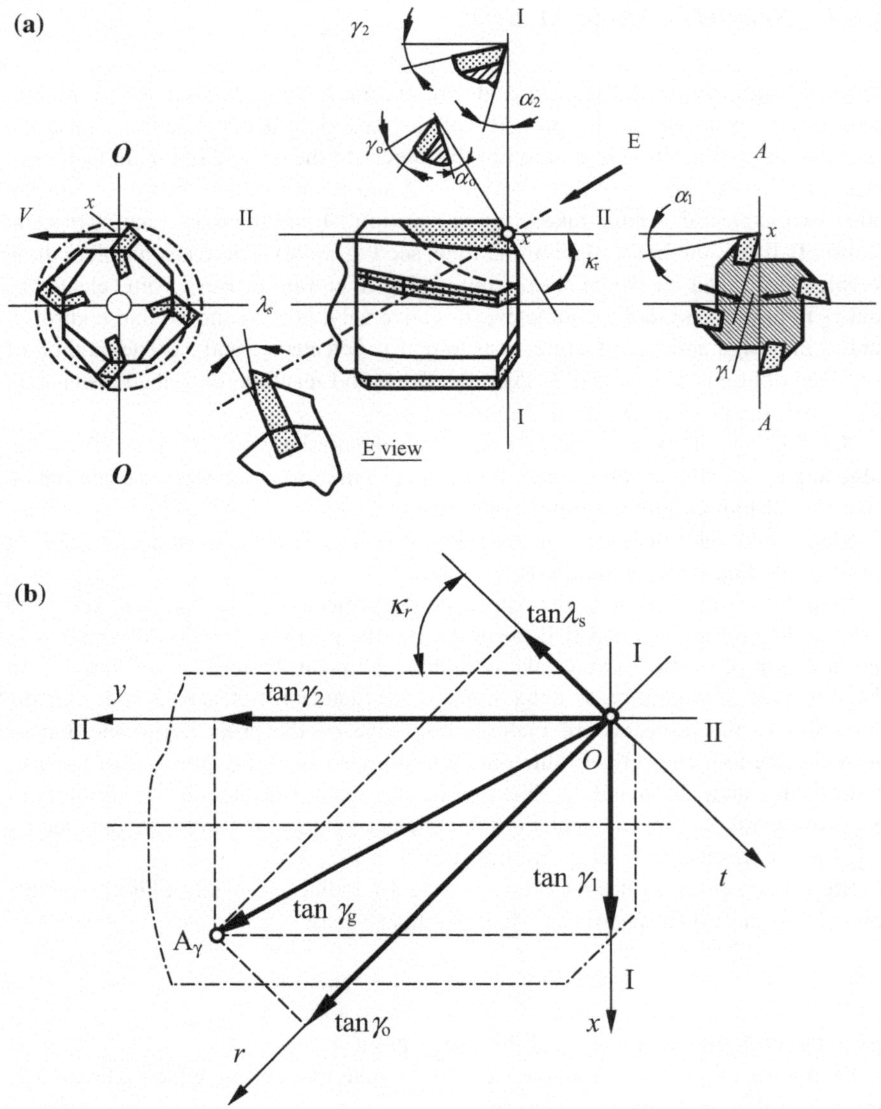

Fig. 3.9 Geometric analysis for reamers on projective plane. **a** The working drawing of cutting tool. **b** The diagram of plane vectors

Substituting $\gamma_1 = 10°$, $\gamma_2 = 20°$ and $\kappa_r = 45°$ into above two equations results in $\lambda_s = -7.5°$, $\gamma_o = 20.9°$. In this example the reason why the value of tool cutting edge inclination λ_s is negative is that the projection of $\boldsymbol{A}_\gamma$ on axis t is at the negative direction.

By means of similar method we also can calculate γ_1 and γ_2 from given λ_s, γ_o and κ_r.

3.2.2 Twist Drill's Angle Analysis

The method of analyzing tool angles through planar vector graphic is not limited to the cases that the tool face is a plane but also can be used in the cases when the face is a curved surface, such as the formula of twist drill's rake is also can be deduced by means of this method.

Let's see Fig. 3.10a, the main cutting edge is in plane T, which is parallel to drill axis and apart from the axis by R_o, the radius of drill web. The reference plane P_r of a point x on the major edge is the radial plane of the point. The angle λ_{TX} between planes T and P_r is called cutting edge inclination of twist drill at end face,

$$\text{Sin}\,\lambda_{TX} = R_o/R_x, \quad (3.17)$$

where R_x is the radius at the point x.

With the P_r of point x as base plane, we can draw plane image diagram on the projection as follows. The obliquity vector of plane T is perpendicular to the drill axis, and its length is tan λ_{TX}. Because the main cutting edge is on the plane T, so projecting obliquity vector of plane T on the direction t of tool cutting edge plane gives the tool cutting edge inclination at x point of the main cutting edge

$$\tan\lambda_x = \tan\lambda_{TX}\mathbf{j}\cdot\mathbf{t} = \tan\lambda_{TX}\,\sin\kappa_{rx}, \quad (3.18)$$

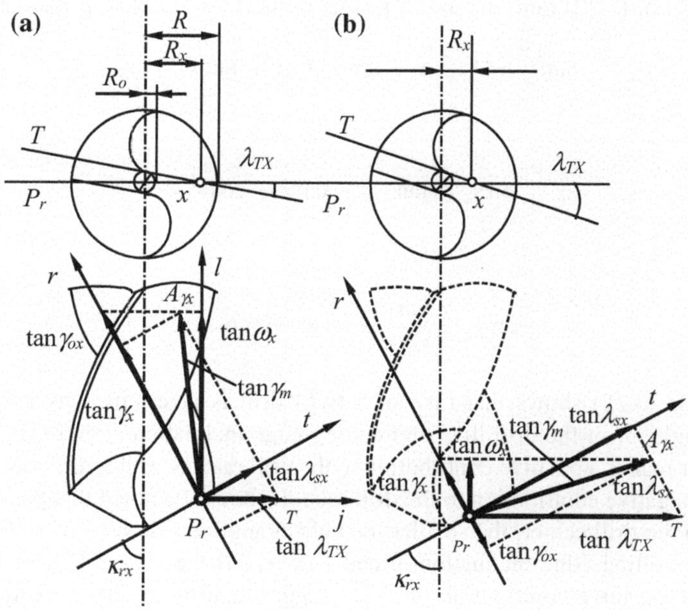

Fig. 3.10 The analysis of geometric angles of a twist drill. **a** The angles close to the periphery. **b** The angles close to the web

where **j** and **t** are the unit vectors in direction of j and t, respectively, which are not shown in Fig. 3.10, and κ_{rx} is tool cutting edge angle at x point.

Because major cutting edge is on the tool face, so tan λ_{sx} is also the obliquity of tool face in direction of t; on the other hand, the obliquity of tool face in direction of axial section l is namely tan ω_x, and ω_x is the helical angle of the drill flute at point x.

$$\tan \omega_x = \tan \omega (R_x/R), \tag{3.19}$$

where ω is helical angle of the drill flute at periphery, and R is the radius of the drill. Passing through the end points of vectors tan $\lambda_{Tx}\mathbf{t}$ and tan $\omega_x\mathbf{l}$ draw their normals, and they intersect at point $A_{\gamma x}$, which is the image point of tangent plane to the tool face at point x, where **l** are the unit vectors in direction of l (it is not shown in Fig. 3.10). From Fig. 3.10 we can see that the obliquity vector $\boldsymbol{A}_{\gamma x}$ of face $A_{\gamma x}$ at point x is equal to the sum of two vectors,

$$\boldsymbol{A}_{\gamma x} = \tan \gamma_{ox}\mathbf{r} + \tan \lambda_{TX}\mathbf{j}, \tag{3.20}$$

where **r** is the unit vector in direction r, which is not shown in Fig. 3.10. The term tan γ_{ox} reflects the face obliquity caused by helical angle of the drill flute,

$$\tan \gamma_{ox} = \tan \omega_x / \sin \kappa_{rx}; \tag{3.21}$$

The term tan λ_{TX} reflects the variation in face obliquity caused by drill web. Projecting Eq. (3.20) onto the axis r results in the formula of drill rake at point x,

$$\tan \gamma_x = \boldsymbol{A}_{\gamma X} \cdot \mathbf{r} = \tan \gamma_{ox}\mathbf{r} \cdot \mathbf{r} + \tan \lambda_{TX}\mathbf{j} \cdot \mathbf{r},$$

or,

$$\tan \gamma_x = \tan \gamma_{ox} - \tan \lambda_{TX} \cos \kappa_{rx}$$

so,

$$\tan \gamma_x = \frac{\tan \omega_x}{\sin \kappa_{rx}} - \tan \lambda_{TX} \cos \kappa_{rx}. \tag{3.22}$$

Formula (3.22) shows that rake of a twist drill is determined by two factors: helical angle ω_x of the drill flute and cutting edge inclination at end face λ_{TX}. The first factor makes a positive contribution to the drill rake as well as the second factor makes a negative contribution to the drill rake. It should be noted that the closer the point x to the drill center, the smaller the helical angle ω_x is (see Eq. 3.19) and the larger the cutting edge inclination at end face λ_{TX} is (see Eq. 3.17). To a certain extent, the negative contribution of λ_{TX} is bigger than the positive contribution of ω_x, and the drill rake γ_x at point x thus become negative. By a comparison of Fig. 3.10b with Fig. 3.10a you can intuitively get the above conclusion. This is why the part of the main cutting edge close to the drill web of a standard twist drill has a worse cutting ability.

What we described above is of the situation for standard grinding of standard twist drills. However the method introduced above can be also used for the situation of non-standard grinding of standard twist drills, such as Qun-Drill. Because of its non-standard grinding the main cutting edge is no longer two straight lines and its rake and tool cutting edge inclination may be analyzed as follows.

Take the reference plane P_r at point x on the major cutting edge as the base plane, under standard grinding the working cutting edge angle is κ_{rx} and under arbitrary grinding the working cutting edge angle is κ_{rx}'. In Fig. 3.11 the two situations are overlapped for a comparison. Under the both conditions, the reference plane and the face are the same (because only the flank is ground), and so the obliquity vector of the face with respect to the reference plane $\boldsymbol{A}_{\gamma x}$ does not change. What are changed are the coordinate axes t (cutting edge plane) and r (orthogonal plane) become t' and r', and this is a problem of rotation transformation of coordinate axes. Paying attention to the fact that new axes (t', r') differ from the old axes (t, r) by a rotation angle $-(\kappa_{rx}' - \kappa_{rx})$, we can promptly write out the formulae for calculating cutting edge inclination λ_x' and orthogonal rake γ_x' under arbitrary grinding,

$$\begin{pmatrix} \tan\lambda_x' \\ \tan\gamma_x' \end{pmatrix} = \begin{pmatrix} \cos(\kappa_{rx}' - \kappa_{rx}) & -\sin(\kappa_{rx}' - \kappa_{rx}) \\ \sin(\kappa_{rx}' - \kappa_{rx}) & \cos(\kappa_{rx}' - \kappa_{rx}) \end{pmatrix} \begin{pmatrix} \tan\lambda_x \\ \tan\gamma_x \end{pmatrix},$$

Expansion the above equation and substitution of Eqs. (3.18) and (3.22) into above equation result in

$$\tan\lambda_x' = \tan\omega_x\left(\cos\kappa_{rx}' - \frac{\sin\kappa_{rx}'}{\tan\kappa_{rx}}\right) + \tan\lambda_{TX}\sin\kappa_{rx}', \tag{3.23}$$

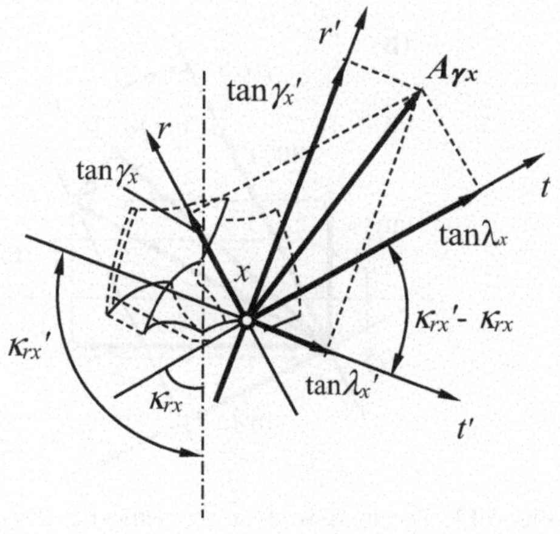

Fig. 3.11 The analysis of geometric angles of Qun-Drill

$$\tan \gamma_x' = \tan \omega_x \left(\sin \kappa_{rx}' + \frac{\cos \kappa_{rx}'}{\tan \kappa_{rx}}\right) - \tan \lambda_{TX} \cos \kappa_{rx}'. \tag{3.24}$$

It is obvious from Fig. 3.11 that in the part of round cutting edge the orthogonal plane r' of Qun-Drill is closer to the obliquity vector of the face $\boldsymbol{A}_{\gamma x}$ than the standardly ground drill, so $\tan \gamma_x' > \tan \gamma_x$. Actually, this intuitively explained why cutting performance of Qun-Drill is better than a standardly ground drill.

3.2.3 Screw Tap's Angle Analyses

Take the reference plane of x point on the starting taper as the base plane as shown in Fig. 3.12a. The obliquity of the face, i.e. chip-ejection flute of the screw tap, in radial section I-I is $\tan \gamma_1$, which is determined by the cross section profile at point x;

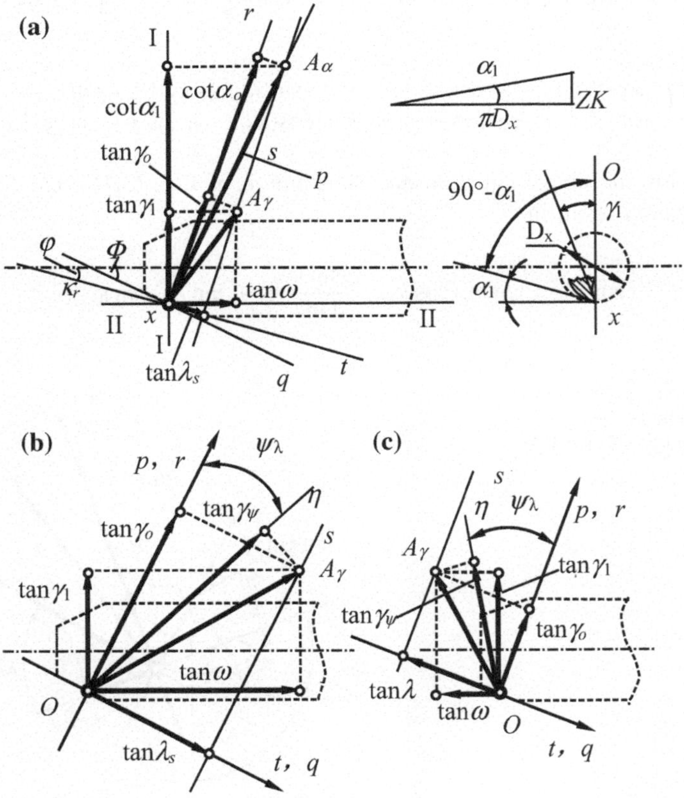

Fig. 3.12 The analysis of the geometrical angles on screw taps. **a** Exact solution. **b** Approximate solution, right-handed rotated flute. **c** Approximate solution, left-handed rotated flute

the obliquity in axial section II-II is tan ω, where ω is the helical angle at point x on the flute (it is supposed that flute is right-handed rotated). The sum of above two terms determine the image point A_γ of the face.

The flank in the starting taper is formed by radial back-off (or relieving turning), the forming process may be imaged as a generating line q in the axial section and intersects with tap's axis into an angle of Φ (the half cone angle of the starting taper), and line q is rotating about tap's axis and proportionally moving towards the tap's axis. Thus the tap's flank is formed. The radial clearance α_1, the back-off per tooth K and the diameter at point x have the relation as shown by a triangle in Fig. 3.12a. The obliquity of the flank in the radial direction is tan $(90 - \alpha_1) = \cot \alpha_1$. On the other hand, q is obviously a contour of the flank, and so the line p, on which the obliquity vector of flank $\boldsymbol{A}_\alpha$ is lying, should perpendicular to the line q. Passing through the end point of $\cot \alpha_1 \mathbf{I}$ draw its normal intersecting with p at point A_α, which is the image point of flank. Where $\mathbf{I}$ is the unit vector in the direction I-I, which is not shown in Fig. 3.12a.

The connecting line of image point A_γ and A_α is the image line s of the top edge at starting taper. Passing through origin draw line t perpendicular to s, we get the direction of tool cutting edge plane, the angle φ between line t and tap's axis is tool cutting edge angle at point x. Drawing a line perpendicular to t results in the direction r.

Projecting $\boldsymbol{A}_\gamma$, $\boldsymbol{A}_\alpha$ onto directions r and t, respectively, leads to rake tan γ_o, tool cutting edge inclination tan λ_s and clearance cot α_o.

As a matter of fact, because radial clearance α_1 is relatively small, and thus $\cot \alpha_1 \gg \tan \gamma_1$, $\tan \omega$, compared with A_γ, A_α may be considered at the infinite point of line p. If only rake and cutting edge inclination to be analyzed, s may be regards as a line passing through point A_γ and parallel to p as shown in Fig. 3.12b, c. In this case, p and r coincide with each other, and $\varphi \approx \Phi$. In the figures, the chip-ejection direction η and the rake in direction γ_ψ are also shown. Figure 3.12b is of right-handed rotated flute and Fig. 3.12c is of left-handed rotated flute. The angle ψ_λ between η and r is chip-ejection angle. The determination of ψ_λ is a physical problem but not a purely geometrical problem. Such a problem will be discussed in the second part of this book.

What's shown in Fig. 3.12 are static angles without consideration of the influence of feed motion. The feed per revolution of screw taps is the pitch of screw thread, the influence of which is sometimes not ignorable. Here is an analysis of screw taps' working angles taking the feed motion into consideration. As shown in Fig. 3.13, suppose the angle between resultant cutting speed vector $\boldsymbol{V_e}$ and cutting speed vector $\boldsymbol{V}$ is μ (it is actually the angle of spiral at point x), and thus the working reference P_{re} plane rotates over an angle μ with relative to static reference plane P_r. The image point of P_{re} is at the negative part of axis II-II. If the screw tap rotates about axis I-I by an angle μ the image of working reference P_{re} plane would move to the origin O and become new base plane P^*_{re}. Correspondingly, the image of tool face A_γ along a hyperbola moves over μ longitudes to its new position $A_\gamma{}^*$. Also, the intersection point B of cutting edge image s with axis II-II also across μ longitudes

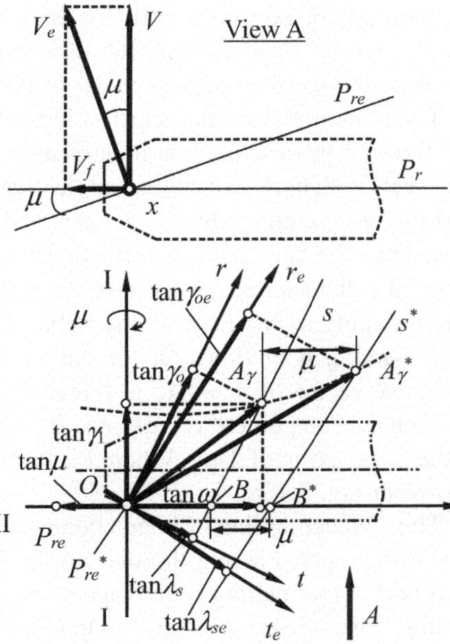

Fig. 3.13 The working angles of screw taps

to its new position B^*. The connecting line of $A_\gamma{}^*$ and B^* is the image s^* of cutting edge in working coordinate system. Setting the directions of working orthogonal plane $r_e \perp s^*$ and working cutting edge plane $t_e \| s^*$ and projecting $A_\gamma{}^*$ onto axes r_e and t_e, respectively, result in working rake tan γ_{oe} and working cutting edge inclination angle λ_{se} immediately.

3.2.4 Gear Shaper Cutter's Angle Analyses

Shaper tool's reference plane is its end section, its face is a conical surface, and its side flank is an involute helicoid (see Fig. 3.14a).

In the strict sense of the word, the projection of the side edge (the intersection line of involute helicoid with conical surface) on end section is not an involute, and however, it's accurate enough to regard the projection of the side edge on end section as an involute for the purpose of geometrical angle analysis. Hence for the point x on the side edge, the axis of tool orthogonal plane r is in the direction of the tangent to the base circle and the axis t of tool cutting edge plane is the tangent to the side edge profile and perpendicular to axis r.

The contour of the tool face at point x is a circle and so the obliquity vector $\boldsymbol{A}_\gamma$ of face is in the radial direction (the direction I-I) and the length is tan γ_1, where γ_1 is the cone base Angle. Projecting $\boldsymbol{A}_\gamma$ onto t and r axes results in tool cutting edge inclination tan λ_s and tool orthogonal rake tan γ_o, respectively.

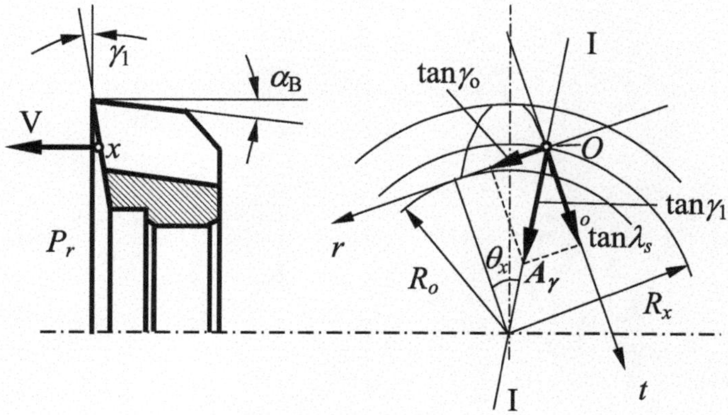

Fig. 3.14 Gear shaper cutter's angle analyses

In this chapter we have explained how to analyze and synthesize the geometric angles of various cutting tools by means of their images on a projective plane. If these are simpler, more intuitive methods than conventional ones for getting the same results as in the conventional way, then the next chapter, a research on the curve-edged twits drill, some results would be very hard to get in the conventional ways.

Chapter 4
A Study on Curved Edge Drills

Standard twist drill, under the standard grinding its flank, has a pair of straight major edges. Various new re-sharpening methods, including Qun-drill, however, break the "straight edge" restrictions, re-sharpening the major cutting edges of standard twist drill into a variety of curves to improve its drilling performance. This chapter, with the help of previously introduced method of mapping space angles onto a projective plane, made an in depth study on the cutting edge shape of twist drill, cutting angle distribution along the curved cutting edge and the relationship between cutting edge shape and drilling performance. Study reveals: there is a special cutting edge allowing the rake at each point on it reaches the possible maximum value at the point, thus enabling significantly improved machining ability. This chapter provides the technology for optimizing cutting edge shape to obtain desired rake angle distribution, as well as the technology through the drill flank design and grinding in order to control distribution of drill clearance. Theoretical analysis and experimental research have been performed. Computer software was developed, so that the cutting edge shape realizing required rake angle distribution is calculated.

4.1 Twist Drill's Geometry and Its Images on Projective Plane

4.1.1 Rake Surface and Its Obliquity Vector Field

The flute helicoids are the rake surface of a twist drill which is defined by the following parameter equations in the coordinate system *O-xyz* (see Fig. 4.1).

H. Shi, *Metal Cutting Theory*, Springer Series in Advanced Manufacturing,
https://doi.org/10.1007/978-3-319-73561-0_4

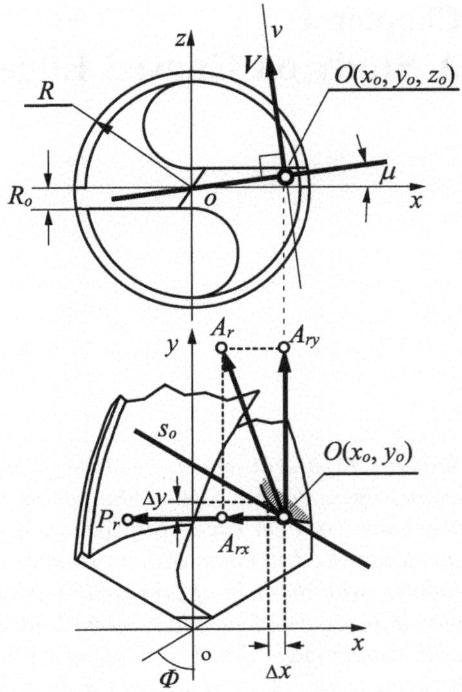

Fig. 4.1 Rake surface of a twist drill and its obliquity vector field

$$\begin{cases} x = t \sin \Phi \cos \theta + R_o \sin \theta, \\ y = t \cos \theta + H\theta/(2\pi), \\ z = -t \sin \Phi \sin \theta + R_o \cos \theta. \end{cases} \quad (4.1)$$

where 2Φ—point angle of the drill, $2R_o$—web thickness, H—pitch of the drill flute helicoid and θ, t—the variable parameters. Let O (x_o, y_o, z_o) be a given point on the rake surface and O (x_o, y_o) its projection in plane xoy, which is taken as the base plane.

The obliquity of the surface at point O (i.e., the obliquity of the tangential plane to the flute surface at point O) is expressed by vector $\mathbf{A}_\gamma$ $(A_{\gamma x}, A_{\gamma y})$, and

$$A_{\gamma x} = -\frac{\partial z}{\partial x} = \frac{H \tan \Phi \tan \theta - 2\pi t \sin \Phi - 2\pi R_o \tan \theta}{H \tan \Phi + 2\pi t \sin \Phi \tan \theta - 2\pi R_o}, \quad (4.2)$$

$$A_{\gamma y} = -\frac{\partial z}{\partial y} = \frac{2\pi t \sin^2 \Phi/(\cos \Phi \cos \theta)}{H \tan \Phi + 2\pi t \sin \Phi \tan \theta - 2\pi R_o}. \quad (4.3)$$

where $A_{\gamma x}$, $A_{\gamma y}$ are the plane Cartesian coordinates of image point A_γ and they have a relationship with the homogeneous coordinates $A_{\gamma 1}$, $A_{\gamma 2}$ and $A_{\gamma 3}$,

$$A_{\gamma x} = A_{\gamma 1}/A_{\gamma 3}, \tag{4.4}$$

$$A_{\gamma y} = A_{\gamma 2}/A_{\gamma 3}. \tag{4.5}$$

From this,

$$\mathbf{A}_\gamma = \begin{pmatrix} A_{\gamma 1} \\ A_{\gamma 2} \\ A_{\gamma 3} \end{pmatrix} = \begin{pmatrix} H \tan \Phi \tan \theta - 2\pi t \sin \Phi - 2\pi R_o \tan \theta \\ 2\pi t \sin^2 \Phi/(\cos \Phi \cos \theta) \\ H \tan \Phi + 2\pi t \sin \Phi \tan \theta - 2\pi R_o \end{pmatrix}, \tag{4.6}$$

From the middle one of Eq. (4.1),

$$t = (y - H\theta/2\pi)/\cos \Phi, \tag{4.7}$$

Substituting Eq. (4.7) into the upper one of Eq. (4.1) results in

$$x = (y - H\theta/2\pi) \tan \Phi \cos \theta + R_o \sin \theta. \tag{4.8}$$

For the given point O (x_o, y_o), the parameter θ can be calculated form Eq. (4.8) through numerical approach. Then parameter t can be worked out by substituting the value of obtained and y_o into Eq. (4.7). Knowing t and θ we can calculate $A_{\gamma x}$ (x_o, y_o) and $A_{\gamma y}$ (x_o, y_o) by Eqs. (4.2) and (4.3), so as to determine the obliquity vector $\mathbf{A}_\gamma$ (x_o, y_o) = $\mathbf{OA}_\gamma$ (x_o, y_o) and the image point A_γ related to point O (x_o, y_o). The whole of the obliquity vectors related to all the points on the rake surface forms a vector field $\mathbf{A}_\gamma$ (x, y) on the projective plane xoy (see Appendix 4.1).

4.1.2 *Cutting Edge*

The small segment of the cutting edge at point O is denoted in Fig. 4.1 by curved line and shade (here, we study arbitrary, non-standard edge shape). The line denoted by s_o is the projection of the tangent to the edge at point O onto plane xoy.

In Fig. 4.2 point O, line s_o and vector $\mathbf{A}_\gamma$ = $\mathbf{OA}_\gamma$ are shown together with some other geometrical elements which are to be introduced and explained later on. The coordinate axes x and y are translated to point O.

An edge element at point O is described by its three resolutions Δx, Δy and Δz. The projection of the segment and its tangent s_o may be determined by the choice of Δx and Δy (Fig. 4.1). Once s_o is determined drawing a straight line through point A_γ and perpendicular to s_o gives the elementary edge image s and its obliquity $\boldsymbol{p}_s$ (p_{sx}, p_{sy}) as shown in Fig. 4.2. The equation of the image line s in point-slope form is

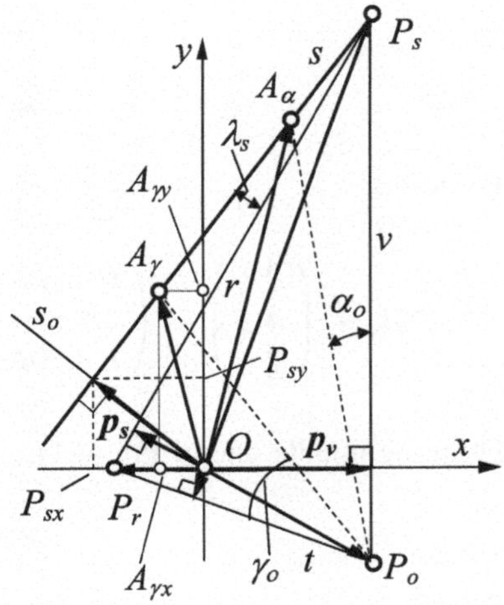

Fig. 4.2 The projection of the geometric elements related to a twist drill onto projective plane

$$\frac{A_{\gamma y} - y}{A_{\gamma x} - x} = -\frac{\Delta x}{\Delta y},$$

It is rearranged as

$$\Delta x \cdot x + \Delta y \cdot y - (\Delta x A_{\gamma x} + \Delta y A_{\gamma y}) = 0. \tag{4.9}$$

A comparison of the above equation with Eqs. 1.6 and 1.7 results in the

$$s = (s_1, s_2, s_3)^T = [\Delta x, \Delta y, \ -(\Delta x \cdot A_{\gamma x} + \Delta y \cdot A_{\gamma y})]^T, \tag{4.10}$$

From Eqs. 1.5 and 1.7 we know that the homogeneous coordinates of s can be also expressed as

$$s = (s_1, s_2, s_3)^T = (\Delta x, \Delta y, \Delta z)^T$$

And from this we know

$$\Delta z = -(\Delta x \cdot A_{\gamma x} + \Delta y \cdot A_{\gamma y}). \tag{4.11}$$

Changing Eq. (4.9) into normal form and taking Eq. (4.11) into consideration, we have

$$\frac{\Delta x}{\pm\sqrt{(\Delta x)^2 + (\Delta y)^2}} x + \frac{\Delta y}{\pm\sqrt{(\Delta x)^2 + (\Delta y)^2}} y + \frac{\Delta z}{\pm\sqrt{(\Delta x)^2 + (\Delta y)^2}} = 0,$$

And this leads to length and the components p_{sx}, p_{sy} of the obliquity vector of the cutting edge $\boldsymbol{p_s}$ are

$$P_s = |\Delta z|/\sqrt{(\Delta x)^2 + (\Delta y)^2}, \tag{4.12}$$

$$P_{sx} = -\Delta x \Delta z/[(\Delta x)^2 + (\Delta y^2)], \tag{4.13}$$

$$P_{sy} = -\Delta y \Delta z/[(\Delta x)^2 + (\Delta y^2)]. \tag{4.14}$$

Knowing the obliquity vector $\boldsymbol{A_\gamma}$ $(x_o, y_o) = \boldsymbol{OA_\gamma}$ and the image point A_γ related to point O (x_o, y_o), and defining the orientation of cutting edge on the projective plane at point O by resolutions Δx and Δy, then we can calculate Δz through Eq. (4.11), and determine the homogeneous coordinates and obliquity vector of the tangent to edge at point O by means of Eqs. (4.10)–(4.14) and thus determine the image line of edge.

4.1.3 Flank Surface

The image of the tangential plane to the flank surface at point O is denoted by A_α which must be on image line s (Fig. 4.2). Image line s with the two image point A_γ and A_α on it describes the shape and the orientation of an "elementary cutting tool" at point O.

For a given point O (x_o, y_o), the obliquity vector of rake $\boldsymbol{A_\gamma}$ (A_x, A_y) is determined by the design of the drill flute, while obliquity vector of cutting edge $\boldsymbol{p_s}$ (p_{sx}, p_{sy}) and the obliquity vector of flank $\boldsymbol{A_\alpha}$ $(A_{\alpha x}, A_{\alpha y})$, however, must be determined from an optimum procedure which will be explained later on.

4.1.4 Cutting Speed

The cutting speed $\boldsymbol{V}$ of point O is perpendicular to radius oO as shown in Fig. 4.1 (see the upper one), and the image of space line v in the direction of $\boldsymbol{V}$ together with its obliquity vector $\boldsymbol{p_v}$ is shown in Fig. 4.2. From Fig. 4.2 we have

$$p_v = \cot \mu = x_o/z_o, \tag{4.15}$$

The equation of image v is

$$x = p_v = x_o/z_o,$$

Or written in the form of

$$z_o x + 0 \cdot y - x_o = 0,$$

Then from Eqs. (1.6) and (1.7),

$$\boldsymbol{v} = (v_1, v_2, v_3)^T = (z_o, 0, -x_o)^T. \tag{4.16}$$

4.1.5 Local Cutting Coordinate System

The local cutting coordinate system is used to determine the cutting angles of the elementary tool at point O.

4.1.5.1 Tool Reference Plane P_r

P_r is perpendicular to $\boldsymbol{V}$ in the space, the image point P_r is shown in Fig. 4.2, and

$$\boldsymbol{P_r} = (P_{r1}, P_{r2}, P_{r3})^{\mathrm{T}} = \perp(v) = v = (z_o, 0, -x_o)^{\mathrm{T}}, \tag{4.17}$$

And from Eq. (1.3),

$$P_r = P_{rx} = P_{r1}/P_{r3} = -z_0/x_0, P_{ry} = 0. \tag{4.18}$$

4.1.5.2 Tool Cutting Edge Plane P_s

P_s is determined by s and v in the space, thus the image of P_s is the intersection point of image lines s and v as shown in Fig. 4.2. From composite operators Eq. (1.30),

$$\boldsymbol{P_s} = (P_{s1}, P_{s2}, P_{s3})^{\mathrm{T}} = \boldsymbol{s} \times \boldsymbol{v} = \left(\begin{vmatrix} s_2 & s_3 \\ 0 & -x_o \end{vmatrix}, \begin{vmatrix} s_3 & s_1 \\ -x_o & z_o \end{vmatrix}, \begin{vmatrix} s_1 & s_2 \\ z_o & 0 \end{vmatrix} \right)^{\mathrm{T}},$$

i.e.

$$\boldsymbol{P_s} = (P_{s1}, P_{s2}, P_{s3})^{\mathrm{T}} = (-x_0 s_2, z_0 s_3 + x_0 s_1, -z_0 s_2)^{\mathrm{T}} \tag{4.19}$$

Then

$$P_{sx} = P_{s1}/P_{s3} = x_0/z_o, \quad P_{sy} = P_{s2}/P_{s3} = -s_3/s_2 - x_0 s_1/z_0 s_2. \tag{4.20}$$

4.1.5.3 The Projection r of s onto P_r

Image line r is the connecting line of image points P_r and P_s, and from composite operators Eq. (1.30), or projection operator Eq. (1.34) and Eqs. (4.17) and (4.19),

$$\boldsymbol{r} = (r_1, r_2, r_3)^{\mathrm{T}} = \boldsymbol{P_r} \times \boldsymbol{P_s} = \left[x_0(s_1x_0 + s_3z_0), s_2(x_0^2 + z_0^2), z_0(s_1x_0 + s_3z_0)\right]^{\mathrm{T}} \tag{4.21}$$

4.1.5.4 Tool Orthogonal Plane P_o

P_o is normal to r as shown in Fig. 4.2, and from orthogonality operator Equation (1.29),

$$\boldsymbol{P_o} = (P_{o1}, P_{o2}, P_{o3})^T = \perp(r) = r = (r_1, r_2, r_3)^T, \tag{4.22}$$

And from Eq. (4.21), we have

$$P_{ox} = P_{o1}/P_{o3} = x_0/z_0, \quad P_{oy} = P_{o2}/P_{o3} = s_2(x_0^2 + z_0^2)/[z_0(s_1x_0 + s_3z_0)] \tag{4.23}$$

The three planes P_r, P_s and P_o, which are perpendicular to each other, form an orthogonal coordinate system and their images construct a triangle with its orthocenter at origin O as shown in Fig. 4.2.

4.1.6 The Cutting Angles on the Elementary Tool at Point O

Angles γ_o, λ_s and α_o indicated in Fig. 4.2 (and some other angles not shown in the figure) defined by ISO/IDS-3002 can be calculated from the homogeneous coordinates of relevant image points. In this paper, however, another approach to calculating those angles by means of moving images is introduced.

In Fig. 4.3 images A_γ, s and P_r are shown. A rotation of the drill body about its axis y over an angle of μ moves the image point A_γ to the origin O becoming $P_r{}^*$, and $P_r{}^*$ is now the new place of reference plane P_r which coincides with the projection plane xOy and becomes new base plane. Because of the rotation the image A_γ ($A_{\gamma x}$, $A_{\gamma y}$) moves to $A_\gamma{}^*$ ($A_{\gamma x}{}^*$, $A_{\gamma y}{}^*$) along a hyperbola a-a, and image s moves to s^* accordingly. $A_{\gamma x}{}^*$, $A_{\gamma y}{}^*$ are determined by Eqs. (2.24) and (2.25),

$$A_{\gamma x}{}^* = \tan\alpha^* = \tan(\alpha + \mu), \tag{4.24}$$

$$A_{\gamma y}{}^* = A_{\gamma y}\cos\alpha/\cos\alpha^*, \tag{4.25}$$

Fig. 4.3 Calculating tool orthogonal rake and tool cutting edge inclination by means of the method of moving images

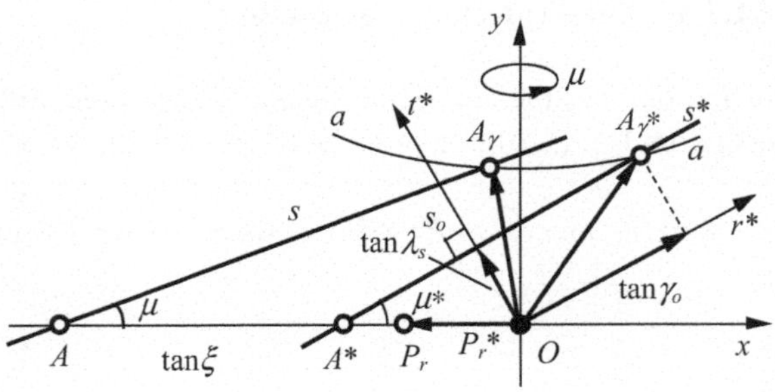

where

$$\alpha = \arctan A_{\gamma x}, \tag{4.26}$$

and

$$\mathbf{A}_{\gamma}{}^{*} = (A^{*}_{\gamma 1}, A^{*}_{\gamma 2}, A^{*}_{\gamma 3})^{\mathrm{T}} = (A_{\gamma x}{}^{*}, A_{\gamma y}{}^{*}, 1)^{\mathrm{T}}. \tag{4.27}$$

The image of the cutting edge s^* after rotation is determined in the following way: the intersection point of s with axis x is A (Fig. 4.3), and suppose $A = A_x = \tan\xi$. With the rotation A moves to A^*, $A^* = A_x{}^* = \tan\xi^*$ and $\xi^* = \xi + \mu$, $A^* = (A_x{}^*, 0, 1)^{\mathrm{T}}$. Once $A_\gamma{}^*$ and A^* are known, s^* can be determined by composition operator $s^* = A_\gamma{}^* \times A^*$ according to Eq. (2.31).

$A_\gamma{}^*$ and s^* may be also calculated by moving operator defined by Eq. (2.38),

$$A_\gamma{}^* = \mathrm{Y}(\mu)A_\gamma, \quad s^* = \mathrm{Y}(\mu)s$$

where

$$\boldsymbol{Y}(\mu) = \begin{pmatrix} \cos\mu & 0 & \sin\mu \\ 0 & 1 & 0 \\ -\sin\mu & 0 & \cos\mu \end{pmatrix}. \tag{4.28}$$

Drawing from origin O the two axes r^* and t^* perpendicular to and parallel with s^*, respectively (Fig. 4.3), where r^* and t^* are the intersections of tool orthogonal plane $P_o{}^*$ and tool cutting edge plane $P_s{}^*$ with tool reference plane $P_r{}^*$, respectively. The projections of $\mathbf{A}_\gamma{}^*$ onto r^* and t^* give, respectively, the orthogonal rake $\tan\gamma_o$ and tool cutting edge inclination $\tan\lambda_s$.

Generally speaking, for a given point O on the flute the cutting angles γ_o and λ_s related to the tool face depend on only the orientation of the cutting edge at the point. As an example, in Fig. 4.4 the values of γ_o (thick line) and λ_s (dotted line) are plotted versus the orientation of cutting edge s_o. It is clear in the figure that if s_o is oriented in the direction normal to $\mathbf{P}_{\mathrm{r}}\,\mathbf{A}_\gamma$, then γ_o reaches its maximum and λ_s

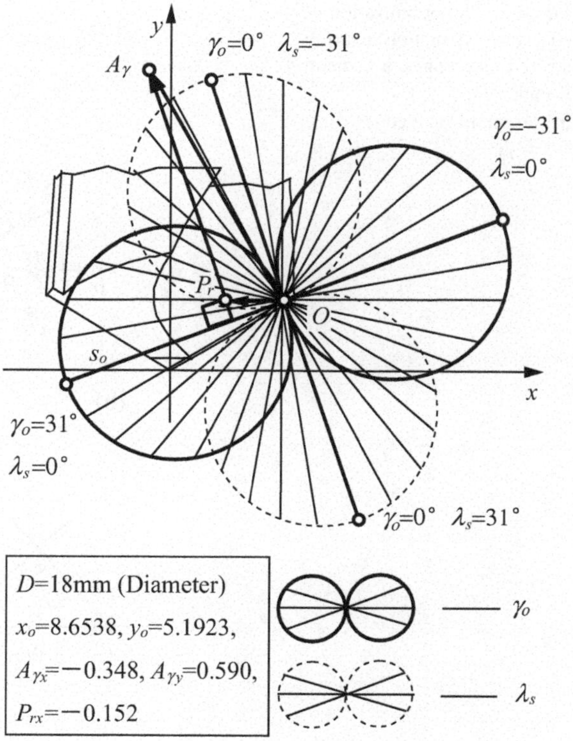

Fig. 4.4 The values of γ_o (thick line) and λ_s (dotted line) are plotted versus the orientation of cutting edge s_o

reduces to 0°; and if s_o is oriented in the direction parallel with $\boldsymbol{P}_r\, \boldsymbol{A}_\gamma$, then λ_s reaches its maximum and γ_o reduces to 0°.

4.2 The Optimization of Cutting Edge Shape Based on Given Rake Angle Distributions

4.2.1 The Edge Orientation Insuring Desired Rake Angle

The obliquity of tool face $\boldsymbol{A}_\gamma(x_o, y_o)$ and reference plane $\boldsymbol{P}_r(x_o, y_o)$ related to a given point $O(x_o, y_o)$ in rake surface are shown in Fig. 4.5a. A rotation moves P_r to the origin O, becoming P_r^*, and A_γ to A_γ^* along a hyperbola *b-b*, accordingly. The coordinates of $A_{\gamma x}^*$ and $A_{\gamma y}^*$ of A_γ^* can be determined by Eqs. (4.24)–(4.26). Circle *c* is the obliquity circle of $\boldsymbol{A}_\gamma^*$. Drawing another circle *a-a* with its center at point A_γ^* and the radius equal to tan γ_o, where γ_o is the desired rake angle at point O, results in two intersection points J_1 and J_2 of the two circles *c* and *a-a*. The connecting lines s_1^* and s_2^* of A_γ^* and J_1, A_γ^* and J_2, respectively, are the two possible images of the cutting edge insuring the desired rake angle γ_o. The choice between them is

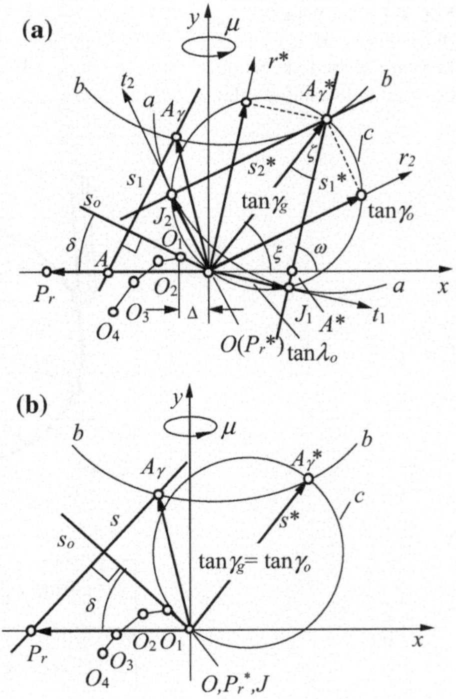

Fig. 4.5 The optimization of edge line orientation insuring desired rake angle. **a** General situation ($\gamma_o \leq \gamma_g$). **b** Maximum-rake-edge ($\gamma_o = \gamma_g$)

made on the principle of keeping the continuity of the edge line. Suppose here $s_1{}^*$ is chosen, then an opposite rotation moves $A_\gamma{}^*$, $P_r{}^*$ back to A_γ, P_r, respectively, and $s_1{}^*$ to s_1 accordingly (we may first rotate the intersection point A^* of $s_1{}^*$ and axis x back to A and then connect points A and A_γ so as to get s_1).

Based on the plane geometry described above it is easy to write the "point-slope" equation of $s_1{}^*$,

$$(A_{\gamma y}{}^* - y)/(A_{\gamma x}{}^* - x) = \tan\omega, \tag{4.29}$$

And from $\Delta P_r{}^* A_\gamma{}^* A^*$ in Fig. 4.5a,

$$\omega = \xi + \zeta, \tag{4.30}$$

where

$$\xi = \arctan A_{\gamma y}{}^*/A_{\gamma x}{}^*, \tag{4.31}$$

and

$$\zeta = \arccos(\tan\gamma_o/\tan\gamma_g), \tag{4.32}$$

where γ_g is the maximum rake which is defined as

$$A_\gamma{}^* = \tan\gamma_g = \sqrt{(A_{\gamma x}{}^*)^2 + (A_{\gamma y}{}^*)^2}. \tag{4.33}$$

A rearranging Eq. (4.29) leads to

$$\tan\omega \cdot x - y + (A_{\gamma y}{}^* - A_{\gamma x}{}^* \tan\omega) = 0,$$

And so the homogeneous coordinates of edge $\boldsymbol{s}_1^*$ are

$$\boldsymbol{s}_1^* = (\tan\omega,\ -1,\ A_{\gamma y}{}^* - A_{\gamma x}{}^* \tan\omega)^{\mathrm{T}}. \tag{4.34}$$

The homogeneous coordinates $\boldsymbol{s}_1$ of the edge before rotation may be calculated by moving operator

$$\boldsymbol{s}_1 = \boldsymbol{Y}(-\boldsymbol{\mu})\boldsymbol{s}_1^* = (s_{11}, s_{12}, s_{13})^{\mathrm{T}} \tag{4.35}$$

Substituting Eq. (4.28) into above equation leads to

$$\boldsymbol{s}_1 = \begin{pmatrix} s_{11} \\ s_{12} \\ s_{13} \end{pmatrix} = \begin{pmatrix} \sin\mu \tan\omega A_{\gamma x}{}^* - \sin\mu A_{\gamma y}{}^* + \cos\mu \tan\omega \\ -1 \\ -\cos\mu \tan\omega A_{\gamma x}{}^* + \cos\mu A_{\gamma y}{}^* + \sin\mu \tan\omega \end{pmatrix}. \tag{4.36}$$

The line s_o normal to s_1 (Fig. 4.5a) is the proper direction the elementary edge at point O should orient in so as to insure the desired rake angle γ_o. The slope $\tan\delta$ can be determined from Eq. (4.36),

$$\tan\delta = -s_{12}/s_{11} = 1/(\sin\mu \tan\omega A_{\gamma x}{}^* - \sin\mu A_{\gamma y} * + \cos\mu \tan\omega). \tag{4.37}$$

Taking a segment OO_1 on s_o we get an elementary edge and reach a new point $O_1(x_1, y_1)$,

$$x_1 = x_0 + \Delta, \tag{4.38}$$

$$y_1 = y_0 - \Delta \tan\delta, \tag{4.39}$$

where Δ is the length of the projection of elementary edge OO_1 (in the case of Fig. 4.5a we have $\Delta < 0$).

The above described procedure can be repeatedly continued so as to find a series of points $O_2(x_2, y_2), O_3(x_3, y_3)\ldots$. The corresponding coordinates z_i (i = 1, 2, 3…) can be determined from Eqs. (4.7) and (4.8) and the last one of (4.1). All the points $O_i(x_i, y_i, z_i)$ (i = 1, 2, 3…) may be connected to form the optimized cutting edge which insures the desired rake angle distribution.

4.2.2 The Maximum-Rake-Edge

There exists a special cutting edge line for a twist drill which insures the rake angle at each point reaching its maximum γ_g at the point. It is clear from Fig. 4.5a that when $\gamma_o < \gamma_g$ there are two intersection points of circles *c* and *a-a*, i.e., J_1 and J_2. If, however, $\gamma_o = \gamma_g$, i.e. the desired rake angle is the maximum rake at the point, then the intersections reduce themselves to only one point *J* which coincides with the origin *O*, i.e., P_r^* as shown in Fig. 4.5b. Such being the case, the edge image s^* (after rotation) becomes the connecting line of points A_γ^* and P_r^*. It must be noticed that the incidence relation of images are kept without any change while they are driven by the same moving operator. Bearing this point in mint it is easy to understand that before rotation the edge image *s* must be also the connecting line of A_γ, P_r, i.e., $s = A_\gamma \times P_r$. Thus, the way of finding the maximum-rake-edge is simply to keep the tangent to the edge at each point of the edge line always normal to the vector $\boldsymbol{P_r A_\gamma}$ related to the same point as shown in Fig. 4.5b. On the bases of this the slope of s_o is determined as follows:

$$\tan \delta = (A_{\gamma x} - P_{rx})/(A_{\gamma y} - P_{ry}) = (A_{\gamma x} + z_0/x_0)/A_{\gamma y} \tag{4.40}$$

All the points $O_i(x_i, y_i, z_i)$ (i = 1, 2…) on the edge may be worked out in the same way as mentioned above.

4.2.3 Various Edge Line Shapes Corresponding to Different Rake Distributions

Computer software is developed to compute edge shapes corresponding to various desired rake angle distributions, or to compute the real rake angle distributions from designed edge shapes. Some of the computing results are shown in Fig. 4.6. Figure 4.6a is the rake distribution of standard twist drill under standard grinding. Figure 4.6b shows the edge shape with a step rake distribution. Note that a sudden change in rake angle distribution results a sudden change in the orientation of edge line. Figure 4.6c is the maximum-rake-edge, which bends upward and the rake angle goes to π/2 when the point moving towards the drill axis. The central part of maximum-rake-edge is, of cause, not feasible because of the weakness at the web and the wedge. Figure 4.6d is a combination of Fig. 4.6b, c which insures maximum rake in the outer part (a-b) of the edge and keeps the rake constant in the central part (b-c). Figure 4.6e resents the same edge shape as Fig. 4.6d except with a point at drill axis so as to centralize the drill during a drilling operation. As shown in Fig. 4.6e it is clear that the central point results in negative rake.

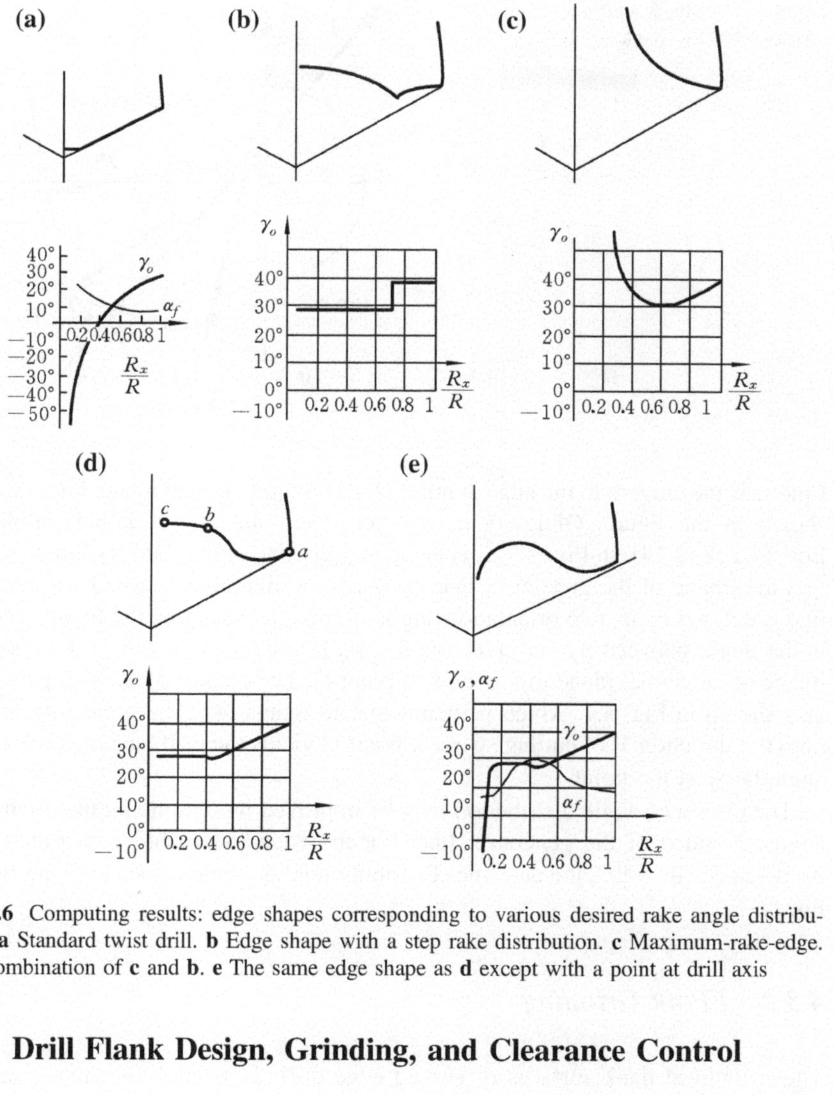

Fig. 4.6 Computing results: edge shapes corresponding to various desired rake angle distributions. **a** Standard twist drill. **b** Edge shape with a step rake distribution. **c** Maximum-rake-edge. **d** A combination of **c** and **b**. **e** The same edge shape as **d** except with a point at drill axis

4.3 Drill Flank Design, Grinding, and Clearance Control

The optimized cutting edge shape of a twist drill is realized by properly grinding its flank. A kind of cylindroid surface is chosen to be the flank in this study.

4.3.1 The Mapping of Flank and Clearance Control

In Fig. 4.7, the end view of a drill is shown with corresponding system *xoz*. Here plane *xoz* is taken as base plane and O (x_o, z_o) is the projection of a given point O (x_o, y_o, z_o) on plane *xoz* and O (x_o, y_o, z_o) is a point on the cutting edge designed.

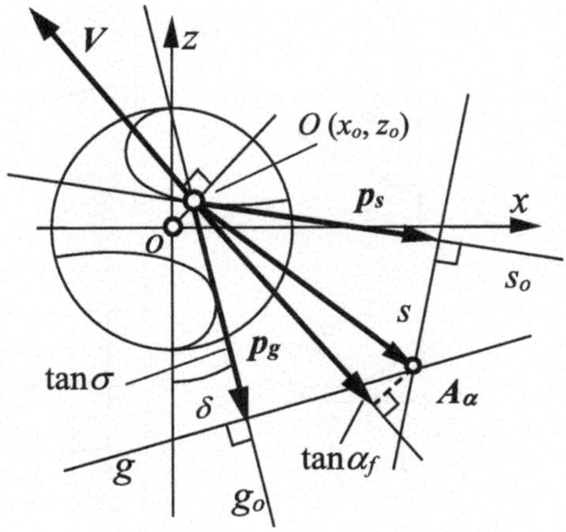

Fig. 4.7 Drill flank and the images related to flank

Line s_o is the tangent to the edge at point O, its obliquity $\boldsymbol{p}_s$ and image line s are also shown in the figure. Obliquity $\boldsymbol{p}_s$ may be determined by equations similar to Eqs. (4.12)–(4.14). In Fig. 4.7 g_o is the generatrix line of the flank cylinder surface, g is the image of the generatrix line and $\boldsymbol{p}_g$ is its obliquity vector. The generatrix line is defined by its two orientation angles δ and σ as shown in the figure, where δ is the angle between $\boldsymbol{p}_g$ and axis z and $\left|\boldsymbol{p}_g\right| = \tan\sigma$. $\boldsymbol{A}_\alpha = \boldsymbol{OA}_\alpha = \boldsymbol{s} \times \boldsymbol{g}$ is the image of tangential plane to the flank at point O. The cutting speed $\boldsymbol{V}$ at point O is also shown in Fig. 4.7, which perpendicular to radius oO. The projecting of $\boldsymbol{OA}_\alpha$ onto the direction $\boldsymbol{V}$ of cutting speed at point O yields the drill tangential clearance angle $\tan\alpha_f$ at the point.

The clearance angle distribution may be improved by optimizing the orientation angles δ and σ of the generatrix line. For the edge shape shown in Fig. 4.6e if $\delta = -24.5°$, $\sigma = 17°$, the clearance distribution is shown in the same figure by fine line.

4.3.2 Flank Grinding

The cylindroid flank surfaces of curved edge drills is ground on a universal tool grinder by means of form grinding. The orientation angle δ and σ of the flank generator are adjusted by a universal vice in which the ground drill is fixed. The grinding wheel is dressed by a diamond dresser according to the optimized edge shape and checked by a template. The curve of the template is calculated from the points $O_i(x_i, y_i, z_i)$, $i = 1, 2, \ldots, n$, on the optimized edge line by the following coordinate translation:

$$(u_i, v_i, w_i)^T = \boldsymbol{X}(-\sigma)\boldsymbol{Y}(-\delta)(x_i, y_i, z)^T, \quad i = 0, 1, 2, \ldots, n, \tag{4.41}$$

where $X(-\sigma)$ and $Y(-\delta)$ are the moving operators similar to Eqs. 2.39 and 2.40, and (u_i, v_i), $i = 0, 1, 2, \ldots,$ n, give the template curve.

4.3.3 *The Inspection of Edge Line Shape*

It is necessary to check if the grinding operation has really realized the optimized edge line and the desired rake distribution. For this reason, the projection of the edge line on *xoy* plane is measured by a projector machine so as to get a series of data point (x_i, y_i); $i = 0, 1, 2 \ldots n$, and from the data the rake distribution can be calculated by the procedure and equations above presented.

Figure 4.8a presents desired rake distribution (dotted line) and calculated rake distribution from the really measured edge line data (full line). Figure 4.8b presents the designed edge line (full line). Figure 4.8c presents really ground drill bit and its edge shape. The figure also shows the central point, the chisel edge and Chisel edge thinning. In Fig. 4.8a, the really ground rake turns negative when the edge approaches the drill center (full line). Besides, some deviation between the full line and the dotted line is caused by grinding error.

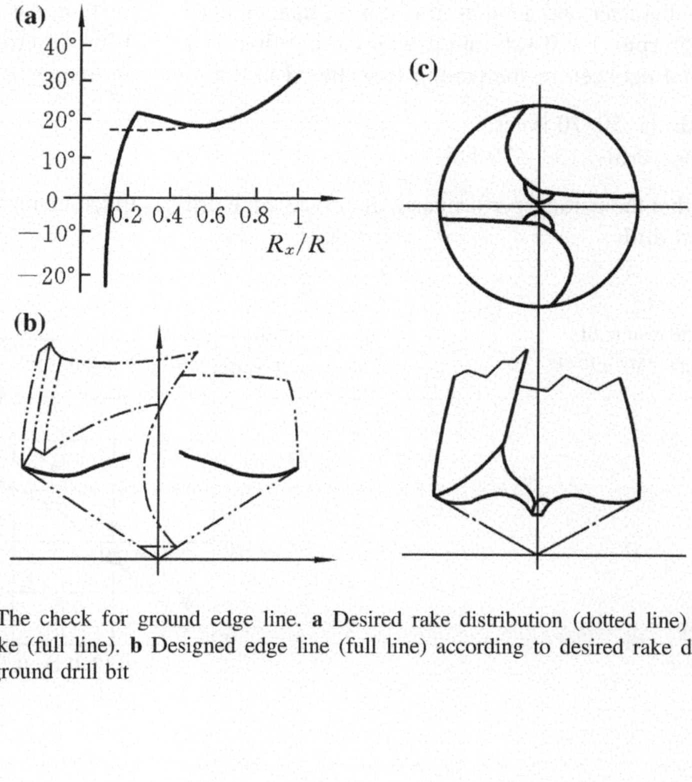

Fig. 4.8 The check for ground edge line. **a** Desired rake distribution (dotted line) and really ground rake (full line). **b** Designed edge line (full line) according to desired rake distribution. **c** Really ground drill bit

4.4 The Drilling Experiments on Curved Edge Drills

4.4.1 Drilling Force

The drilling trust and the torque are measured for both standard drills and for curved edge drills while drilling a work piece of medium carbon steel 45#. The drill diameter is 18 mm. The cutting force is measured by a drilling dynamometer and strain gauge.

Figure 4.9 shows the torque of standard drill and the curved edge drills. As shown in Fig. 4.9, a great deal of experimental date shows that under experimental conditions compared with standard drills the drilling torque of curved edge drills is reduced by about a quarter. If the edge line is more accurately ground the reduction in torque of curved edge drill is expected more notable.

The experiments show that the thrust of curved edge drills is also lower than that of standard drills, but the reduction is not very notable.

4.4.2 Tool Life

Standard drills and curved edge drills were used in workshop for drilling holes of 18 mm in diameter and 18 mm in depth on flanges made of medium carbon steel A_3, n = 355 rpm, f = 0.126 mm/r, with cutting fluid. The tool life in terms of the holes drilled between re-sharpening is scattered in the following ranges:

Standard drills: 50–70 holes,
Curved edge drills: 132–520 holes.

It is clear that the cutting performance of curved edge drills is much better than that of standard drills.

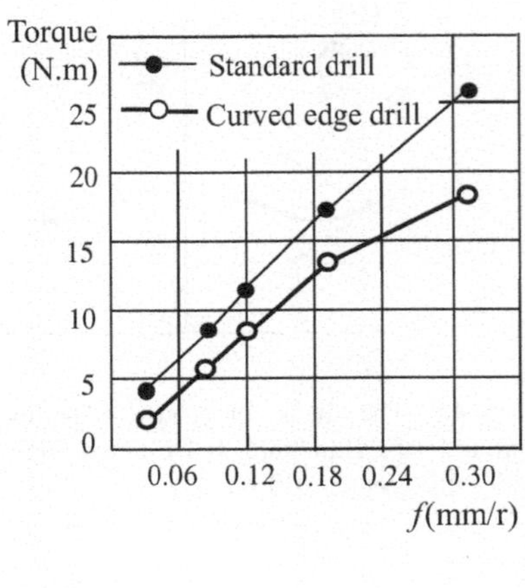

Fig. 4.9 The results of drilling torque experiments

4.5 Remarks

1. It is possible to realize a desired rake angle distribution along the cutting edge of a twist drill by means of controlling the orientation of the edge line point by point, provided the following requirements for the desired rake distribution are fulfilled:

   (1) The distribution is continuous on the whole edge line or continues piecewise without large sudden changes in the distribution which would result in a very sharp and weak point in the edge line;

   (2) $|\gamma_o| \leq \gamma_g$ For all the points, where γ_o is desired rake angle for a given point and γ_g is the maximum rake for the same point. Besides, γ_o should be within a reasonable limit set by machining operation requirements.

2. The cutting edge design may be also done on the bases of a given cutting edge inclination λ_s distribution in a similar way as described in this chapter.
3. The flank of a drill may be ground into a cylindroid surface by means of form grinding. In this way the desired edge line shape can be realized in a simplified grinding operation, and the clearance (α_f) distribution can be controlled by means of optimizing the orientation angles of the cylindroid surface generatrix (δ, σ).
4. The experimental results showed that the author's work on displaying and analyzing space angles in a projective plane is proved to be effective in analyzing angles of complex cutting tools, such as twist drills.

Appendix 4.1: The Numerical Table of Obliquity Field on Rake Surface of Standard Drill Flute

The following table provides the face obliquities of standard twist drill $A_\gamma(A_{\gamma x}, A_{\gamma y})$ and the obliquity of reference plane $P_r = P_{rx}$ at grid points on drill flute as shown in Fig. 4.10. Such data may be used for understanding the incline conditions or for a rough estimation the cutting angle distributions of the various edge line shape and their cutting performs as introduced in this book above (Table 4.1).

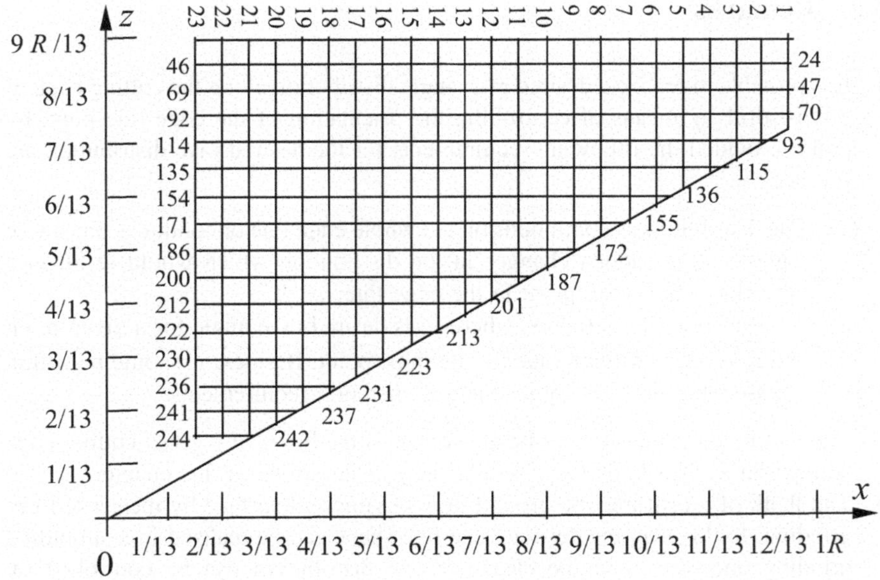

Fig. 4.10 Grid points on the face of standard twist drill

Table 4.1 Face obliquities of standard twist drill A_γ ($A_{\gamma x}$, $A_{\gamma y}$) and the obliquity of reference plane $P_r = P_{rx}$ at grid points on drill flute (the positions of grid points are shown in Fig. 4.10)

No	$A_{\gamma x}$	$A_{\gamma z}$	P_{rx}	No	$A_{\gamma x}$	$A_{\gamma z}$	P_{rx}	No	$A_{\gamma x}$	$A_{\gamma z}$	P_{rx}
1	−0.269	0.573	−0.083	2	−0.240	0.548	−0.076	3	−0.212	0.523	−0.069
4	−0.184	0.499	−0.063	5	−0.156	0.476	−0.058	6	−0.129	0.452	−0.054
7	−0.101	0.429	−0.051	8	−0.074	0.406	−0.049	9	−0.046	0.383	−0.058
10	−0.019	0.360	−0.049	11	0.009	0.338	−0.052	12	0.037	0.315	−0.057
13	0.065	0.292	−0.066	14	0.093	0.269	−0.078	15	0.122	0.246	−0.095
16	0.151	0.223	−0.118	17	0.181	0.199	−0.149	18	0.211	0.175	−0.193
19	0.242	0.150	−0.252	20	0.273	0.125	−0.337	21	0.306	0.100	−0.463
22	0.339	0.074	−0.659	23	0.374	0.046	−0.998	24	−0.294	0.578	−0.106
25	−0.265	0.552	−0.098	26	−0.237	0.527	−0.092	27	−0.208	0.502	−0.086
28	−0.180	0.478	−0.081	29	−0.152	0.454	−0.076	30	−0.124	0.431	−0.073
31	−0.097	0.407	−0.071	32	−0.069	0.384	−0.070	33	−0.041	0.361	−0.071
34	−0.014	0.338	−0.074	35	0.014	0.315	−0.080	36	0.042	0.292	−0.088
37	0.070	0.269	−0.100	38	0.099	0.246	−0.117	39	0.128	0.223	−0.140
40	0.157	0.199	−0.171	41	0.187	0.175	−0.214	42	0.217	0.151	−0.273
43	0.248	0.126	−0.358	44	0.280	0.101	−0.482	45	0.313	0.075	−0.677
46	0.346	0.048	−1.012	47	−0.321	0.584	−0.129	48	−0.291	0.557	−0.121
49	−0.262	0.531	−0.115	50	−0.233	0.506	−0.109	51	−0.204	0.481	−0.103
52	−0.176	−0.457	−0.099	53	−0.148	0.433	−0.096	54	−0.012	0.409	−0.094
55	−0.092	0.386	−0.093	56	−0.065	0.362	−0.094	57	−0.037	0.339	−0.097
58	−0.009	0.316	−0.102	59	0.019	0.293	−0.110	60	0.047	0.270	−0.122

(continued)

Table 4.1 (continued)

No	$A_{\gamma x}$	$A_{\gamma z}$	P_{rx}	No	$A_{\gamma x}$	$A_{\gamma z}$	P_{rx}	No	$A_{\gamma x}$	$A_{\gamma z}$	P_{rx}
61	0.076	0.246	−0.139	62	0.104	0.223	−0.161	63	0.133	0.199	−0.193
64	0.163	0.175	−0.235	65	0.193	0.151	−0.294	66	0.223	0.127	−0.378
67	0.255	0.101	−0.501	68	0.287	0.076	−0.694	69	0.319	0.049	−1.027
70	−0.348	0.590	−0.152	71	−0.318	0.563	−0.144	72	−0.288	0.536	−0.138
73	−0.258	0.510	−0.131	74	−0.229	0.485	−0.126	75	−0.201	0.460	−0.122
76	−0.172	0.436	−0.118	77	−0.154	0.411	−0.116	78	−0.116	0.387	−0.115
79	−0.088	0.364	−0.116	80	−0.060	0.340	−0.119	81	−0.032	0.317	−0.124
82	−0.004	0.294	−0.132	83	0.024	0.270	−0.144	84	0.052	0.247	−0.161
85	0.081	0.223	−0.183	86	0.110	0.200	−0.214	87	0.139	0.176	0.257
88	0.168	0.152	−0.315	89	0.199	0.127	−0.388	90	0.229	0.102	−0.521
91	0.261	0.077	−0.712	92	0.293	0.570	−0.168	93	−0.345	0.051	−1.042
94	−0.314	0.542	−0.161	95	−0.284	0.515	−0.155	96	−0.255	0.489	−0.149
97	−0.226	0.464	−0.145	98	−0.197	0.439	−0.141	99	−0.168	0.414	−0.139
100	−0.140	0.390	−0.138	101	−0.112	0.366	−0.139	102	−0.084	0.342	−0.141
103	−0.056	0.318	−0.146	104	−0.028	0.295	−0.154	105	0.001	0.271	−0.166
106	0.029	0.248	−0.183	107	0.057	0.224	−0.205	108	0.086	0.200	−0.236
109	0.115	0.177	−0.278	110	0.144	0.153	−0.336	111	0.174	0.128	−0.419
112	0.204	0.103	−0.540	113	0.235	0.078	−0.730	114	0.267	0.053	−i.058
115	−0.342	0.549	−0.184	116	−0.311	0.521	−0.178	117	−0.281	0.494	−0.172
118	−0.251	0.468	−0.168	119	−0.222	0.443	−0.164	120	−0.193	0.417	−0.162
121	−0.164	0.393	−0.161	122	−0.136	0.368	−0.161	123	−0.108	0.344	−0.164
124	−0.079	0.320	−0.169	125	−0.051	0.296	−0.177	126	−0.023	0.272	−0.189
127	0.005	0.249	−0.205	128	0.034	0.225	−0.227	129	0.062	0.201	−0.258
130	0.091	0.177	−0.300	131	0.120	0.153	−0.357	132	0.150	0.129	−0.440
133	0.179	0.105	−0.560	134	0.210	0.08	−0.749	135	0.241	0.054	−1.074
136	−0.308	0.500	−0.196	137	−0.277	0.473	−0.191	138	−0.248	0.447	−0.187
139	−0.218	0.421	−0.185	140	−0.189	0.396	−0.183	141	−0.160	0.371	−0.184
142	−0.132	0.346	−0.187	143	−0.103	0.322	−0.191	144	−0.075	0.298	−0.199
145	−0.047	0.274	−0.211	146	−0.018	0.250	−0.227	147	0.010	0.260	−0.249
148	0.038	0.202	−0.280	149	0.067	0.179	−0.321	150	0.096	0.154	−0.379
151	0.125	0.130	−0.460	152	0.155	0.106	−0.580	153	0.185	0.081	−0.768
154	0.215	0.056	−1.091	155	−0.274	0.452	−0.211	156	−0.244	0.425	−0.208
157	−0.214	0.399	−0.207	158	−0.185	0.374	−0.207	159	−0.156	0.349	−0.209
160	−0.128	0.324	−0.214	161	−0.099	0.300	−0.222	162	−0.071	0.276	−0.233
163	−0.042	0.252	−0.249	164	−0.014	0.228	−0.272	165	0.014	0.204	−0.302
166	0.043	0.180	−0.343	167	0.072	0.156	−0.400	168	0.101	0.131	−0.481
169	0.130	0.107	−0.601	170	0.160	0.082	−0.781	171	0.190	0.057	−1.108
172	−0.240	0.404	−0.230	173	−0.211	0.378	−0.230	174	−0.181	0.352	−0.232
175	−0.152	0.327	−0.237	176	−0.124	0.302	−0.245	177	−0.059	0.278	−0.256
178	−0.067	0.254	−0.272	179	−0.038	0.229	−0.294	180	−0.010	0.205	−0.324
181	0.019	0.181	−0.365	182	0.048	0.157	−0.422	183	0.077	0.133	−0.503
184	0.106	0.109	−0.622	185	0.135	0.084	−0.817	186	0.165	0.059	−1.126
187	−0.237	0.382	−0.254	188	−0.207	0.356	−0.256	189	−0.178	0.330	−0.260
190	−0.148	0.305	−0.268	191	−0.120	0.280	−0.279	192	−0.091	0.256	−0.295

(continued)

Table 4.1 (continued)

No	$A_{\gamma x}$	$A_{\gamma z}$	P_{rx}	No	$A_{\gamma x}$	$A_{\gamma z}$	P_{rx}	No	$A_{\gamma x}$	$A_{\gamma z}$	P_{rx}
193	−0.062	0.231	−0.317	194	−0.034	0.207	−0.346	195	−0.005	0.183	−0.387
196	0.023	0.159	−0.444	197	0.052	0.134	−0.524	198	0.081	0.110	−0.642
199	0.110	0.086	−0.827	200	0.140	0.061	−1.144	201	−0.203	0.334	−0.284
202	−0.174	0.308	−0.291	203	−0.144	0.283	−0.302	204	−0.116	0.258	−0.318
205	−0.087	0.234	−0.339	206	−0.058	0.209	−0.369	207	−0.029	−0.185	−0.410
208	−0.001	0.160	−0.466	209	0.028	0.136	−0.546	210	0.057	0.112	−0.664
211	0.086	0.087	−0.847	212	0.116	0.063	−1.163	213	−0.170	0.286	−0.325
214	−0.140	0.261	−0.341	215	−0.111	0.236	−0.362	216	−0.083	0.211	−0.392
217	−0.540	0.187	−0.432	218	−0.205	0.162	−0.488	219	0.004	0.138	−0.568
220	0.032	0.113	−0.685	221	0.061	0.089	−0.868	222	0.090	0.064	−1.183
223	−0.136	0.239	−0.386	224	−0.107	0.214	−0.415	225	−0.078	0.189	−0.455
226	−0.050	0.164	−0.511	227	−0.021	0.140	−0.590	228	0.008	0.115	−0.707
229	0.037	0.091	−0.870	230	0.066	0.066	−1.203	231	−0.103	0.191	−0.478
232	−0.074	0.167	−0.534	233	−0.045	0.142	−0.613	234	−0.017	0.117	−0.730
235	0.012	0.093	−0.911	236	0.041	0.068	−1.224	237	−0.099	0.169	−0.559
238	−0.070	0.144	−0.636	239	−0.041	0.119	−0.752	240	−0.012	0.095	−0.934
241	0.016	0.070	−1.246	242	−0.066	0.122	−0.776	243	−0.037	0.097	−0.957
244	−0.008	0.072	−1.268	245				246			
247				248				249			

Chapter 5
Approximate Solutions for Space Angle Problems

In the vicinity of original points, the geometrical property of an elliptic plane is very close to that of Euclidian plane, so, when the obliquities of all the planes and lines involved are relatively small we can use Euclidian geometry to approximately replace non-Euclidian geometry and thus greatly simplify the procedure of analyses and calculation or graphic determination of space angle problem with necessary precision; or at least, this approximate approach method can display the varying trend of various angles in mutual effects and it is convenient for optimization tool design by means of error and try.

5.1 The Theoretical Basis for the Approximate Solution of Space Angles

5.1.1 The Euclidean Approximation of the Non-Euclidean Metric Properties in the Origin Vicinity on Elliptical Plane

Substituting Eq. 2.4 into Eq. 2.12 enables the distance between points A and B on projective plane O to be expressed in terms of their Cartesian coordinates (see Fig. 5.1a)

$$\tan S_{AB} = \overline{AB} \cdot K, \tag{5.1}$$

where

$$\overline{AB} = \sqrt{(B_x - A_x)^2 + (B_y - A_y)^2}, \tag{5.2}$$

H. Shi, *Metal Cutting Theory*, Springer Series in Advanced Manufacturing,
https://doi.org/10.1007/978-3-319-73561-0_5

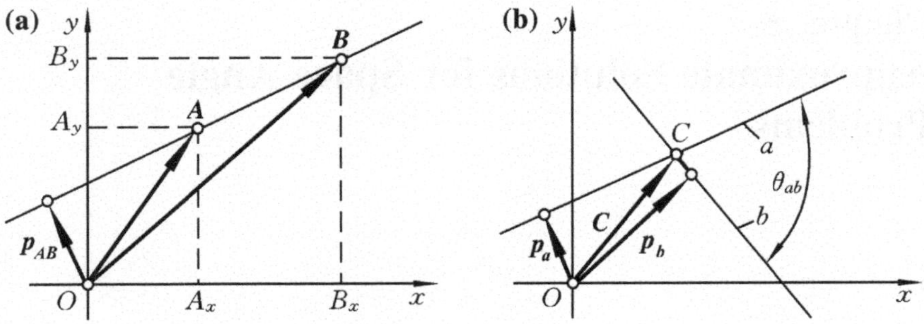

Fig. 5.1 The Euclidean approximation of non-Euclidean metric properties in the origin vicinity on elliptical plane. **a** The Euclidean approximation of non-Euclidean distance. **b** The Euclidean approximation of non-Euclidean angle

$$K = \sqrt{1 + p_{AB}^2} \Big/ (1 + \boldsymbol{A} \cdot \boldsymbol{B}), \tag{5.3}$$

$$p_{AB} = \frac{|\boldsymbol{A} \times \boldsymbol{B}|}{\overline{AB}}. \tag{5.4}$$

where, $\overline{AB}$ is the Euclidian distance between points A and B, K is the length correction factor caused by the deviation of geometrical property of elliptic plane from that of Euclidian plane, $\boldsymbol{p}_{\boldsymbol{AB}}$ is the obliquity of image line AB, $\boldsymbol{A} \times \boldsymbol{B}$ is the cross production and $\boldsymbol{A} \cdot \boldsymbol{B} = A_x B_x + A_y B_y$, i.e. the dot production of obliquity vectors of image points A and B.

In the vicinity of origin, $\boldsymbol{p}_{AB}^2 \ll 1$ and $\boldsymbol{A} \cdot \boldsymbol{B} \ll 1$, neglecting the second order micro amount $\boldsymbol{p}_{AB}^2$ and $\boldsymbol{A} \cdot \boldsymbol{B}$, we have $K \approx 1$, and because $S_{AB} \ll 1$, so tan $S_{AB} \approx S_{AB}$, therefore

$$S_{AB} \approx \overline{AB}, \tag{5.5}$$

This means that in the vicinity of origin non-Euclidian distance S_{AB} maybe approximately replaced by Euclidian distance $\overline{AB}$.

To move forward a single step, since in the vicinity of origin we have $A_x = \tan \alpha_A \approx \alpha_A$, $A_y = \tan \beta_A \approx \beta_A$, thus Eq. (5.2) may be simplified as

$$\overline{AB} = \sqrt{(\alpha_B - \alpha_A)^2 + (\beta_B - \beta_A)^2}, \tag{5.6}$$

This means that in the vicinity of origin the coordinates of image points can be directly expressed by the obliquity angle instead of the tangent of the angle.

Substituting Eqs. (2.7) and (2.9) into Eq. (2.16) we can express the non-Euclidian angle between two image lines by their obliquity vectors (see Fig. 5.1b),

$$\tan\theta_{ab} = \tan\angle ab \cdot k, \tag{5.7}$$

where,

$$\tan\angle ab = \frac{|\boldsymbol{p}_a \times \boldsymbol{p}_b|}{|\boldsymbol{p}_a \cdot \boldsymbol{p}_b|}, \tag{5.8}$$

$$k = \frac{\sqrt{1+C^2}}{1+\boldsymbol{p}_a \cdot \boldsymbol{p}_b/\cos\angle ab}, \tag{5.9}$$

$$C = \boldsymbol{p}_a \cdot \boldsymbol{p}_b |\boldsymbol{p}_a - \boldsymbol{p}_b| / |\boldsymbol{p}_a \times \boldsymbol{p}_b|. \tag{5.10}$$

where, $\angle ab$ is the Euclidian angle between two image lines a and b, k is the angle correction factor caused by the deviation of geometrical property of elliptic plane from that of Euclidian plane. C is the obliquity of the intersection point of image lines a, b.

In the vicinity of origin, $C^2 \ll 1$ and $\boldsymbol{p}_a \cdot \boldsymbol{p}_b \ll 1$, neglecting the second order micro amount C^2 and $\boldsymbol{p}_a \cdot \boldsymbol{p}_b$, we have $k \approx 1$, therefore

$$\theta_{ab} \approx \angle ab. \tag{5.11}$$

This means that that in the vicinity of origin, non-Euclidian angle θ_{ab} may be approximately replaced by Euclidian angle $\angle ab$.

5.1.2 "Small Inclination" Hypothesis and the Approximate Solution for Space Angle Problems

If a problem involves planes and lines which all have relatively smaller obliquities, then based on the approximate Eqs. (5.5), (5.6) and (5.11) the resolution procedure of the problem can be greatly simplified.

Based on "Small inclination" hypothesis, the scope of calculation and analysis is limited in the vicinity of origin, and those plane or lines having relatively larger obliquities, in approximate calculation cannot be taken into consideration. If the planes or lines have very large obliquities, however, we can analyze and calculate their normal lines or normal planes, which will certainly fall into the vicinity of origin. In a word, if a problem involves whether small obliquity angles ($\gamma \ll 1$ rad) or extraordinary large obliquity angles ($\gamma \approx \pi/2$ rad), approximate solution can be considered.

Under this approximation, the angle between two space incline planes appears as the Euclidian distance between their images on projective plane O, which can be calculated through Eq. (5.6), or just measured on projective plane by means of a ruler. The angle between two space incline lines appears as the Euclidian angle between their images on projective plane O, which can be calculated through Eq. (5.8), or just measured on projective plane by means of a protractor.

Under this approximation, Eqs. (2.22)–(2.25) become

$$\begin{cases} A_{2x} = A_{1x}, \\ A_{2y} = A_{1y} - \theta. \end{cases} \tag{5.12}$$

$$\begin{cases} A_{2x} = A_{1x} + \theta, \\ A_{2y} = A_{1y}. \end{cases} \tag{5.13}$$

This means that when a space plane rotating about axis x by a positive angle, its image moves along axis x and towards the direction of negative axis y; when a space plane rotating about axis y by a positive angle, its image moves along axis y and towards the direction of positive axis x.

5.2 Approximate Calculation and Graphical Solution of Cutting Tool Angles

The following examples show how to use above described approximate methods (including approximate graphics and approximate calculation) to solve cutting tool angles.

5.2.1 The Approximate Solution of Working Angles on a Mechanical Clamping Lathe Tools

Example I The working angles of an insert while a pad is installed underneath it. The insert is supposed to be a regular triangle ABC as shown in Fig. 5.2a, in its grinding coordinate system, each cutting edge has a set of grinding angles: tool cutting edge inclination $\lambda_s = 0°$, tool orthogonal rake $\gamma_o = 10°$, tool orthogonal clearance $\alpha_o = 11°$. A pad is also of regular triangle shape and has an incline angle of its top surface as shown in Fig. 5.2a. The pad is installed on the cutter body with its installing base O', which is the working reference plane of the insert, the insert is installed on the pad with its installing base O, which is the grinding reference plane. Obviously, after installed on pad the three edges got a set of working angles. The pad has three sides 1, 2, 3 and different side turned beneath the major cutting edge of the insert results in different working angles. Thus we may have three sets of working angles: $\lambda_1, \gamma_1, \alpha_1$; $\lambda_2, \gamma_2, \alpha_2$; $\lambda_3, \gamma_3, \alpha_3$. Find these 9 working angles through approximate approach.

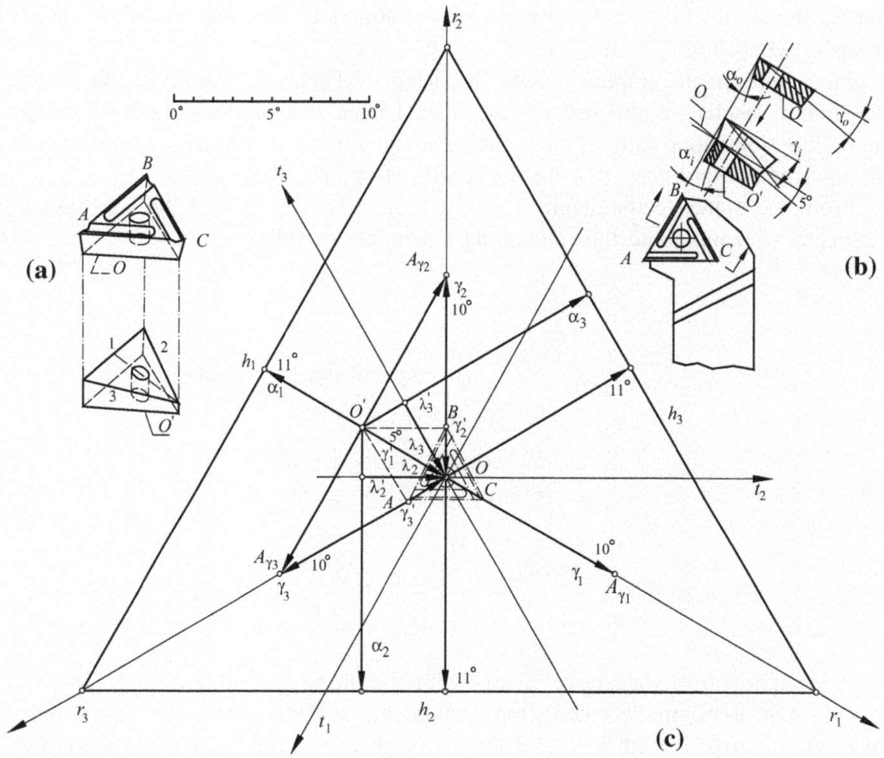

Fig. 5.2 Approximate solution of working angles for mechanical clamping lathe tools when a pad is installed underneath the insert. **a** The insert and the pad. **b** The installation of the insert and the pad on the shank. **c** The graphic determination of working angles by means of approximate method

Solution

Take the grinding reference plane O as the base plane (Fig. 5.2c), draw the profile of the insert with its three cutting edges AB, BC and CA, and draw t_1, t_2, t_3, the axes in the directions of tool cutting edge planes (each is parallel with the corresponding edge and pointing from the tool nose towards the rear part of the insert), draw r_1, r_2, r_3, the axes in the directions of tool orthogonal planes (each is perpendicular to the corresponding edge and pointing from the tool edge to the inside of the insert), draw the images of tool face $A_{\gamma1}$, $A_{\gamma2}$, $A_{\gamma3}$, draw the normals h_1, h_2, h_3 to the flanks, and at last, draw image point of working reference plane O' on grinding reference plane O. In the figure the tangents of the angles are directly replaced by the angles themselves e.g. tan γ_1 is replaced by γ_1, etc.

To work out working angles what we need to do is just to move origin from O to O'. Then projecting $\boldsymbol{O'A_{\gamma i}}(i = 1, 2, 3)$, the obliquities of faces $A_{\gamma1}$, $A_{\gamma2}$, $A_{\gamma3}$ with respect to working reference O', onto the axes t_i, $r_i(i = 1, 2, 3)$, respectively, result in angle λ_i and γ_i; and from O' drawing normals to h_i, respectively, result in $\alpha_i(i = 1, 2, 3)$. Besides, projecting $\boldsymbol{O'O}$ onto the axes t_i, r_i, respectively, result in λ_i'

and γ_i', the cutting edge inclination and the orthogonal rake of the working sides of the pad, where $i = 1, 2, 3$.

The approximate graphic results are listed in Tables 5.1 and 5.2, the precise calculation results are also listed in the same Tables. It is obvious that even though including both approximate error and graphic error the results of approximate graphic method are very close to the results of precise calculation.

From the plane vector diagram in Fig. 5.2c it is easy to find the approximate calculating formula, and the calculating results are as follows:

$$\begin{aligned}
\lambda_1' &= 0^\circ \\
\gamma_1' &= 5^\circ \\
\lambda_2' &= 5^\circ \cdot \cos 30^\circ = 4^\circ\, 19'\, 48'' \\
\gamma_2' &= -5^\circ \cdot \sin 30^\circ = -2^\circ\, 30'\, 0'' \\
\lambda_3' &= -5^\circ \cdot \cos 30^\circ = -4^\circ\, 19'\, 48'' \\
\gamma_3' &= -5^\circ \cdot \sin 30^\circ = -2^\circ\, 30'\, 0'' \\
\lambda_i &= \lambda_i', \quad i = 1, 2, 3 \\
\gamma_i &= 10^\circ + \lambda_i', \quad i = 1, 2, 3 \\
\alpha_i &= 11^\circ - \lambda_i', \quad i = 1, 2, 3
\end{aligned}$$

A comparison of the above results with the data listed in Tables 5.1 and 5.2 reveals that approximate calculation method has a higher precision than graphic method, the error is within 3′, and for the purpose of cutting tool angles analyses and calculation it is perfectly capable of meeting the accuracy requirement.

Example II The working angles of an insert after installed on the shank.
The insert is supposed to be an equilateral five-side polygon indicated by OABCD as shown in Fig. 5.3. Each edge of the insert has a set of grinding angles: rake $\gamma_o = 20^\circ$, edge inclination $\lambda_s = 0^\circ$, clearance $\alpha_o = 0^\circ$ and tool included angle $\varepsilon_r = 108^\circ$ with respect to its grinding base O. Insert is installed on the shank, the installing base P, i.e. the working reference plane, is $\phi = 10^\circ$ inclined with respect to the grinding base O in the direction of the bisector of tool included angle as shown in Fig. 5.3. The tool cutting edge angle is $\kappa_r = 60^\circ$. Determine the working

Table 5.1 The working angles when a pad is installed beneath the insert

i	λ_i		γ_i		α_i	
	Approximate graphic method	Precise calculation	Approximate graphic method	Precise calculation	Approximate graphic method	Precise calculation
1	0°	0°	15°	15°	6°	6°
2	4° 22′	4° 19′ 48″	7° 32′	7° 31′ 48″	13° 34′	13° 27′ 36″
3	−4°19′	−4° 19′ 48″	7° 28′	7° 31′ 48″	13° 34′	13° 27′ 36″

Table 5.2 The working angles of the pad sides

i	λ'_i		γ'_i	
	Approximate graphic method	Precise calculation	Approximate graphic method	Precise calculation
1	0°	0°	5°	5°
2	4° 22′	4° 19′ 48″	−2° 32′	−2° 30′ 0″
3	−4° 19′	−4° 19′ 48″	−2° 36′	−2° 30′ 0″

angle of the insert with respect to the working base P: λ_{se}, γ_{oe}, α_{oe} related to the major cutting edge and λ'_{se}, γ'_{oe}, α'_{oe} related to the minor cutting edge.

Solution

Take the grinding base O as the base plane (its image is at the origin of the projective plane), draw axes of various directions and obliquity vectors with respect to O, such as the directions of feed f, the tool orthogonal plane r, the tool cutting edge t, the tool minor orthogonal plane r' and the tool minor orthogonal plane t', as shown in Fig. 5.3.

As stated above, the tool cutting edge angle is considered as $\kappa_r = 60°$, however, this should be measured in working reference plane P as the angle between the projections of major cutting edge and the feed direction f on plane P, in grinding

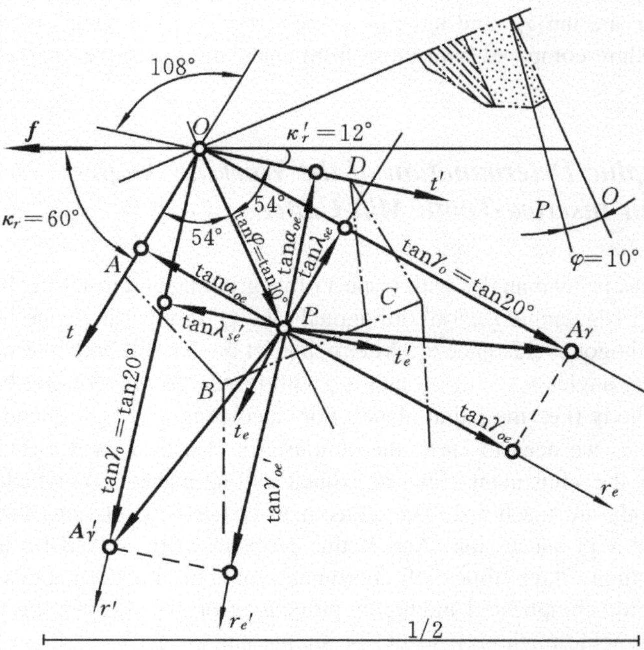

Fig. 5.3 Approximate graphic determination of working angles of a mechanical clamping insert tool

Table 5.3 Working angles of an insert after installed on a shank

Methods	Angles					
	Edge inclination		Rake		Clearance	
	$\tan \lambda_{se}$ $\tan \lambda'_{se}$	λ_{se} λ'_{se}	$\tan \gamma_{oe}$ $\tan \gamma_{oe}'$	γ_{oe} γ_{oe}'	$\tan \alpha_{oe}$ $\tan \alpha_{oe}'$	γ_{oe} γ_{oe}'
Graphics	−0.1025	−5° 51′	0.222	12° 31′	0.1425	8° 7′
Calculation		5° 51′		12° 44′		8° 7′

base O (Fig. 5.3), the angle between the projections of major cutting edge and the feed direction is, strictly speaking, not the κ_r, but the error is negligible for an approximate solution, and we just regard the angle in grinding base O is $\kappa_r = 60°$. According to $\gamma_o = 20°$ and $\lambda_s = 0°$ determine the positions of images of A_γ, A_γ', According to ϕ = 10° determine the position of image P on bisector of tool included angle. Because $\alpha_o = 0°$, so the image points A_α and A_α' are at the infinite points of axes r and r' and the obliquities of their normal become zero, and image lines become axes t and t', respectively.

The remaining steps is just translating the origin of the orthogonal coordinates (t, r) and (t', r') from O to P and this step results in new working systems (t_e, r_e) and (t_e', r_e'), and then projecting the obliquities $\boldsymbol{PA}_\gamma$ and $\boldsymbol{PA}'_\gamma$ of A_γ and A_γ' with respect to plane P onto t_e, r_e 和 t_e', r_e', respectively, and this results in working angles λ_{se}, γ_{oe}, λ'_{se} and γ_{oe}', and the length of the sections on the negative parts of axes r_e, r_e' cut by t and t' are $\tan \alpha_{oe}$ and $\tan \alpha_{oe}'$, respectively. All the results of the above graphic procedure compared with those from calculation are given in Table 5.3.

5.2.2 *Graphic Determination of the Working Angles of an Inserted-Tooth Mill Cutter*

Every insert has its own angles with respect of its grinding and installing base plane O: tool cutting edge angle κ_r, tool orthogonal rake γ_o, tool cutting edge inclination λ_s and tool orthogonal clearance α_o. After being set on the mill body the whole mill has its working angles $\kappa_{\gamma e}$, γ_{oe}, λ_{se} and α_{oe} with respect to the working base plane P of the mill body (i.e., the radial plane). For calculating $\kappa_{\gamma e}$, γ_{oe}, λ_{se} and α_{oe} from κ_r, γ_o, λ_s and α_o we need to know the obliquity of P with respect to O, which is determined by the adjustment of the horizontal milling machine on which the slots of the mill body are machined. The adjustment consists of rotations thrice about axes z, y, and x in succession. And if this problem to be solved by traditional analytical method, three times of coordinate rotation transformations well be involved, it is too complicated and highly prone to error. If, however, it is solved by approximate graphic method, it would be simple and clear.

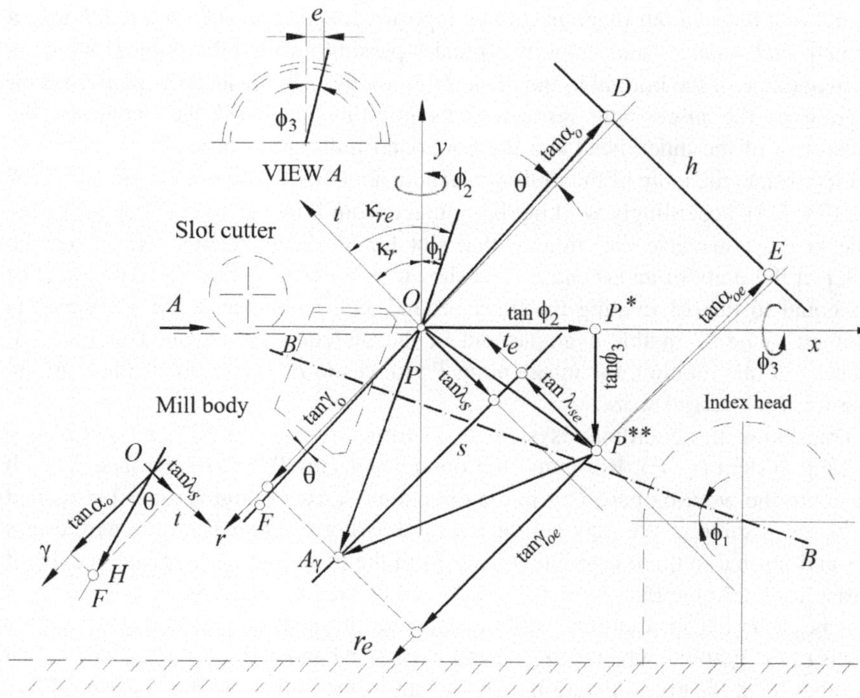

Fig. 5.4 Graphic determination of the working angles of an inserted-tooth mill cutter

As shown in Fig. 5.4, while machined the mill body is installed on the arbor of an index head, and the latter is mounted on the table of a horizontal milling machine. The arbor (i.e., the axis of the mill body *BB*) of the index head is adjusted to tilt the angle of $\phi_1 = 20°$ in relation to the table surface. The grinding angles of the inserts are $\gamma_o = 12°$, $\lambda_s = -6°$, $\alpha_o = 15°$, $\kappa_\gamma = 45°$ (normally, rake γ_o and edge inclination λ_s of the inserts are set to zero, here, for sake of generality, they are supposed not zero).

Because of the adjustment of ϕ_1, tool cutting edge angle κ_r changes to $\kappa_{re} = \kappa_r + \varphi_1 = 45° + 20° = 65°$.

It is supposed that at this moment the grinding base *O* and the working base plane *P* of the mill body are coincide with each other and both located at the origin as shown in Fig. 5.4. Set axis *t* in the direction of cutting edge and axis *r* in its normal direction. Determine the position of image A_γ and its obliquity $\boldsymbol{OA_\gamma}$ with respect to grinding base *O* by its two coordinates $\tan\lambda_s$ and $\tan\gamma_o$. The obliquity of cutting edge is $\tan\lambda_s$, so passing through the end point of the obliquity draw its normal line *s* which is the image of cutting edge.

The image of tool flank A_α must at line *OF*, the angle between line *OF* and axis *r* is $\theta = \arctan(\tan\lambda_s/\cot\alpha_o) = \arctan(\tan\lambda_s\tan\alpha_o) = 1.6°$ (see the small figure at lower left quarter of Fig. 5.4). Because $\cot\alpha_o$ is very large we have to use the normal h of flank in our diagram. In the opposite direction of ***OF*** on line OF take a segment OD = tan α_o, and draw its normal h passing through the point D which is the image line of the normal to the flank. Now we already finished the planar vector diagram of the insert with respect to its grinding base and let's continue the adjustment of the index head and the horizontal milling machine.

First, rotate the table of the milling machine about axis y over an angle $\phi_2 = 10°$ (see Fig. 5.4), accordingly working base image from P moves to P^*, then move the table in the transverse direction so that slot for the insert moves away from the center of the body of an eccentric e, as shown in View A of Fig. 5.4. The effect of e is equal to that of rotating the body around axis x over an angle ϕ_3, which is assumed to be 7° in this example and in the direction as indicated in Fig. 5.4. Because of this rotation, the image point P^* moves to P^{**}, and this is the working base for the insert to work.

Translating the coordinate system (t, r) from origin O to P^{**}, we get a new working system (t_e, r_e). Resolving the obliquity vector $\boldsymbol{P^{**}A_\gamma}$ of the face A_γ with respect to the working base P^{**} in the directions of axes t_e and r_e gives tan λ_{se} and tan γ_{oe}, respectively. We may get tan α_{oe} by drawing a vector from the new origin P^{**} and normal to the image line h. And from the magnitude of relevant vectors, it is easy to determine that $\lambda_{se} = 6.1°$, $\gamma_{oe} = 14.0°$ and $\alpha_{oe} = 13.5°$.

It is clear that in a similar way an opposite problem can be easily solved by means of the graphic: when γ_o, λ_s, α_o and κ_r on one hand and γ_{oe}, λ_{se}, α_{oe} and κ_{re} on the other hand are given, determine how to adjust the miller and the index head, i.e. to determine the adjustment ϕ_1, ϕ_2 and ϕ_3.

It is also easy to understand that from the geometrical relations shown in Fig. 5.4, it's easy to write the simple approximate mathematical formulae of grinding angles, working angles and the adjustment angles and it is possible to intuitively estimate the direction and the intensity of interactive effect between various angles, and it is convenient to design and to analyze the cutting angles of the inserted miller.

5.2.3 *Dynamic Cutting Angles Under Complex Cases*

Under the situation of multifactor concurrency, such as tools installation error, the effect of feed motion and vibration occurrence etc. analysis and calculation of the working angle or the dynamic angle of the tool, is a very complex issue.

It is supposed that the tool point is installed δ mm above the work piece center (see Fig. 5.5a); the longitudinal feed rate is f mm/revolution (see Fig. 5.5b) and turning tool is vibrating relative to the work piece along the y axis. Try to find the dynamic angles. It is almost impossible to solve such a problem by traditional precise methods, this approximate approach is, however, very simple and neat. The analysis procedure is as follows:

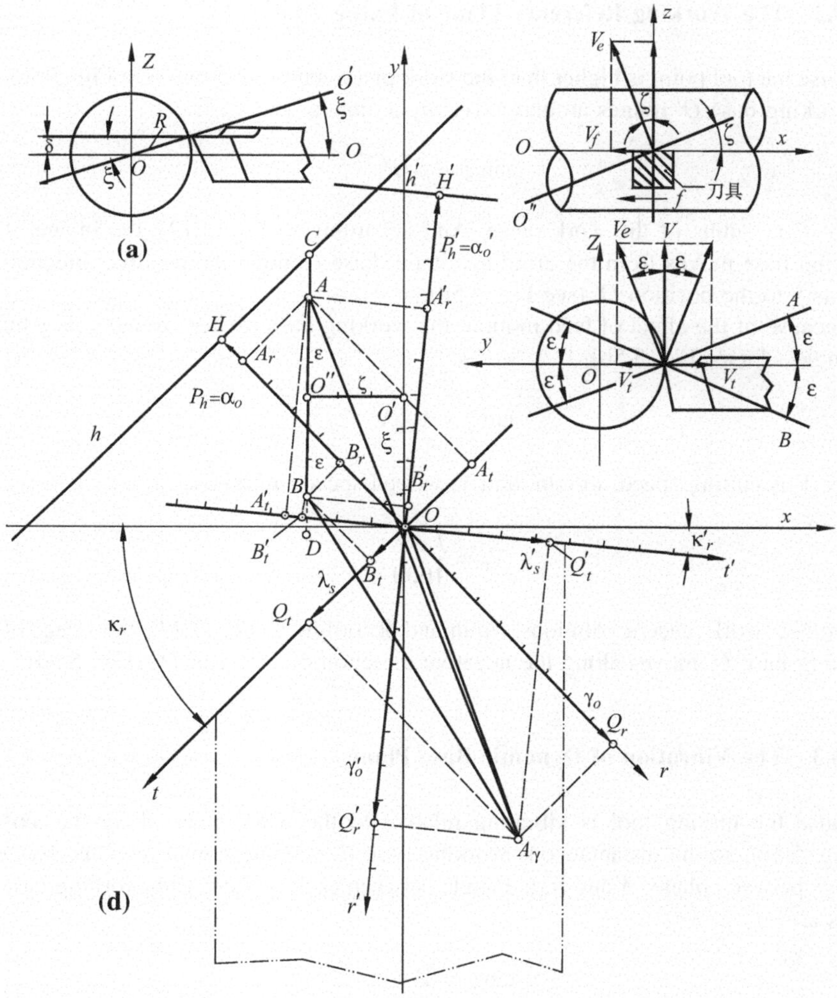

Fig. 5.5 Dynamic cutting angles under complex cases. **a** The effect of tool point installed higher than work piece center. **b** The effect of feed motion. **c** The effect of the cutter vibration in radial direction of work piece. **d** The graphic determination of working angles or the dynamic angles

5.2.3.1 The Static Angle of Lathe Tools (Grinding Angles)

On the bases of grinding angles of the lathe tool κ_r, κ_r', λ_s, λ_s', γ_o, γ_o', α_o and α_o', in Fig. 5.5d draw the profile of the lathe tool, the axes in the directions of tool cutting edge plane, tool orthogonal plane, tool cutting edge plane for minor cutting edge and tool orthogonal plane for minor cutting edge *t*, *r*, *t'* and *r'*, and the images of the rake $A\gamma$, the normal to major flank and the normal to minor flank *h* and *h'* with respect to the grinding (static) base *O*.

5.2.3.2 The Working Reference Plane of Lathe Tool

Because the tool point is higher than the work piece center by δ mm (see Fig. 5.5a), so working base O' rotates around axis x by an angle—ξ,

$$\sin \xi = \delta / R,$$

where R is radius of the work piece. And according to Eq. (5.12), the image of working base moves from the grinding (static) base O along the positive direction of axis y to the position O' (see Fig. 5.5d).

Because of the effect of feed motion, the working base rotates around axis y by an angle—ζ (see Fig. 5.5b)

$$\tan \zeta = V_f / V$$

where V is cutting speed, m/min, and V_f is feed speed, m/min,

$$V_f = \frac{f \cdot N}{1000},$$

where N is work piece revolutions, r/min and according to Eq. (5.13), the image of working base O' moves along the negative direction of axis x to O'' (Fig. 5.5d).

5.2.3.3 The Vibration of Dynamic Base Plane

Because the turning tool is vibrating relative to the work piece along the axis y (Fig. 5.5c), so the instantaneous working base (it become now a dynamic base) swings between planes A and B, the angle between each of them with working base O'' is ε,

$$\tan \varepsilon = V_1 / V.$$

According to Eq. (5.12), the image of dynamic base is vibrating in the direction of axis y with the image O'' as center and ε as amplitude, and its extreme positions are the images of A and B.

5.2.3.4 The Instantaneous Dynamic Working Angles of Lathe Tool

The obliquity of the tool face with respect to the dynamic working base is varying within $\boldsymbol{AA}_\gamma$ and $\boldsymbol{BA}_\gamma$. Projecting the above two vectors onto t, r, t', r' and the obliquity $\boldsymbol{p}_h$, $\boldsymbol{p}_h{}'$ of h, h' results in the varying range of dynamic cutting edge inclination λ_{se}, dynamic orthogonal rake γ_{oe}, dynamic minor cutting edge inclination λ'_{se}, dynamic orthogonal rake for minor cutting edge $\gamma_{oe}{}'$, dynamic orthogonal

Table 5.4 The varying ranges of various dynamic angles

Angles	λ_{se}	γ_{oe}	α_{oe}	λ'_{se}	$\gamma_{oe}{}'$	$\alpha_{oe}{}'$
Corresponding line segments	$\overline{B_tQ_t} \sim \overline{A_tQ_t}$	$\overline{B_rQ_r} \sim \overline{A_rQ_r}$	$\overline{A_rH} \sim \overline{B_rH}$	$\overline{B'_tQ'_t} \sim \overline{A'_tQ'_t}$	$\overline{B'_rQ'_r} \sim \overline{A'_rQ'_r}$	$\overline{A'_rH'} \sim \overline{B'_rH'}$
Varying range of angles	2° 36′ ~6° 48′	11° 48′ ~16° 0′	1° 0′ ~6° 12′	8° 0′ ~7° 24′	9° 30′ ~15° 30′	3° 24′ ~9° 24′

clearance α_{oe} and dynamic orthogonal clearance for minor cutting edge $\alpha_{oe}{}'$. Under the assumed condition of this example, the varying range of various dynamic cutting angles may be directly measured from Fig. 5.5d, and the results are listed in Table 5.4.

It is clear from Fig. 5.5d, that when the working base vibrates to the point A, the minimum clearance is $\alpha_{omin} = \overline{A_rH} = 1°$. Obviously, from the "clearance limitation" point of view, the maximum vibration amplitude of instantaneous dynamic base is $\varepsilon_{\max} = O''C$, once the vibration amplitude exceeds this threshold the clearance would turn to negative and cutting become impossible. It is also clear that if the tool point is installed a litter bit higher than the work piece center, the dynamic base points O', O'' would rise and $\varepsilon_{\max}$ will be reduced. This is what many people know: tool point installed a litter bit higher then work piece center is beneficial for chatter suppression.

From the above described examples we can understand that the essence of approximate method is to replace coordinate rotations with coordinate translations in the vicinity of origin, and this is actually a kind of linearization of the issue.

Part II
Non-Free Cutting Mechanics

Chapter 6
The Principle of Minimum Energy Dissipation in Metal Cutting

In this chapter based on experimental and theoretical studies, a principle of minimum energy dissipation is suggested as a basic physical law restricting the process of metal cutting. The principle may be summarized as that the state parameters of a cutting process, under the precondition of satisfying the constraints set by controlling (technological) parameters, always minimize cutting power consumption. On account of this principle, the basic law of cutting mechanics has been deeply studied. The results of theoretical analysis and predictions are verified and supported by experimental facts, and agree with people's traditional understanding and knowledge about cutting process. Besides giving a general description of the principle of minimum energy dissipation, this chapter focuses on the principle of minimum energy dissipation for the cutting process of an elementary cutting tool, as a basis for further study of the cutting performance of modern complex shaped cutting tools.

6.1 The Meaning of the Principle of Minimum Energy Dissipation

The parameters describing a cutting process may be categorized into two types: the technological parameters, such as cutting parameter (including cutting speed V, revolutions of the machine spindle N, thickness of cut a_c, feed rate f, width of cut a_w and depth of cut a_p, etc.), geometrical parameters on cutting tool (including rake γ_o, edge inclination λ_s, etc.), and the property parameters of the work piece materials, they can be artificially controlled from outside the cutting process, so they are also called controlling parameters; other type of parameters are state parameters, such as chip-ejection vector $\boldsymbol{U}$, which is defined by chip ejection angle ψ_λ (the angle between chip ejection direction and the normal to the cutting edge see Fig. 6.2) and chip-ejection speed U, and the latter is related to shear angle φ, cutting ratio

H. Shi, *Metal Cutting Theory*, Springer Series in Advanced Manufacturing,
https://doi.org/10.1007/978-3-319-73561-0_6

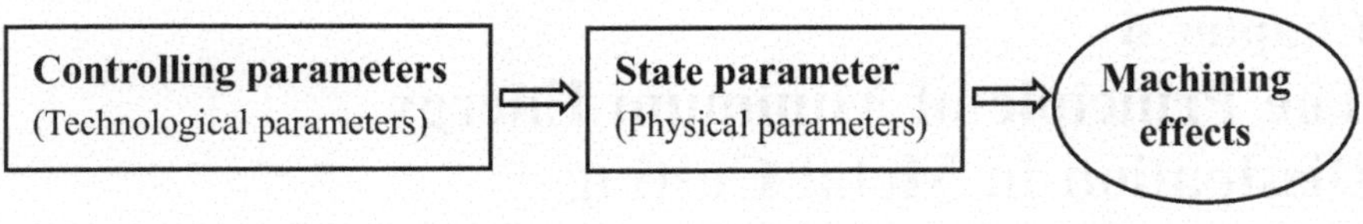

Fig. 6.1 The relations of controlling parameters, state parameters and machining effects

$r = U/V$, in addition, there are cutting force F_c, cutting power A and cutting temperature etc. such type of parameters describes the inner physical state of a cutting process, so it is also called physical parameters. They are restricted by physical law and are determined from inside of a cutting process. They cannot be directly selected according to people's aspiration. What people can do is to select proper technological (controlling) parameter so as to indirectly influence the values of state parameters. People pursuant machining effects, such as machining precision and surface quality, are directly determined by the state parameters but indirectly controlled by technological parameters. The relations of them are shown in Fig. 6.1.

Cutting power A is determined by both the controlling parameters and the state parameters. If denote state parameters as $a, b, c, \ldots$ and controlling parameters as $\alpha, \beta, \gamma, \ldots$, then we have

$$A = A(a, b, c, \ldots; \alpha, \beta, \gamma, \ldots). \tag{6.1}$$

Under a set of given controlling parameters $\alpha, \beta, \gamma, \ldots$ state parameters have many sets of values which all satisfy the constraint conditions set by controlling parameters. The "principle of minimum energy dissipation" asserts that among all the values satisfying constraint conditions, only the set of values $a_o, b_o, c_o, \ldots$, which minimize cutting power $A_{\min}$, is really going to happen, i.e.,

$$A_{\min} = A(a_o, b_o, c_o, \ldots; \alpha, \beta, \gamma, \ldots). \tag{6.2}$$

$a_o, b_o, c_o, \ldots$ are called natural values of state parameters. If the state parameters are by any means forced to deviate from their natural values then this will certainly causes an increase in cutting power. This principle comes from the induction of a lot of test data and observe facts, and based on a kind of philosophy: a natural process under the condition of satisfying constraints always carry on in the least labor-saving way.

In the near-equilibrium thermodynamics there is a "principle of minimum dissipation of energy" [67–69], it means that on the premise of satisfying constraints, the entropy production or energy dissipation of a near-equilibrium states tends to a minimum and any additional forces or human intervention will result in a process of entropy production or energy dissipation increases. This is the physical basis of the above principle of minimum energy dissipation.

In history, the earliest person who used principle of minimum energy dissipation to the research of metal cutting process, was Merchant, M. E. [66]. He thought that during right angle free cutting, the value of shear angle φ must minimize cutting power A. This law has caused many researchers interested in and large amount of follow-up work.

The author as early as in the early 60s last century had been conducted the study of the minimum energy dissipation principle and the smallest resistance chip-ejection direction.

This chapter has broadened the scope of this principle in the research of cutting theory. It is regarded as a basic physical law restricting cutting process. As a matter of fact, the principle of minimum energy dissipation has imposed a very strong limitation upon cutting process and values of state parameters. It is a key to understanding the cutting process of modern tools, and the basis for modeling the complex cutting process of modern tools.

6.2 The Principle of Minimum Energy Dissipation for the Cutting Process of an Elementary Cutting Tool

An elementary cutting tool consists of rake, flank and intersection line of them—the cutting edge, as shown in Fig. 6.2. It is the elements and basis composing various modern complex shaped cutting tools. An elementary cutting tool normally performs oblique cutting. Because the cutting edge of an elementary cutting tool is very short, it may be treated as a segment of a straight line, and the interference among the chip-ejections of different segments on its edge can be neglected. Therefore, while cutting alone, an elementary cutting tool performs free-cutting.

6.2.1 A Theoretical Model of Elementary Cutting Tool Performing Oblique Cutting

In the coordinate system shown in Fig. 6.2, the unit vector in the direction of cutting edge is

$$\mathbf{b} = \cos\lambda_s \cdot \mathbf{i} + 0 \cdot \mathbf{j} - \sin\lambda_s \cdot \mathbf{k}, \tag{6.3}$$

The unit vector in the direction normal to cutting edge is

$$\mathbf{a} = -\sin\gamma_n \cdot \sin\lambda_s \cdot \mathbf{i} + \cos\gamma_n \cdot \mathbf{j} - \sin\gamma_n \cdot \cos\lambda_s \cdot \mathbf{k}, \tag{6.4}$$

where γ_n is cutting edge normal rake,

$$\tan\gamma_n = \tan\gamma_o \cdot \cos\lambda_s. \tag{6.5}$$

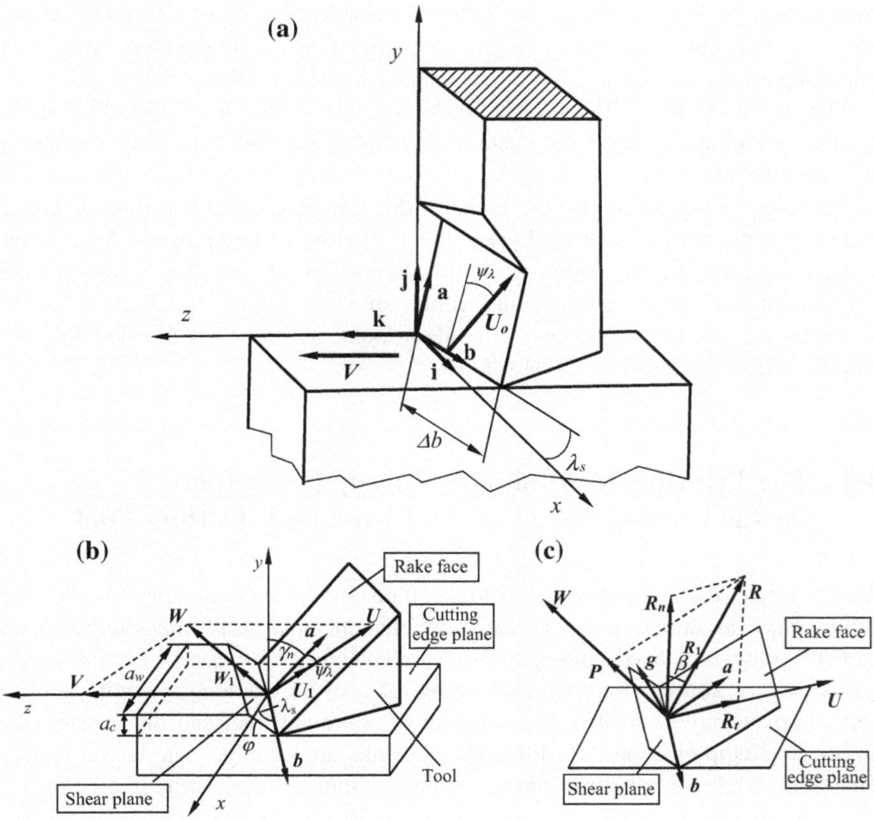

Fig. 6.2 Elementary cutting tool performing oblique cutting and relevant parameters. **a** The elementary cutting tool. **b** The speed vector's relationship. **c** The force vector's relationship

The unit vector perpendicular to the rake face

$$\mathbf{g} = \mathbf{b} \times \mathbf{a}, \tag{6.6}$$

Chip-ejection vector

$$\boldsymbol{U} = rV(\cos\psi_\lambda \mathbf{a} + \sin\psi_\lambda \mathbf{b}). \tag{6.7}$$

Cutting speed vector is $\boldsymbol{V} = V\mathbf{k}$, shear speed vector is $\boldsymbol{W} = \boldsymbol{U} + \boldsymbol{V}$. The unit vector normal to shear plane is

$$\mathbf{e} = \frac{\boldsymbol{W} \times \mathbf{b}}{|\boldsymbol{W} \times \mathbf{b}|}, \tag{6.8}$$

The area of shear plane is

$$s = \frac{a_c a_w}{|\mathbf{e} \cdot \mathbf{k}|}. \tag{6.9}$$

Shear force is $P = s\tau$, where τ is shear stress on the shear plane. The unit vector in the direction of the resultant force imposed on chips by the rake face is

$$\mathbf{R}_1 = \cos\beta\,\mathbf{g} + \sin\beta\,\boldsymbol{U}/U,$$

Here, it is supposed that friction force on the contacting area between the rake face and the bottom face of chips is coincide with the chip-ejection direction. Where β is the friction angle between the rake face and the bottom face of chips. This resultant force is

$$\boldsymbol{R} = R\mathbf{R}_1. \tag{6.10}$$

The component of the force $\boldsymbol{R}$ in the shear direction $\boldsymbol{W}$ is

$$\boldsymbol{R} \cdot \boldsymbol{W}/W = R\mathbf{R}_1 \cdot \boldsymbol{W}/W = P,$$

From this,

$$R = P/(\mathbf{R}_1 \cdot \boldsymbol{W}/W),$$

Substituting the above equation into Eq. (6.10) results in $\boldsymbol{R}$, and finally cutting power $A = \boldsymbol{R} \cdot \boldsymbol{V}$ can be expressed as

$$A = A(r, \psi_\lambda; \gamma_o, \lambda_s, V, a_c, a_w, \tau, \beta), \tag{6.11}$$

where γ_o, λ_s, V, a_c, a_w, τ and β are the controlling parameters and r, ψ_λ are the state parameters.

6.2.2 Computer Simulation for Free Oblique Cutting of Elementary Cutting Tools

Based on the model above described, for different values of $\gamma_o = -20°–40°$, $\lambda_s = 0°–40°$ and $\beta = 20°–50°$, a large number of computer simulation experiments have been done. In the simulation, through numerical approach the values of natural cutting ratio r_o and natural chip-ejection angle $\psi_{\lambda o}$ which minimize cutting power $A = A_{\min}$ are obtained. Under different γ_o, λ_s and β, the values of r_o and $\psi_{\lambda o}$ are listed in Appendices 6.1 and 6.2, respectively.

The values in Appendices 6.1 and 6.2 indicate that γ_o, λ_s and β, all influence the values of r_o and $\psi_{\lambda o}$, however, the main factor affecting r_o is γ_o as well as the main factor affecting $\psi_{\lambda o}$ is λ_s.

6.2.3 The Extremum Characteristics of Cutting Power

In the vicinity of r_o, $\psi_{\lambda o}$ the chip-ejection vector is disturbed, the pictures of cutting power A versus variation of chip-ejection vector are shown in Fig. 6.3. In the figure, ζ and ξ are the relative variation of the chip-ejection vector $\boldsymbol{U}$ in the directions parallel with or perpendicular to cutting edge, respectively (Fig. 6.4),

$$\begin{aligned} \zeta &= \Delta U_t/U_o = (U_t - U_{ot})/U_o \\ \xi &= \Delta U_n/U_o = (U_n - U_{on})/U_o \end{aligned}, \tag{6.12}$$

As shown in Fig. 6.3, while an elementary cutting tool is cutting, the cutting power does exist a minimum value and in the vicinity of this minimum, the relation $A = f(\zeta, \xi)$ may be approximately regarded as a elliptic paraboloid,

$$A = A_{\min}(1 + \mu_1\zeta^2 + \mu_2\xi^2), \tag{6.13}$$

where μ_1 and μ_2 indicate the strength of influence of ζ and ξ on cutting power, respectively. Rearranging the above equation results in

$$\frac{\zeta^2}{\left(\sqrt{\alpha/\mu_1}\right)^2} + \frac{\xi^2}{\left(\sqrt{\alpha/\mu_2}\right)^2} = 1, \tag{6.14}$$

where,

$$\alpha = (A - A_{\min})/A_{\min}$$

When A takes a particular value Eq. (6.14) is a standard equation of ellipse, and μ_1 and μ_2 are related to the long axis and the short axis of the ellipse, respectively.

Equation (6.13) and Fig. 6.3 clearly show that if and only if $\zeta = \xi = 0$, i.e. $\boldsymbol{U} = \boldsymbol{U}_o$, $A = A_{\min}$, otherwise, $A > A_{\min}$. This fully shows that extremum characteristics of cutting power in the vicinity of natural chip-ejection vector $\boldsymbol{U}_o$. Figure 6.3 gives only three examples from large number of the results of numerical experiments. Under various conditions, perturbation of chip-ejection vector $\boldsymbol{U}$ in the vicinity of r_o and $\psi_{\lambda o}$ and fitting the data according to Eq. (6.13) lead to coefficients μ_1 and μ_2, which are listed in Appendix 6.3, and the relation of μ_1–γ_o–λ_s and μ_2–γ_o–λ_s are shown in Fig. 6.5.

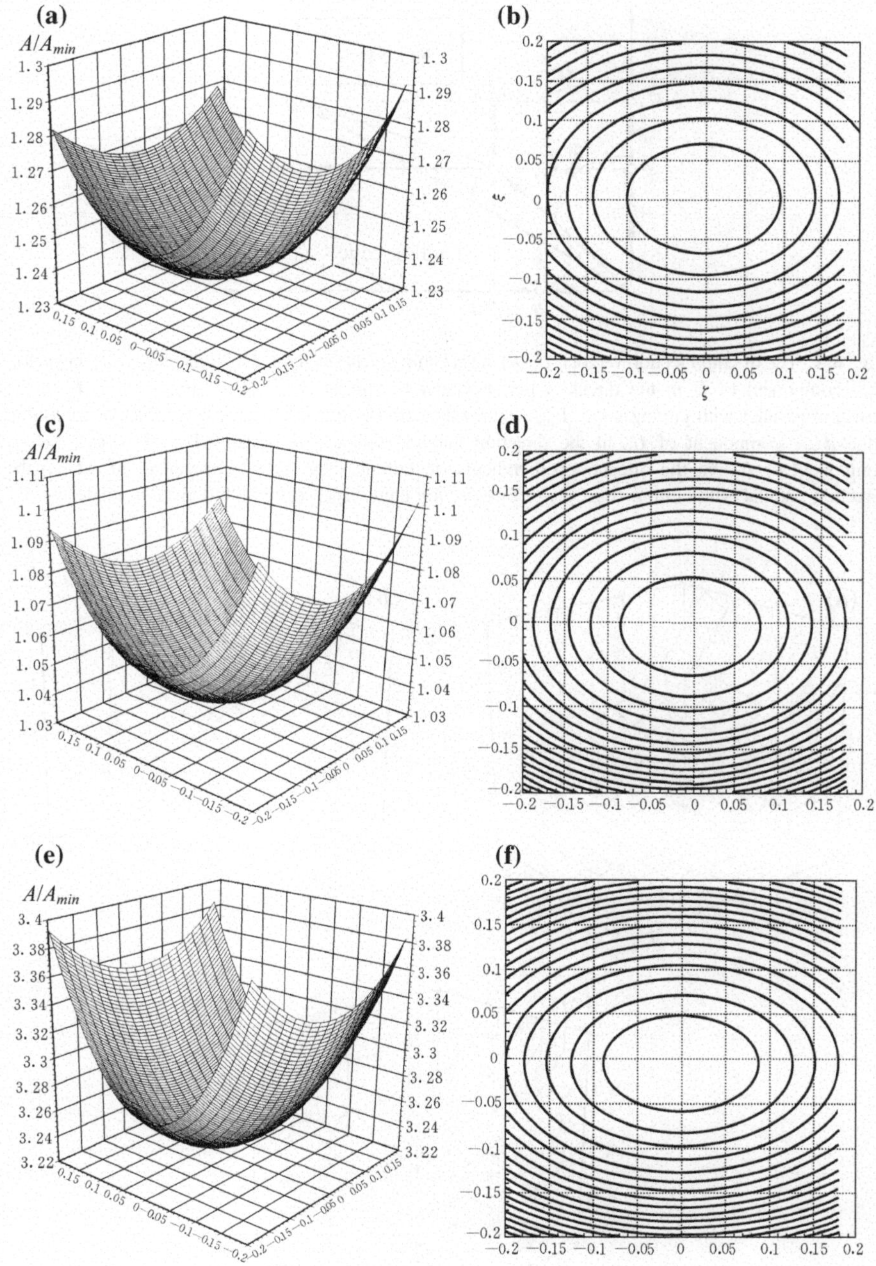

Fig. 6.3 Cutting power versus chip-ejection vector $\boldsymbol{U}$ in the vicinity of natural chip-ejection vector $\boldsymbol{U_o}$. **a**, **b** $\gamma_o = \lambda_s = 0°$, $\beta = 20°$; **c**, **d** $\gamma_o = 15°$, $\lambda_s = 10°$, $\beta = 25°$; **e**, **f** $\gamma_o = -10°$, $\lambda_s = 10°$, $\beta = 50°$; **a**, **c**, **e** The elliptic paraboloid in A–ζ–ξ space; **b**, **d**, **f** The contours in ζ–ξ plane

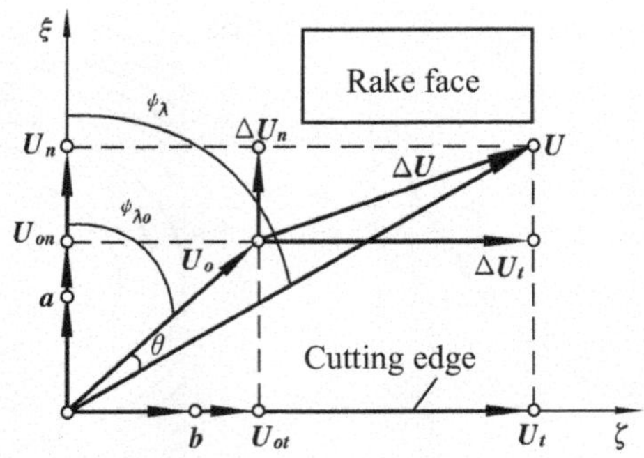

Fig. 6.4 The of chip-ejection vectors. $\boldsymbol{U_o}$—Natural chip-ejection vector. $\boldsymbol{U}$—Real chip-ejection vector. $\boldsymbol{U_n}$—component of U in the direction perpendicular to cutting edge. $\boldsymbol{U_t}$—component of $\boldsymbol{U}$ in the direction parallel with cutting edge. $\boldsymbol{U_{on}}$—component of $\boldsymbol{U_o}$ in the direction perpendicular to cutting edge. $\boldsymbol{U_{ot}}$—component of $\boldsymbol{U_o}$ in the direction parallel with cutting edge. $\boldsymbol{\Delta U} = \boldsymbol{U} - \boldsymbol{U_o}$. $\boldsymbol{\Delta U_n}$—component of $\boldsymbol{\Delta U}$ in the direction perpendicular to cutting edge. $\boldsymbol{\Delta U_t}$—component of $\boldsymbol{\Delta U}$ in the direction parallel with cutting edge. ψ_λ—chip ejection angle. ψ_λ—natural chip ejection angle

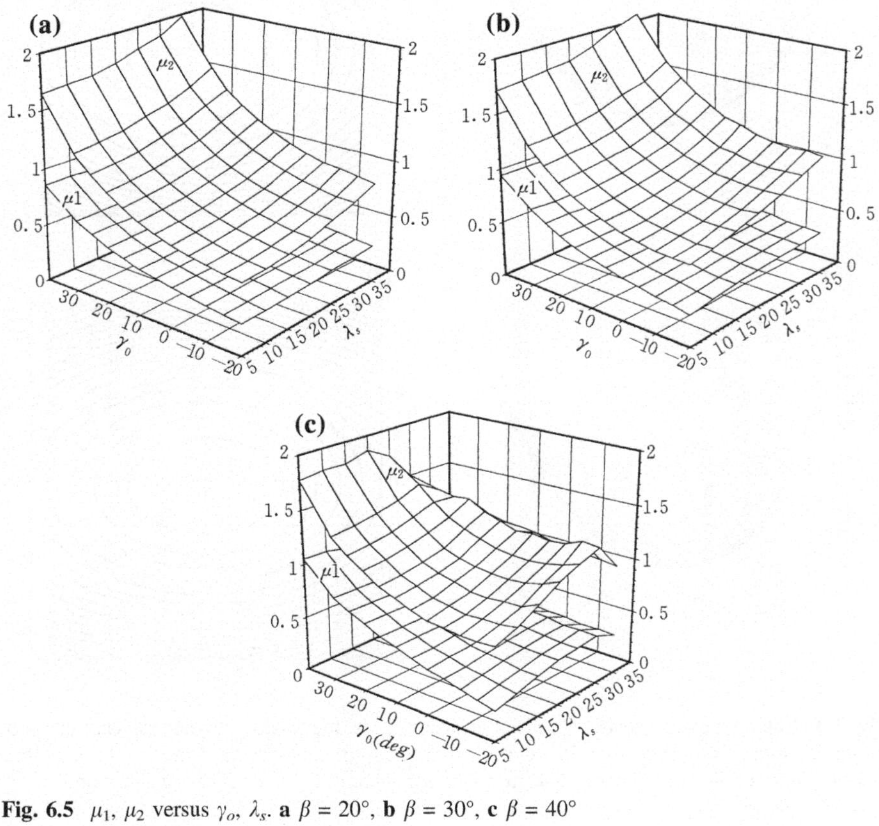

Fig. 6.5 μ_1, μ_2 versus γ_o, λ_s. **a** $\beta = 20°$, **b** $\beta = 30°$, **c** $\beta = 40°$

The data in Appendix 6.3 and Fig. 6.5 show that in a fairly wide technological conditions, $\mu_1 \approx 0.4$–0.6, $\mu_2 \approx 0.8$–1.2, and $\mu_2 \approx 2\mu_1$, i.e. deviation of $\boldsymbol{U}$ from $\boldsymbol{U_o}$ in the direction perpendicular to the cutting edge has a greater influence on the cutting power than that in the direction parallel with the cutting edge.

Obviously, in Eq. (6.13) $A_{\min}$ is the function of controlling (technological) parameters, so

$$A_{\min} = A_{\min}(\gamma_o, \lambda_s; a_w, a_c, V; \ldots).$$

The previous researchers have made great efforts in experimental work and theoretical analysis about the above functional relationship. This book, however, lays emphasis on the part of brackets in Eq. (6.13), which reflect the principle of minimum energy dissipation while cutting with an elementary cutting tool and provides basis or clew for the further research of non-free cutting process of complex shaped tools.

It is should be explained that in the about model the shear stress τ and friction angle β are regarded as controlling parameter and treated as a constant in the simulation. This is, of course, approximate way, and, as a matter of fact, τ and β are both variable according to the cutting conditions and join the process of minimizing the cutting power, i.e. they are really the state parameters. One of my Ph.D. students modified the above model by taking the variation of τ and β during simulation into consideration, however his results had no substantial changes [73].

6.2.4 The Experimental Studies on the Extremum Characteristics of Cutting Power

For experimentally verifying the extremum characteristics of cutting power, it is necessary to force chip-ejection direction and speed to change in cutting process of an elementary cutting tool, and observe the variation of cutting power. Here it should be explained that by means of changing cutting conditions, such as cutting tool angles or cutting parameters, so as to cause chip-ejection vector and cutting power to change is not the purpose of this study. This way only leads a change of a natural chip ejection to another, our purpose, however, is to force a natural chip-ejection vector to deviate from its natural value and observe the change in cutting power caused by the deviation, so as to verify the extremum characteristics of cutting power. For this purpose, in specially designed experiments several elementary cutting tools are working concurrently, the natural chip-ejection vectors of them are different, and because of the chip-ejection interference their chip-ejection directions or speeds were forced to deviate from their natural values, and the variations in cutting force or cutting power caused by the deviations are measured.

6.2.4.1 The Experiments in Which the Chip-Ejection Directions Are Forced to Change

In this group of experiments, double-edged tools were used as shown in Fig. 6.6. Because of the chip-ejection interference of the two edges, their chip-ejection directions are forced to change. The author designed and performed this experiments as early as in 1960s [10–11], because the conditions were limited, he did not get precise results. In early 90s last century he directed Ph.D. students to re-performed such experiments. Steel work piece was cut by double-edged cutters, with transverse feeding. Experimental conditions are shown in Fig. 6.6. Tools with different cutting edge angles θ were used in the experiments, thickness of cut $a_c = 0.07$ mm and width of cut $2a_w = 4$ mm were kept unchanged in the experiments, but transverse feeding f (mm/rev) and width of work piece b (mm) were adjusted according to angle θ,

$$\begin{aligned} b &= 2a_w \cos\theta = 4\cos\theta \\ f &= a_c/\cos\theta = 0.07/\cos\theta \end{aligned} \tag{6.15}$$

Because the tool cutting edge inclination of both the left and right edges are zero, so if they are cutting separately, the chip-ejection directions should be perpendicular to each edge, as shown by the interrupted vectors $\boldsymbol{U}_o$ and $\boldsymbol{U}'_o$ in Fig. 6.6. When

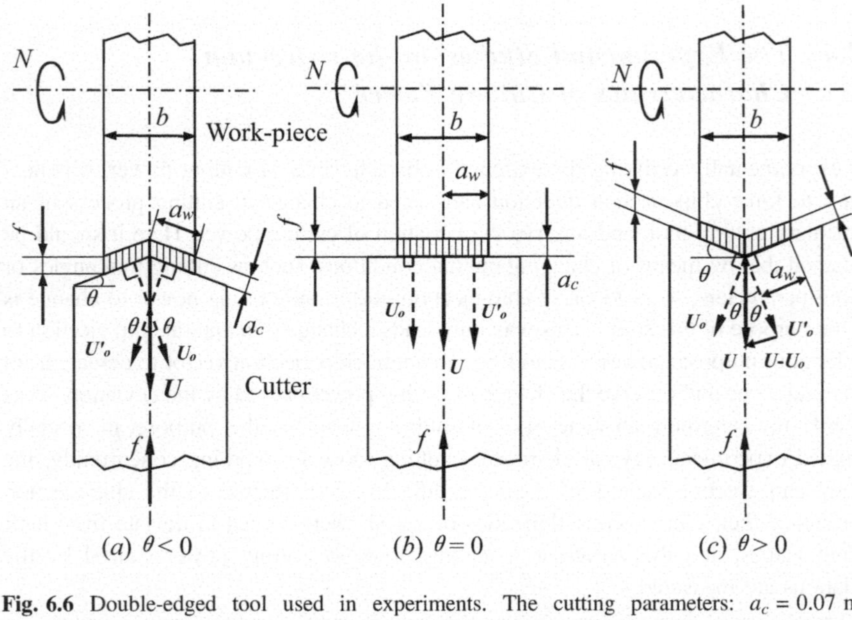

Fig. 6.6 Double-edged tool used in experiments. The cutting parameters: $a_c = 0.07$ mm, $a_w = 2$ mm, $V = 20$ m/min; $\gamma_o = \lambda_s = 0°$, $\alpha_o = 10°$, transverse feeding. The tool: cutting edge angles $\theta = -45°–45°$, HSS cutting tool, **a** $\theta < 0°$, **b** $\theta = 0°$, **c** $\theta > 0°$; The work-piece: mild steel, diameter: 101 mm

the two edges are working concurrently, because of the integrality of the chip and the symmetry of the cut, the common chip of the two edges can only be ejected in the direction of the shank axis, as shown by vector $\boldsymbol{U}$. Therefore, the chip-ejection direction were forced to deviate from their natural direction by an angle θ, i.e. the cutting edge angle. Under various angle θ the cutting forces F_c were measured and the results are given in Table 6.1.

Figure 6.7 shows the relations of cutting force F_c versus cutting edge angle θ. The figure shows that: first, cutting force F_c (and therefore cutting power) is really related to chip-ejection direction; second, when $\theta = 0°$, i.e. $\psi_\lambda = \psi_{\lambda o}$, two edges all ejects their chips in their natural chip-ejection direction (no chip-ejection interference), cutting force takes its minimum, and when absolute values of angle θ increase (no matter θ is positive or negative) cutting force or cutting power will increase. This is to say that any deviation of the chip-ejection direction from its natural direction would leads to an increase in cutting force or power (the interrupted line part in Fig. 6.7 or the data when $\theta \geq 25°$ in Table 6.1 will be explained in the next chapter).

From the data in Table 6.1 an Empirical Formula (6.16) can be fitted,

$$F_c = F_{c\,\min}(1 + \mu\theta^2) \tag{6.16}$$

where $F_{c\,\min} = 581.4$ (N), $\mu = 5.45 \times 10^{-4}$.

Besides, another group of experiments of cutting bronze with HSS double edged tools (also as shown in Fig. 6.6) were conducted, transverse feeding. The work-piece is turned into ring-like shape beforehand, the width of the ring is b. The cutting edge angle $\theta = 0°$–$-44.5°$ (only experiments of $\theta \leq 0°$ are performed). While the cutting edge angle θ is changing, $a_c = 0.0375$ mm, $2a_w = 4$ mm are kept unchanged and the transverse feeding f (mm/rev) and the width of the ring b is adjusted accordingly [also as shown by Eq. (6.15)]. The results are listed in Table 6.2.

From the data in Table 6.2 an empirical formula (6.17) also can be fitted (it is assumed that the figure of relation $F_c \sim \theta$ is symmetrical in fitting),

$$F_c = F_{c\,\min}(1 + \mu\theta^2) \tag{6.17}$$

where min = 850.8 (N), $\mu = 0.76$, and the values of θ should be in radian.

As stated previously, $F_{c\,\min}$ is a function of cutting parameters,

$$F_{c\,\min} = F_{c\,\min}(a_c, a_w),$$

For fitting the relation between $F_{c\,\min}$ and a_c, a set of experiments were conducted with the same work-piece material, cutter material and the way of feeding, take width of cut $2a_w = 4$ mm, and $\theta = 0°$, i.e. performing free cutting. Under various thickness of cut a_c the cutting force (free cutting) measured are given in Table 6.3.

Table 6.1 The results of experiments of cutting steel work-piece with double edged tools (the experimental conditions are in Fig. 6.6)

θ (°)	0	15	20	25	30	35	40	45
f (mm/rev)	0.0703	0.0731	0.0750	0.0780	0.0805	0.0863	0.0920	0.1000
b (mm)	4.000	3.864	3.759	3.625	3.464	3.277	3.064	2.828
Fc (N)	581.4	649.8	746.5	707.5	673.8	601.2	614.3	630.4
θ (°)	0	−15	−20	−25	−30	−35	−40	−45
f (mm/rev)	0.0703	0.0731	0.0750	0.0780	0.0805	0.0863	0.0920	0.1000
b (mm)	4.000	3.864	3.759	3.625	3.464	3.277	3.064	2.828
F_c (N)	581.4	678.0	735.6	793.4	892.6	1011.2	1088.9	1177.0

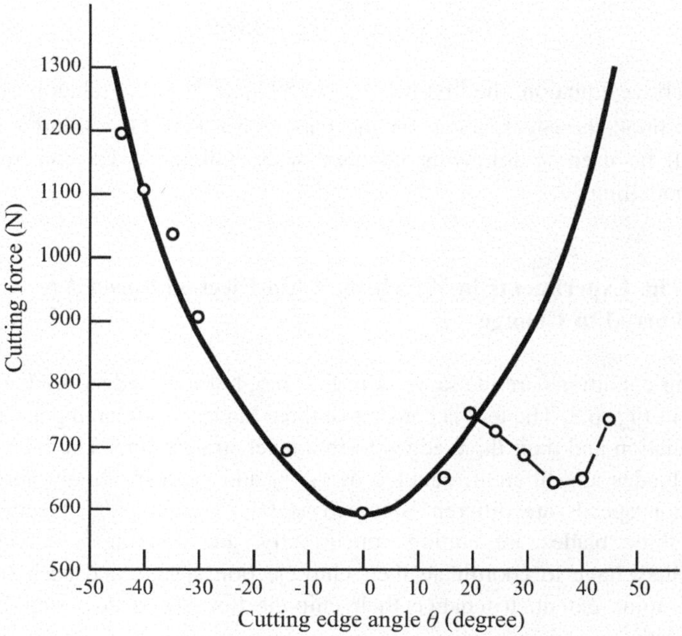

Fig. 6.7 The extreme characteristic of the cutting force

Table 6.2 Experimental results of cutting bronze with HSS double edged tools ($\gamma_o = \lambda_s = 0°$, $\alpha_o = 10°$, $a_c = 0.037$ mm, $2a_w = 4$ mm)

θ (°)	0	−13.1	−29.0	−38.5	−44.5
f (mm/rev)	0.0375	0.0385	0.0430	0.0480	0.0525
b (mm)	4.000	3.896	3.488	3.320	2.860
Fc (N)	859.8	887.0	1005.4	1164.8	1275.8

Table 6.3 Cutting force $F_{c\min}$ versus thickness of cut a_c, free cutting of bronze (HSS tool, $\gamma_o = \lambda_s = 0°$, $a_o = 10°$, $q = 0°$, $2a_w = 4$ mm)

a_c (mm)	0.0375	0.0385	0.0675	0.1535	0.195	0.250
$F_{c\min}$ (N)	859.8	833.0	1234.8	2087.4	2401.0	2508.8

From data in above table, the following formula may be fitted,

$$F_{c\min} = 6372a_c^{0.61}, \tag{6.18}$$

Substitution Eqs. (6.18) into (6.17), and noticing that above cutting force is of $2a_w = 4$ mm, and cutting force is normally in direct proportion to width of cut a_w, leads to,

$$F_c = 1593a_w a_c^{0.61}(1 + 0.76\theta^2). \tag{6.19}$$

In the above equation, the first part $F_c = 1593a_w a_c^{0.61}$ is the relation every body talks, this book, however, focus on the part in brackets $(1 + 0.76\theta^2)$. Equation (6.19) will be used in following chapters when talking about non-free cutting process modelling.

6.2.4.2 The Experiments in Which the Chip-Ejection Speed Are Forced to Change

The cutting conditions are the same as in Fig. 6.6, but triple edged tools were used as shown in Fig. 6.8. The tool is consists of three blades, each having a zero cutting edge inclination and their three edges form a level straight line. The rake angles of the three blades are different, denoted as γ_1, γ_2 and γ_3, respectively, and so their chip-ejection speeds are different, denoted as U_{o1}, U_{o2} and U_{o3}, respectively. In fact, the three blades are cutting concurrently, and ejecting a common chip, therefore they have to coordinate their chip-ejection speeds and each of them is forced, to some extent, to change their chip-ejection speed deviating from their natural speed.

In the experiments, γ_1, γ_2 and γ_3 are all changing though, γ_1 and γ_3 are kept equal, and because of symmetry, the chip-ejection direction only can be in length wise direction of the three blades. Therefore, the chip-ejection direction did not change, and only the ejection speed changed.

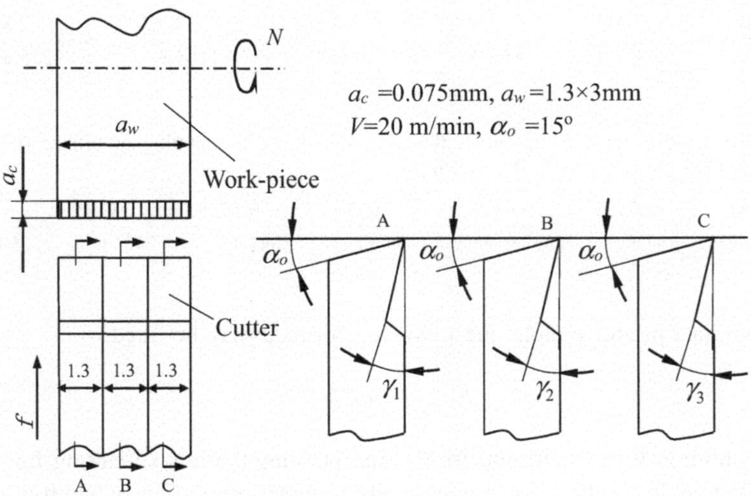

Fig. 6.8 Triple edged tool used in experiments. Tools: HSS; work-piece: mild steel; diameter: 101 mm; transverse feeding

Table 6.4 Single blade cutting experiments

No.	γ_i (°)	F_c (N)	r
1	0	433.2	0.481
2	5	416.4	0.492
3	10	387.5	0.514

Table 6.5 Triple blades combination experiments

Group		Rake	Cutting force F_c (N)				Cutting ratio r		
		$\gamma_1/\gamma_2/\gamma_3$	Sum	Measured	Δ F_c	Δ%	Predicted	Measured	Δ%
I	1	0/0/0	1299.6	1314.5	14.0	1.15	0.481	0.472	1.90
	2	5/5/5	1249.2	1267.0	17.8	1.42	0.492	0.486	1.23
	3	10/10/10	1162.5	1180.7	18.2	1.36	0.514	0.510	0.78
II	1	0/5/0	1282.8	1487.2	204.4	15.93	0.484	0.479	1.04
	2	5/0/5	1266.0	1445.2	179.3	14.16	0.488	0.487	0.21
	3	0/10/0	1253.9	1446.7	192.8	15.37	0.491	0.485	1.24
	4	10/0/10	1208.2	1408.7	200.5	16.59	0.502	0.502	0.00
	5	10/5/10	1191.4	1378.3	186.9	15.69	0.506	0.505	0.20
	6	5/10/5	1415.6	1220.3	195.3	16.00	0.498	0.495	0.60

In the experiments, the cutting force of single blade and the combination of blades are measured, and the chip length before and after cutting L_o and L also measured and cutting ratios were calculated as $r = L/L_o$. All the results are listed in Tables 6.4 and 6.5.

Table 6.4 gives the experimental results while each blade cutting alone.

The data in Group I, Table 6.5 present the experimental results with the combinations of blades of the same rake angles. Since the chip-ejection speeds are the same for the three blades in a combination, there is no interference among the chip ejections of the blades, and the common chip flows out at their natural speed. The cutting ratio of the common chip is approximately equal to that of each blade while cutting alone. The total cutting force is equal to the sum of that of all the blades in the combination. This confirms a well-known conclusion in traditional metal-cutting theory: the working length of a cutting edge, i.e. the width of cut, has little influence on the cutting process, but proportionally (linearly) controls the amount of cut. The conclusion, however, should be treated with care, because it is valid only when the interference among the chip-ejections from different segments of the cutting edge is negligible.

The data in Group II, Table 6.5 present the experimental results with the combination of blades of different rake angles. It can be seen from the data that the total cutting force is larger than the sum of that of the three blades while each of them is cutting separately. This is, of course, due to the forced changes in chip ejection speeds of each blade. The changes may be verified by the differences between the common cutting ratio of the whole tool and the natural cutting ratios of the three blades as presented in Table 6.4. The experimental results also verified that the deviation of cutting ratio from its natural value, no matter bigger or smaller, would leads to an increase in cutting force.

The predicted values and the prediction method in Table 6.5 will be explained in next chapter.

6.2.4.3 A Comparison of Coefficients μ_1 and μ_2 Obtained Experimentally with that Obtained Through Theoretical Model Simulation

Equation (6.13) gives the relation of cutting power A with ζ, ξ. Figure 6.5 and Appendix 6.3 present the values of coefficients μ_1 and μ_2 in Eq. (6.13), which are obtained from numerical simulations.

Equation (6.16) gives the relation of cutting force and the chip-ejection angle obtained in experiments of changing chip-ejection direction as described in Sect. 6.2.4.1,

$$F_c = 581.4(1 + 5.45 \times 10^{-4}\theta^2),$$

where the unit of θ is in degree. When θ varies from 0° to 45°,

$$\begin{aligned} \zeta^2 &= \Delta U_{\mathrm{t}}^2/U_0^2 = \sin^2\theta = 0\text{–}0.500 \\ \xi^2 &= \Delta U_{\mathrm{n}}^2/U_0^2 = (1 - \cos^2\theta) = 0\text{–}0.086 \end{aligned}.$$

Compared with ζ^2, ξ^2 is negligible, therefore the experiment is mainly showing the influence of chip-ejection vector changes in the direction parallel to the cutting edge, and from the data, we have

$$F_c = 581.4(1 + \mu_1\zeta^2) \tag{6.20}$$

where μ_1 is fitted as 2.17, which is bigger than the parameter μ_1 obtained through numerical simulation.

In experiments of changing chip-ejection speed described in Sect. 6.2.4.2 there is only ξ changing, and an empirical formula may be fitted,

$$F_c = 581.4(1 + \mu_2\xi^2) \tag{6.21}$$

where

$$\xi = \frac{r - r_o}{r_o}$$

Since the variation of ξ in the experiment was rather small it was difficult to find out the exact value of parameter μ_2 from the measured data. But one point may be identified that

$$\mu_2 > \mu_1,$$

This means that cutting force, or cutting power is more sensitively affected by the change in $\boldsymbol{U}$ at the direction normal to the cutting edge, i.e. by $\boldsymbol{U}_n$ rather than by $\boldsymbol{U}_t$.

6.2.5 The Natural Values of State Parameters as Functions of Controlling Parameters

Because of the principle of minimum energy dissipation, the natural values r_o and $\psi_{\lambda o}$ of state parameters must minimize the cutting power A. So from Eq. (6.11),

$$\begin{aligned} &\partial A(r_o, \psi_{\lambda o}; \gamma_o, \lambda_s, V, a_c, a_w, \tau, \beta)/\partial r = 0; \\ &\partial A(r_o, \psi_{\lambda o}; \gamma_o, \lambda_s, V, a_c, a_w, \tau, \beta)/\partial \psi_\lambda = 0. \end{aligned} \tag{6.22}$$

The solution of above simultaneous equations will be

$$\begin{aligned} r_o &= r_o(\gamma_o, \lambda_s, V, a_c, a_w, \tau, \beta); \\ \psi_{\lambda o} &= \psi_{\lambda o}(\gamma_o, \lambda_s, V, a_c, a_w, \tau, \beta). \end{aligned} \tag{6.23}$$

Previous researchers conducted a great deal of experiments on the physical substance and the quantitative law of the above relations. Here based on principle of minimum energy dissipation, we will give the problem theoretical summarize and explanation. In the following section we will, based on the data of model simulation, check the consistency of principle of minimum energy dissipation with well-known Merchant model and Stabler law, and check the correctness or rationality of the principle of minimum energy dissipation.

6.3 About the Correctness of the Principle of Minimum Energy Dissipation

6.3.1 The Consistency of Our Model with Merchant Model

Merchant model is about orthogonal cutting (λ_s = 0°) [66], it says that the value of shear angle φ (a state parameter) must minimizing the cutting power, it leads to,

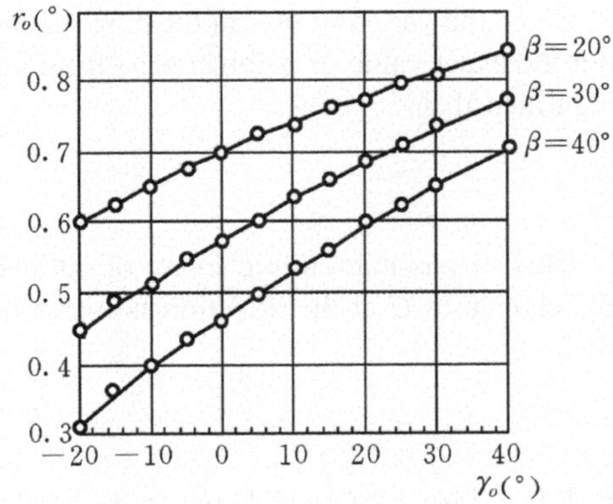

Fig. 6.9 The consistency of our model with Merchant model. Solid line—Merchant model, open circle—our model

$$\begin{aligned} \varphi &= 45^\circ + \gamma_o/2 - \beta/2 \\ \tan\varphi &= r_o\cos\gamma_o/(1 - r_o\sin\gamma_o), \end{aligned} \tag{6.23}$$

Cancelling φ from the above two equations results in the relation of cutting ratio r_o with rake γ_o and friction angle β; on the other hand, our model of Elementary cutting tool and the principle of minimum energy dissipation may be regarded as a generalized Merchant model, let $\lambda_s = 0°$, it will degenerate to Merchant model. Figure 6.9 shows that they are coincide with each other very well.

6.3.2 The Compatibility with Stabler Law

Stabler, G. V. [69] gave a simple empirical relation between chip-ejection angle $\psi_{\lambda o}$ and cutting edge inclination λ_s,

$$\psi_{\lambda o} \approx \lambda_s.$$

Based on our model of Elementary cutting tool and the principle of minimum energy dissipation, deduced $\psi_{\lambda o} \sim \lambda_s$ relation is given in Fig. 6.10, the diagonal line in the figure shows the Stabler law. It is clear that within normal technological scope $\gamma_o = -20°–25°$, $\beta = 20°–40°$, the chip-ejection angle given by Stabler from his experience is approximately the natural chip-ejection angle which minimizes the cutting power. This, on one hand, has given the empirical relation of Stabler law the theoretical explanation, and on the other hand, approved the correctness of the principle of minimum energy dissipation by previous experience accumulate in the past and summarized by Stabler.

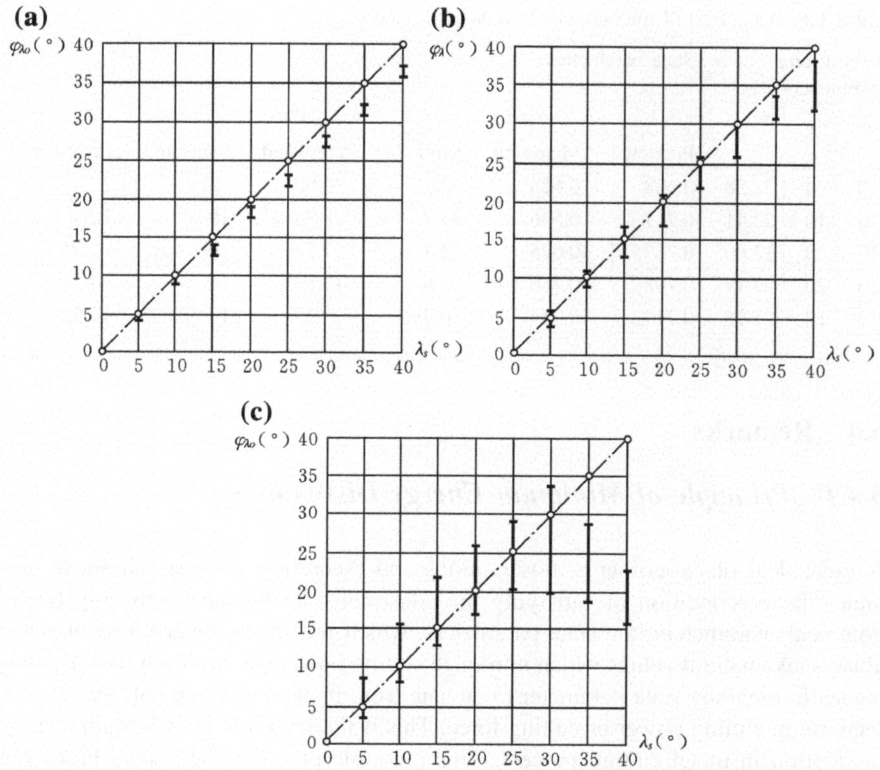

Fig. 6.10 A comparison of natural chip-ejection angle deduced from the principle of minimum energy dissipation with Stabler law. $\gamma_o = -20°–25°$, **a** $\beta = 20°$, **b** $\beta = 30°$, **c** $\beta = 40°$. Line with open circle—Stabler law, I—natural chip-ejection

6.3.3 Verified Experimentally

Besides verified by previous knowledge, our model of elementary cutting tool and the principle of minimum energy dissipation are also verified experimentally. The end face of a seamless steel tube was cut by HSS tools (free cutting). $V = 21.49$ m/min, $a_c = f = 0.075$ mm, $a_w = 2$ mm (the thickness of the tube wall). The friction angle β is calculated from measured cutting force components by a three-phase dynamometer. The measured natural cutting ratio r_o and natural chip-ejection angle $\psi_{\lambda o}$ are compared with theoretically predicted ones and both are listed in Table 6.6. Considering the simplicity of the model and the measurement error, it should be mentioned that the measured and predicted data coincide with each other fairly good, and this shows that the principle of minimum energy dissipation for elementary cutting tools is correct by and large.

Table 6.6 A contrast of measured and predicted r_o and $\psi_{\lambda o}$

Controlling parameters (°)			State parameters					
γ_o	λ_s	β	r_o			$\psi_{\lambda o}$ (°)		
			Predicted	Measured	Error (%)	Predicted	Measured	Error (%)
10	0	32.58	0.608	0.625	2.7	0.00	2	2
10	10	22.41	0.717	0.536	33.7	8.75	10	1.25
10	20	17.60	0.767	0.625	22.7	17.75	16	−1.75
20	20	27.17	0.708	0.750	5.6	17.50	16	−1.50
20	10	25.98	0.717	0.714	0.4	8.75	10	1.25

6.4 Remarks

6.4.1 *Principle of Minimum Energy Dissipation*

A great deal of experiments, observations and theoretical analysis all show that under the precondition of satisfying the constraints set by the controlling (technological) parameters, the state parameters, which reflect the inner physical state, always take natural values which minimize cutting power (or cutting force). By any means forcing any state parameter to deviate from its natural value will leads to an increase in cutting power or cutting force. This is the principle of minimum energy dissipation in metal cutting process. This principle provides theoretical bases for explaining, modelling, predicting and optimizing various complex phenomena in modern machining process.

6.4.2 *Principle of Minimum Energy Dissipation for Elementary Cutting Tool*

An elementary cutting tool has an edge of small segment of straight line and his rake and flank are small facets. Normally it performs oblique free cutting. In the model presented in this chapter the independent state parameter is chip-ejection vector (including chip-ejection angle and chip-ejection speed two parameters, or the vector can be broken into components, parallel with or normal to cutting edge). Both the model analysis and experiments under various conditions verified that the natural values of chip-ejection will minimize the cutting power. This is the principle of minimum energy dissipation of elementary cutting tools. In following chapters, this principle will be applied to modelling cutting properties of various modern complex shaped cutting tools.

Appendix 6.1: r_o Versus γ_o and λ_s (Degree), the Results of Simulation

$\beta = 20°$

λ_s	γ											
	−20	−25	−10	−5	0	5	10	15	20	25	30	40
5	0.6000	0.6250	0.6500	0.6750	0.7000	0.7250	0.7375	0.7625	0.7750	0.8000	0.8125	0.8500
10	0.6000	0.6250	0.6500	0.6750	0.7000	0.7250	0.7375	0.7625	0.7750	0.8000	0.8125	0.8500
15	0.6000	0.6375	0.6625	0.6750	0.7000	0.7250	0.7375	0.7625	0.7750	0.7875	0.8125	0.8375
20	0.6125	0.6375	0.6625	0.6750	0.7000	0.7250	0.7375	0.7625	0.7750	0.7875	0.8125	0.8375
25	0.6250	0.6375	0.6625	0.6875	0.7000	0.7250	0.7375	0.7500	0.7750	0.7875	0.8000	0.8375
30	0.6250	0.6500	0.6625	0.6875	0.7000	0.7250	0.7375	0.7500	0.7750	0.7875	0.8000	0.8250
35	0.6375	0.6500	0.6750	0.6875	0.7000	0.7250	0.7375	0.7500	0.7750	0.7875	0.8000	0.8250
40	0.6500	0.6625	0.6750	0.6875	0.7125	0.7250	0.7375	0.7500	0.7625	0.7750	0.7875	0.8250

$\beta = 30°$

λ_s	γ_o											
	−20	−25	−10	−5	0	5	10	15	20	25	30	40
5	0.4500	0.4875	0.5125	0.5500	0.5750	0.6125	0.6375	0.6625	0.6875	0.7125	0.7375	0.7750
10	0.4625	0.4875	0.5250	0.5500	0.5750	0.6125	0.6375	0.6625	0.6875	0.7125	0.7250	0.7750
15	0.4750	0.5000	0.5750	0.5625	0.5875	0.6125	0.6375	0.6625	0.6875	0.7000	0.7250	0.7750
20	0.4875	0.5125	0.5375	0.5625	0.5875	0.6125	0.6375	0.6625	0.6750	0.7000	0.7250	0.7625
25	0.5000	0.5250	0.5500	0.5750	0.5875	0.6125	0.6375	0.6625	0.6750	0.7000	0.7250	0.7625
30	0.5125	0.5375	0.5500	0.5750	0.6000	0.6125	0.6375	0.6625	0.6750	0.7000	0.7250	0.7625
35	0.5250	0.5500	0.5625	0.5875	0.6000	0.6250	0.6375	0.6625	0.6750	0.7000	0.7250	0.7500
40	0.5375	0.5625	0.5750	0.5875	0.6125	0.6250	0.6375	0.6625	0.6750	0.6875	0.7125	0.6500

$\beta = 40°$

λ_s	γ_o											
	−20	−25	−10	−5	0	5	10	15	20	25	30	40
5	0.3250	0.3625	0.4000	0.4750	0.4625	0.5000	0.5375	0.5625	0.6000	0.6250	0.6500	0.7125
10	0.3375	0.3750	0.4125	0.4375	0.4750	0.5000	0.5375	0.5625	0.6000	0.6250	0.6500	0.7125
15	0.3675	0.3875	0.4250	0.4500	0.4750	0.5125	0.5375	0.5625	0.6000	0.6250	0.6500	0.7000
20	0.3875	0.4125	0.4375	0.4625	0.4875	0.5125	0.5375	0.5625	0.6000	0.6250	0.6500	0.7000
25	0.4000	0.4250	0.4500	0.4750	0.5000	0.5250	0.5500	0.5750	0.6000	0.6250	0.6500	0.6500
30	0.4125	0.4500	0.4625	0.4875	0.5125	0.5250	0.5500	0.5750	0.6000	0.5875	0.5750	0.5375
35	0.3125	0.3500	0.3750	0.4125	0.4250	0.4500	0.4625	0.4625	0.4750	0.4625	0.4625	0.4125
40	0.2000	0.2375	0.2625	0.2875	0.3000	0.3125	0.3250	0.3375	0.3375	0.3375	0.3250	0.3000

$\beta = 50°$

λ_s	γ_o											
	−20	−25	−10	−5	0	5	10	15	20	25	30	40
5	0.2250	0.2625	0.2875	0.3250	0.3625	0.4000	0.4375	0.4750	0.5125	0.5375	0.5750	0.6375
10	0.2500	0.2750	0.3125	0.3375	0.3750	0.4125	0.4375	0.4750	0.5125	0.5375	0.5750	0.6375
15	0.2750	0.3000	0.3250	0.3650	0.3875	0.4125	0.4500	0.4750	0.5125	0.5375	0.5750	0.5875
20	0.2875	0.3250	0.3500	0.3750	0.4000	0.4250	0.4625	0.4875	0.5125	0.2520	0.5125	0.4875
25	0.2125	0.2625	0.3000	0.3250	0.3625	0.3750	0.4000	0.4125	0.4125	0.4125	0.4125	0.3875
30	0.1250	0.1625	0.2000	0.2250	0.2500	0.2750	0.2875	0.3000	0.3000	0.3125	0.3125	0.2875
35	0.0375	0.0625	0.1000	0.1250	0.1375	0.1625	0.1750	0.1875	0.1875	0.2000	0.2000	0.1750
40	0.1625	0.1625	0.1625	0.0125	0.0375	0.0500	0.0625	0.0750	0.0750	0.0875	0.0875	0.0750

Appendix 6.2: $\psi_{\lambda o}$ Versus γ_o and λ_s (Degree), the Results of Simulation

$\beta = 20°$

λ_s	γ_o											
	−20	−25	−10	−5	0	5	10	15	20	25	30	40
5	5	5	5	4	4	4	4	4	4	4	4	4
10	10	9	9	9	9	9	9	9	9	9	9	9
15	14	14	14	13	13	13	13	13	13	13	13	13
20	19	18	18	18	18	18	18	18	18	18	18	18
25	24	23	23	23	22	22	22	22	22	22	22	23
30	28	27	27	27	27	27	27	27	27	27	27	27
35	32	32	32	31	31	31	31	31	31	31	32	32
40	37	36	36	36	36	36	36	36	36	36	36	37

$\beta = 30°$

λ_s	γ_o											
	−20	−25	−10	−5	0	5	10	15	20	25	30	40
5	7	6	5	5	5	5	5	4	4	4	4	4
10	13	11	11	10	9	9	9	9	9	9	9	9
15	18	17	15	15	14	14	14	13	13	13	13	13
20	23	21	20	19	19	18	18	18	17	17	17	17
25	27	26	25	24	23	23	22	22	22	22	22	22
30	31	30	29	28	28	27	27	27	26	26	26	27
35	34	34	33	32	32	32	31	31	31	31	31	31
40	38	38	37	36	36	36	35	36	35	35	36	32

$\beta = 40°$

λ_s	γ_o											
	−20	−25	−10	−5	0	5	10	15	20	25	30	40
5	11	9	7	6	5	5	5	5	5	4	4	4
10	19	16	14	12	11	10	10	9	9	9	9	9
15	25	21	19	17	16	15	14	14	14	13	13	13
20	29	26	24	22	21	20	19	18	18	18	17	17
25	32	30	28	26	25	24	24	23	22	22	22	20
30	34	34	32	30	30	28	28	27	27	25	24	20
35	29	29	29	29	28	28	27	26	25	23	23	19
40	21	22	23	23	23	22	22	22	21	20	19	16

$\beta = 50°$

λ_s	γ_o											
	−20	−25	−10	−5	0	5	10	15	20	25	30	40
5	23	16	10	8	7	6	5	5	5	5	4	4
10	29	23	19	15	13	12	11	10	10	9	9	8
15	33	28	24	21	19	17	16	15	14	13	13	12
20	34	32	29	26	24	22	21	19	18	17	16	13
25	27	27	27	25	24	22	21	20	18	17	16	14
30	18	19	20	20	19	19	18	17	16	16	15	13
35	7	9	12	13	13	13	13	13	12	12	12	10
40	13	13	13	2	4	5	6	7	6	7	6	5

Appendix 6.3: μ_1 and μ_2, the Results of Simulation

μ_1, $\beta = 20°$

λ_s (°)	γ_o (°)											
	−20	−15	−10	−5	0	5	10	15	20	25	30	40
0	0.2576	0.2724	0.2901	0.3113	0.3368	0.3676	0.3956	0.4403	0.4846	0.5521	0.6239	0.8557
5	0.2587	0.2734	0.2910	0.3123	0.3378	0.3685	0.3965	0.4412	0.4853	0.5526	0.6242	0.8551
10	0.2621	0.2765	0.2940	0.3152	0.3407	0.3715	0.3994	0.4441	0.4881	0.5555	0.6267	0.8568
15	0.2673	0.2891	0.3066	0.3199	0.3454	0.3763	0.4040	0.4487	0.4922	0.5465	0.6294	0.8370
20	0.2812	0.2956	0.3132	0.3261	0.3521	0.3834	0.4111	0.4562	0.4995	0.5534	0.6364	0.8418
25	0.2961	0.3025	0.3208	0.3433	0.3607	0.3925	0.4199	0.4536	0.5079	0.5607	0.6270	0.8504
30	0.3036	0.3203	0.3301	0.3537	0.3714	0.4046	0.4323	0.4662	0.5216	0.5742	0.6399	0.8312
35	0.3217	0.3292	0.3510	0.3659	0.3842	0.4188	0.4467	0.4803	0.5212	0.5712	0.6584	0.8474
40	0.3410	0.3516	0.3644	0.3806	0.4135	0.4378	0.4668	0.5014	0.5429	0.5932	0.6549	0.8722

μ_2, $\beta = 20°$

λ_s (°)	γ_o (°)											
	−20	−15	−10	−5	0	5	10	15	20	25	30	40
0	0.5832	0.5999	0.6232	0.6534	0.6913	0.7386	0.7968	0.8691	0.9582	1.0712	1.2127	1.6401
5	0.5887	0.6057	0.6292	0.6574	0.6954	0.7427	0.8009	0.8732	0.9623	1.0756	1.2166	1.6436
10	0.6058	0.6192	0.6430	0.6737	0.7122	0.7599	0.8185	0.8912	0.9807	1.0941	1.2358	1.6636
15	0.6293	0.6481	0.6725	0.6968	0.7357	0.7838	0.8426	0.9155	1.0049	1.1163	1.2597	1.6785
20	0.6720	0.6821	0.7073	0.7392	0.7791	0.8282	0.8880	0.9618	1.0522	1.1646	1.3089	1.7299
25	0.7322	0.7393	0.7660	0.7996	0.8270	0.8767	0.9369	1.0103	1.1014	1.2134	1.3541	1.8034
30	0.7965	0.8024	0.8297	0.8643	0.9064	0.9576	1.0195	1.0945	1.1869	1.3005	1.4426	1.8599
35	0.8798	0.9027	0.9326	0.9468	0.9899	1.0419	1.1042	1.1793	1.2706	1.3828	1.5616	1.9827
40	1.0125	1.0105	1.0418	1.0800	1.1256	1.1798	1.2443	1.3216	1.4149	1.5290	1.6706	2.1468

μ_1, $\beta = 30°$

λ_s (°)	γ_o (°)											
	−20	−15	−10	−5	0	5	10	15	20	25	30	40
0	0.2661	0.2964	0.3160	0.3499	0.3756	0.4060	0.4561	0.5012	0.5566	0.6253	0.7117	0.9417
5	0.2701	0.2989	0.3178	0.3512	0.3765	0.4197	0.4567	0.5020	0.5571	0.6256	0.7117	0.9406
10	0.2916	0.3056	0.3338	0.3547	0.3799	0.4227	0.4591	0.5037	0.5585	0.6267	0.6928	0.9397
15	0.3148	0.3244	0.3410	0.3720	0.3970	0.4262	0.4622	0.5071	0.5611	0.6104	0.6929	0.9357
20	0.3347	0.3455	0.3594	0.3787	0.4013	0.4318	0.4671	0.5107	0.5479	0.6123	0.6931	0.9030
25	0.3527	0.3612	0.3760	0.3962	0.4074	0.4359	0.4730	0.5160	0.5513	0.6153	0.6955	0.9014
30	0.3660	0.3772	0.3794	0.4013	0.4250	0.4418	0.4774	0.5211	0.5560	0.6189	0.6746	0.9023
35	0.3815	0.3907	0.3955	0.4204	0.4298	0.4613	0.4836	0.5278	0.5610	0.6247	0.6788	0.8657
40	0.3878	0.4023	0.4107	0.4224	0.4510	0.4669	0.4897	0.5341	0.5673	0.6069	0.6858	0.6024

μ_2, $\beta = 30°$

λ_s (°)	γ_o (°)											
	−20	−15	−10	−5	0	5	10	15	20	25	30	40
0	0.6869	0.6984	0.7099	0.7398	0.7710	0.8131	0.8760	0.9457	1.0340	1.1462	1.2903	1.7010
5	0.6967	0.7064	0.7162	0.7464	0.7779	0.8287	0.8836	0.9510	1.0394	1.1517	1.2959	1.7068
10	0.7341	0.7267	0.7490	0.9667	0.7946	0.8461	0.9015	0.9719	1.0610	1.1741	1.3030	1.7319
15	0.7830	0.7784	0.7783	0.8111	0.8375	0.8816	0.9382	1.0020	1.0918	1.1920	1.3347	1.7654
20	0.8488	0.8330	0.8385	0.8541	0.8891	0.9249	0.9827	1.0555	1.1233	1.2358	1.3794	1.7856
25	0.9214	0.9151	0.9196	0.9340	0.9481	0.9957	1.0418	1.1159	1.1959	1.3105	1.463	1.8674
30	1.0118	1.0053	0.9978	1.0115	1.0514	1.0737	1.1356	1.2122	1.2741	1.3900	1.5194	1.9776
35	1.1033	1.1178	1.1088	1.1217	1.1529	1.2063	1.2366	1.3153	1.3982	1.5170	1.6488	2.0585
40	1.2378	1.2585	1.2478	1.2465	1.2933	1.3376	1.3637	1.4760	1.5284	1.6322	1.8192	1.8835

μ_1, $\beta = 40°$

λ_s (°)	γ_o (°)											
	−20	−15	−10	−5	0	5	10	15	20	25	30	40
0	0.2270	0.2950	0.3400	0.3833	0.4095	0.4597	0.5127	0.5553	0.6296	0.6952	0.7775	1.0519
5	0.2640	0.3037	0.3444	0.3857	0.4114	0.4589	0.5131	0.5552	0.6293	0.6953	0.7771	1.0504
10	0.3123	0.3383	0.3710	0.3908	0.4327	0.4611	0.5138	0.5567	0.6298	0.6942	0.7752	1.0467
15	0.3657	0.3716	0.3995	0.4161	0.4361	0.4829	0.5171	0.5563	0.6286	0.6938	0.7728	1.0040
20	0.4099	0.4135	0.4208	0.4374	0.4562	0.4825	0.5163	0.5581	0.6290	0.6905	0.7698	0.9938
25	0.4252	0.4273	0.4390	0.4587	0.4781	0.5051	0.5341	0.5771	0.6291	0.6887	0.7640	0.8415
30	0.4434	0.4515	0.4486	0.4742	0.4883	0.5036	0.5326	0.5762	0.6233	0.6119	0.6010	0.5767
35	0.3610	0.3704	0.3672	0.3914	0.3931	0.4093	0.4205	0.4156	0.4346	0.4251	0.4223	0.3779
40	0.2950	0.3063	0.2978	0.3057	0.2979	0.3045	0.3038	0.3062	0.3032	0.3025	0.2871	0.2712

μ_2, $\beta = 40°$

λ_s (°)	γ_o (°)											
	−20	−15	−10	−5	0	5	10	15	20	25	30	40
0	0.7733	0.7989	0.8071	0.8285	0.8408	0.8877	0.9491	1.0051	1.1040	1.2028	1.3312	1.7655
5	0.8222	0.8144	0.8161	0.8365	0.8477	0.8950	0.9569	1.0133	1.1126	1.2096	1.3382	1.7736
10	0.8920	0.8703	0.8710	0.8616	0.8939	0.9174	0.9807	1.0338	1.1345	1.2347	1.3645	1.8033
15	1.0058	0.9366	0.9354	0.9227	0.9326	0.9781	1.0151	1.0743	1.1771	1.2720	1.4033	1.7987
20	1.1182	1.0504	1.0172	1.0025	1.0121	1.0355	1.0727	1.1250	1.2307	1.3349	1.4586	1.8595
25	1.1930	1.1400	1.1077	1.0922	1.1028	1.1268	1.1766	1.2300	1.3018	1.4085	1.5453	1.7496
30	1.2643	1.2808	1.2171	1.2016	1.2291	1.2129	1.2669	1.3207	1.4083	1.3972	1.4288	1.4979
35	1.0120	1.0319	1.0486	1.0857	1.0856	1.1261	1.1428	1.1570	1.1928	1.1979	1.2665	1.3187
40	0.8781	0.8911	0.9181	0.9330	0.9552	0.9635	0.9941	1.0302	1.0540	1.0835	1.1189	1.1838

μ_1, $\beta = 50°$

λ_s (°)	γ_o (°)											
	−20	−15	−10	−5	0	5	10	15	20	25	30	40
0	0.1144	0.2486	0.3270	0.3887	0.4443	0.4995	0.5583	0.6238	0.6996	0.7559	0.8627	1.1203
5	0.2910	0.3163	0.3412	0.3938	0.4456	0.4998	0.5589	0.6233	0.6983	0.7543	0.8621	1.1183
10	0.3987	0.3820	0.4127	0.4301	0.4770	0.5270	0.5554	0.6212	0.6944	0.7522	0.8571	1.1121
15	0.4586	0.4504	0.4503	0.4891	0.5017	0.5254	0.5807	0.6159	0.6923	0.7496	0.8520	0.9302
20	0.4966	0.4931	0.4926	0.5064	0.5206	0.5455	0.6010	0.6445	0.6894	0.7125	0.6813	0.6520
25	0.4422	0.4479	0.4361	0.4428	0.4746	0.4733	0.4957	0.4996	0.4949	0.4968	0.4847	0.4461
30	0.3520	0.3488	0.3486	0.3388	0.3502	0.3554	0.3541	0.3584	0.3451	0.3492	0.3452	0.6065
35	0.1574	0.1868	0.2269	0.2363	0.2262	0.2463	0.2427	0.2426	0.2319	0.2377	0.2260	0.1879

μ_2, $\beta = 50°$

λ_s (°)	γ_o (°)											
	−20	−15	−10	−5	0	5	10	15	20	25	30	40
0	0.8706	0.9015	0.8823	0.8867	0.9080	0.9441	0.9956	1.0646	1.1549	1.2263	1.3723	1.7502
5	1.0730	0.9862	0.8925	0.8958	0.9169	0.9522	1.0028	1.0725	1.1635	1.2356	1.3805	1.7599
10	1.2360	1.0670	1.0275	0.9648	0.9812	1.0168	1.0287	1.0968	1.1897	1.2600	1.4086	1.7895
15	1.3762	1.2102	1.1068	1.0974	1.0643	1.0546	1.1083	1.1387	1.2294	1.3005	1.4526	1.6020
20	1.4061	1.3514	1.2546	1.1911	1.1609	1.1512	1.2096	1.2346	1.2856	1.3142	1.2997	1.3572
25	1.0227	1.0529	1.0710	1.0453	1.0723	1.0458	1.0692	1.0820	1.0783	1.1042	1.1445	1.2416
30	0.8735	0.8738	0.8881	0.8998	0.9025	0.9261	0.9405	0.9609	0.9870	1.0335	1.0716	1.1573
35	0.9430	0.9144	0.8914	0.8895	0.9001	0.9011	0.9194	0.9413	0.9639	0.9923	1.0296	1.0873

Chapter 7
The Chip-Ejection Interference and Compromise in Non-free Cutting

Based on the principle of minimum energy dissipation in metal cutting process this chapter has studied the basic characteristics of non-free cutting, probed into the phenomena of chip-ejection interference and chip-ejection compromise widely existing in modern machining processes. A non-linear synthetic method of elementary cutting tools is proposed for establishing the cutting models of complex tools. The general equation governing chip-ejection motion is deduced and its solutions are studied. On the basis of the general equation some examples of non-free cutting are analyzed and the theoretically analyzing and predicting results are verified by experimental data and facts. The sufficient and necessary condition of eliminating chip-ejection interference and realizing free cutting is explored, the basic ideal and the technique route of "Free Cutting Method" are discussed. The research has provided a feasible way for improving and optimizing modern cutting tools.

7.1 A Brief Introduction

The previous chapter said that the natural values of the chip-ejection vector (including the direction and the speed of the chip-ejection) of an elementary cutting tool always minimizes the power consumption, and this is called "the principle of minimum energy dissipation" for an elementary cutting tool.

A real cutting tool maybe regarded as a combination of a series of elementary cutting tools. Because of the high development in powder metallurgy technology, die making technology and the numerical cutter sharpening technology, the working part of modern tools can be already made into vary complex shape, and their cutting processes are also very complicated.

In cutting process of a real cutting tool, all the elementary cutting tools working concurrently, each of them acts his own way and tries hard to eject chips according

H. Shi, *Metal Cutting Theory*, Springer Series in Advanced Manufacturing,
https://doi.org/10.1007/978-3-319-73561-0_7

to its own natural chip-ejection vector, their chips ejected will certainly mutually interfere and conflict. On the other hand, to keep up the integrity of the common chip ejected by all the elementary cutting tools, they have to compromise with each other on their chip-ejection vectors, so as to determine their overall chip-ejection motion. Researches show that the law restricting the overall chip-ejection motion is also the "principle of minimum energy dissipation": i.e. among all the overall chip-ejection motions satisfying constraint conditions set by controlling parameters, the realizable one must minimize the total cutting power (i.e. the sum of the cutting power consumed by all the elementary cutting tools joining the cutting process).

Those cutting process, in which chip-ejection interference exists and have to be coordinated, is called "non-free cutting". Nearly all the practical machining operations are of non-free cutting. In a non-free cutting process principle of minimum energy dissipation guarantee a minimum cutting power consumed by the whole cutter, but it is impossible to make the real chip-ejection vectors coincide with their natural chip-ejection vectors for all the elementary cutting tools, respectively. Generally speaking, after chip-ejection compromise every elementary cutting tool has to make some concession, and cannot eject chips at its own natural chip-ejection vector. As a result of this the total cutting power of a cutting tool is normally bigger than the sum of the cutting powers of all its elementary cutting tools when each of them cutting alone. This reflect the strong non-linearity of a non-free cutting process. Such a characteristics results in an increase in cutting energy consumption; complicates the chip deformation and deteriorates the machined surface finish; increases the cutting force and deformation of machining system, and decreases machining accuracy; increases cutting temperature, shortens tool life and increase the shut down time for tool changing, and this results in the cost increase. Many problems in modern machining work are often caused by these drawbacks in non-free cutting.

A non-linear synthetic method of elementary cutting tools is proposed in this chapter for explanation, prediction and optimization of complex shaped cutting tools. Compared with the work of other researchers [70–72], the advantage of the method proposed in this book is that it based on the principle of minimum energy dissipation properly dealt with the issue of chip-ejection interference and the non-linear problems caused by the interference.

7.2 The General Pattern of Overall Chip-Ejection Motion

7.2.1 *Dividing Elementary Cutting Tools and Their Description*

Suppose a cutting tool is divided, along its edge, into a series of elementary cutting tools (see Fig. 7.1). The following parameters are necessary for describing every elementary cutting tool: radius vector $\boldsymbol{\rho}_i$, unit vector in the direction of cutting edge $\mathbf{b}_i$, unit vector on the rake and in the direction normal to cutting edge $\mathbf{a}_i$ (Fig. 7.1),

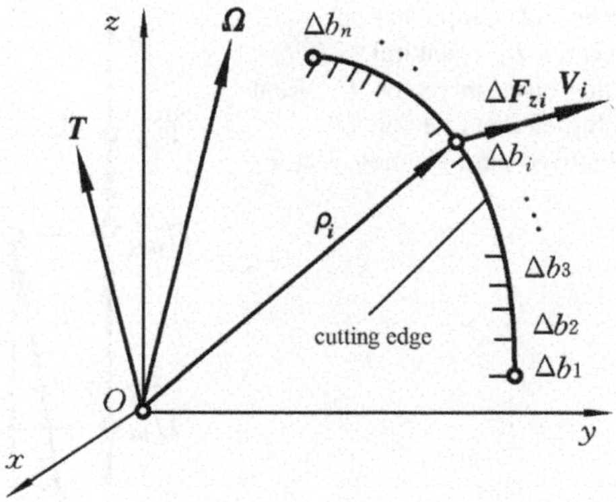

Fig. 7.1 The overall chip-ejection motion and elemental cutting tool dividing

and length of cutting edge Δb_i, geometrical angles γ_{oi}, λ_{si} and cutting parameters a_{ci}, V_i; i = 1,2,..., n. The parameters may be obtained through a geometric and movement analysis. The cutting force on unit length edge of ith elementary cutting tool is

$$\boldsymbol{I}_{coi} = \boldsymbol{I}_{coi}(\gamma_{oi}, \lambda_{si}, a_{ci}, V_i), \quad i = 1, 2 \ldots n, \tag{7.1}$$

And the natural chip-ejection vector is

$$\boldsymbol{U}_{oi} = \boldsymbol{U}_{oi}(\gamma_{oi}, \lambda_{si}, a_{ci}, V_i), \quad i = 1, 2 \ldots n. \tag{7.2}$$

They can be decided through experimental study or theoretical analysis (the materials of work piece and tool and lubrication conditions are also the factors affecting cutting force and natural chip-ejection vector. They are, however, the same for all the elementary cutting tools of a tool, i.e., they are regarded as constants, so they do not appear in the Eqs. 7.1 and 7.2.) $\boldsymbol{U}_{oi}$ can be decomposed into components parallel with and normal to cutting edge,

$$U_{oti} = \boldsymbol{U}_{oi} \cdot \mathbf{b}_i, \quad U_{oni} = \boldsymbol{U}_{oi} \cdot \mathbf{a}_i, \tag{7.3}$$

According to the principle of minimum energy dissipation for elementary cutting tools (see Sect. 6.2.3 and Eq. 6.13), the cutting power of the whole tool is

$$A = \sum_{i=1}^{n} \Delta A_i = \sum_{i=1}^{n} I_{coi} V_i \Delta b_i (1 + \mu_{1i}\zeta_i^2 + \mu_{2i}\xi_i^2), \tag{7.4}$$

And from Eq. (6.12) (see Fig. 7.2),

$$\begin{aligned} \zeta_i &= \Delta U_{ti}/U_{oi} = (U_{ti} - U_{oti})/U_{oi} \\ \xi_i &= \Delta U_{ni}/U_{oi} = (U_{ni} - U_{oni})/U_{oi}. \end{aligned} \tag{7.5}$$

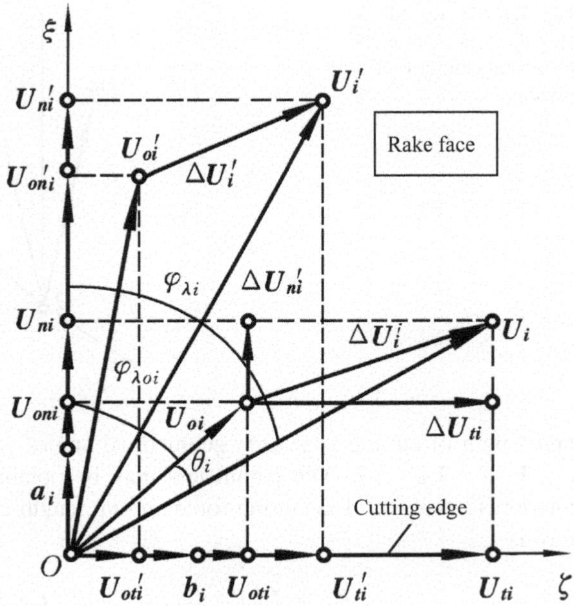

Fig. 7.2 Chip-ejection vector. $\boldsymbol{U}_o$—Natural chip-ejection vector. $\boldsymbol{U}$—Real chip-ejection vector. $\boldsymbol{U}'$—Induced chip-ejection vector

7.2.2 Chip's Overall Motion and the Real Chip-Ejection Vectors for Elementary Cutting Tools

Chip once formed, its motion may be regarded as rigid body motion (compared with cutting power, the elastic deformation energy stored in the chip can be ignored). As everybody knows, the motion of a rigid body is composed by a translation $\boldsymbol{T}$ and a rotation $\boldsymbol{\Omega}$ (Fig. 7.1). It should be noticed that $\boldsymbol{\Omega}$ here refers to the rotation when the chip just leaves the cutting edge, it doesn't include the curl caused by additional elastic or plastic deformation when the chip contacting the tool rake face afterwards.

The real chip-ejection vector of the elementary cutting tool having a radius vector $\boldsymbol{\rho}_i$ is

$$\boldsymbol{U}_i = \boldsymbol{T} + \boldsymbol{\Omega} \times \boldsymbol{\rho}_i, \quad i = 1, 2 \ldots n, \tag{7.6}$$

It can be decomposed into components

$$U_{ti} = \boldsymbol{U}_i \cdot \mathbf{b_i}, U_{ni} = \boldsymbol{U}_i \cdot \mathbf{a_i}. \tag{7.7}$$

7.2.3 The Principle of Minimum Energy Dissipation for the Whole Tool

Substituting Eqs. (7.5)–(6.7) into 7.4 results in

$$A = \sum_{i=1}^{n} I_{coi} V_i \Delta b_i \{1 + \mu_{1i} \frac{[(\boldsymbol{T} + \boldsymbol{\Omega} \times \boldsymbol{\rho}_{\mathrm{i}} - \boldsymbol{U}_{oi}) \cdot \mathbf{b}_{\mathrm{i}}]^2}{\boldsymbol{U}_{oi}^2} + \mu_{2i} \frac{[(\boldsymbol{T} + \boldsymbol{\Omega} \times \boldsymbol{\rho}_{\mathrm{i}} - \boldsymbol{U}_{oi}) \cdot \mathbf{a}_{\mathrm{i}}]^2}{\boldsymbol{U}_{oi}^2}\}, \tag{7.8}$$

where $\boldsymbol{T}$ and $\boldsymbol{\Omega}$ are two state parameters which should be determined according to the principle of minimum energy dissipation, so let $\partial A / \partial \boldsymbol{T} = \mathbf{0}$ and $\partial A / \partial \boldsymbol{\Omega} = \mathbf{0}$, and we can deduce[1]

$$\begin{gathered} \overline{\boldsymbol{U}_i'} = \overline{\boldsymbol{U}_{oi}'} \\ \overline{\boldsymbol{\rho}_i \times \boldsymbol{U}_i'} = \overline{\boldsymbol{\rho}_i \times \boldsymbol{U}_{oi}'}, \end{gathered} \tag{7.9}$$

where $\boldsymbol{U}_i'$ and $\boldsymbol{U}_{oi}'$ are the induced chip-ejection vector and the induced natural chip-ejection vector, respectively (see Fig. 7.2),

$$\begin{aligned} \boldsymbol{U}_i' &= \mu_{1i} U_{ti} \mathbf{b}_{\mathbf{i}} + \mu_{2i} U_{ni} \mathbf{a}_{\mathbf{i}} \\ \boldsymbol{U}_{oi}' &= \mu_{1i} U_{oti} \mathbf{b}_{\mathbf{i}} + \mu_{2i} U_{oni} \mathbf{a}_{\mathbf{i}}. \end{aligned} \tag{7.10}$$

The upper transverse line in Eq. (7.9) means weighted average, i.e.

$$\overline{x} = \sum_{i=1}^{n} w_i x_i, \tag{7.11}$$

where normalized weight coefficients are

$$w_i = \frac{I_{coi} V_i \Delta b_i}{U_{oi}^2} \Bigg/ \sum_{j=1}^{n} \frac{I_{coi} V_i \Delta b_i}{U_{oi}^2}, i = 1, 2, \ldots, n. \tag{7.12}$$

If there are a pair of $\boldsymbol{T}_{\mathbf{m}}$ and $\boldsymbol{\Omega}_{\mathbf{m}}$ really minimize cutting power A, then they must satisfy Eq. (7.9). On the other hand, Eq. (7.9) provides a way to calculate $\boldsymbol{T}_{\mathbf{m}}$ and $\boldsymbol{\Omega}_{\mathbf{m}}$. These two equations show a "law of average" for determining real chip-ejection vector, the law may be described like follows: the weighted average of the induced chip-ejection vectors of all the elementary cutting tools must be equal to the weighted average of the induced natural chip-ejection vectors of all the elementary cutting tools; and the weighted average of the cross product of induced

[1]The deduction of this formulae see the Appendix 7.1 of this chapter.

chip-ejection vectors and the radius vector of all the elementary cutting tools must be equal to the weighted average of the cross product of induced natural chip-ejection vectors and the radius vector of all the elementary cutting tools. And the weight coefficients are in direct proportion to the power consumption and inversely proportional to square of chip-ejection speed. Equation (7.9) is called basic equation of overall chip-ejection motion during non-free cutting.

7.3 The Solution of Basic Equation for Chip-Ejection Motion Under Non-free Cutting

7.3.1 The Property of the Equation and Its Solution

Let's rewrite Eq. (7.9), so as to reflect influence of various factors. Substituting Eqs. (7.6), (7.7), (7.10), (7.11) and (7.12) into (7.9) results in

$$\begin{aligned}&\sum_{i=1}^{n}\frac{I_{coi}V_i\Delta b_i}{U_{oi}^2}\{\mu_{1i}[(\boldsymbol{T}+\boldsymbol{\Omega}\times\boldsymbol{\rho}_i)\cdot\mathbf{b}_i]\mathbf{b}_i+\mu_{2i}[(\boldsymbol{T}+\boldsymbol{\Omega}\times\boldsymbol{\rho}_i)\cdot\mathbf{a}_i]\mathbf{a}_i\}\\&\quad=\sum_{i=1}^{n}\frac{I_{coi}V_i\Delta b_i}{U_{oi}^2}[\mu_{1i}(\boldsymbol{U}_{oi}\cdot\mathbf{b}_i)\mathbf{b}_i+\mu_{2i}(\boldsymbol{U}_{oi}\cdot\mathbf{a}_i)\mathbf{a}_i];\end{aligned}\tag{7.13}$$

$$\begin{aligned}&\sum_{i=1}^{n}\frac{I_{coi}V_i\Delta b_i}{U_{oi}^2}\boldsymbol{\rho}_i\times\{\mu_{1i}[(\boldsymbol{T}+\Omega\times\boldsymbol{\rho}_i)\cdot\mathbf{b}_i]\mathbf{b}_i+\mu_{2i}[(\boldsymbol{T}+\boldsymbol{\Omega}\times\boldsymbol{\rho}_i)\cdot\mathbf{a}_i]\mathbf{a}_i\}\\&\quad=\sum_{i=1}^{n}\frac{I_{coi}V_i\Delta b_i}{U_{oi}^2}\boldsymbol{\rho}_i\boldsymbol{U}_{oi}\times[\mu_{1i}(\boldsymbol{U}_{oi}\cdot\mathbf{b}_i)\mathbf{b}_i+\mu_{2i}(\boldsymbol{U}_{oi}\cdot\mathbf{a}_i)\mathbf{a}_i].\end{aligned}\tag{7.14}$$

The right side of above two equations are two known vectors which can be calculated through known quantities, denoting the two vectors by $\boldsymbol{A}$ and $\boldsymbol{B}$, respectively, we have

$$\sum_{i=1}^{n}\frac{I_{coi}V_i\Delta b_i}{U_{oi}^2}\{\mu_{1i}[(\boldsymbol{T}+\boldsymbol{\Omega}\times\boldsymbol{\rho}_i)\cdot\mathbf{b}_i]\mathbf{b}_i+\mu_{2i}[(\boldsymbol{T}+\boldsymbol{\Omega}\times\boldsymbol{\rho}_i)\cdot\mathbf{a}_i]\mathbf{a}_i\}=\boldsymbol{A},\tag{7.15}$$

$$\sum_{i=1}^{n}\frac{I_{coi}V_i\Delta b_i}{U_{oi}^2}\boldsymbol{\rho}_i\times\{\mu_{1i}[(\boldsymbol{T}+\boldsymbol{\Omega}\times\boldsymbol{\rho}_i)\cdot\mathbf{b}_i]\mathbf{b}_i+\mu_{2i}[(\boldsymbol{T}+\Omega\times\boldsymbol{\rho}_i)\cdot\mathbf{a}_i]\mathbf{a}_i\}=\boldsymbol{B}.\tag{7.16}$$

These are two simultaneous vector equations, which are consisting of six scalar quantity equations. On the other hand, the unknown quantities are two vectors, equivalently six scalar quantities. The number of equations is just equal to the number of unknown quantities, therefore, generally speaking, the group of equations can uniquely determine a pair of $\boldsymbol{T}_{\mathbf{m}}$ and $\boldsymbol{\Omega}_{\mathbf{m}}$, by means of numerical method. Substituting resulted $\boldsymbol{T}_{\mathbf{m}}$ and $\boldsymbol{\Omega}_{\mathbf{m}}$, into Eq. (7.8) we get the real cutting power $A_{\min}$ controlled by the principle of minimum energy dissipation.

7.3.2 Simplification of the Equations and Approximate Solutions

To solve Eqs. (7.15) and (7.16) numerical method has to be used and this is a time-consuming job. From a practical point of view, we might as well simplify the model expressed by Eq. (7.4), the simplification method is to assume that

$$\mu = \mu_{1i} = \mu_{2i} = \frac{1}{2n}\left(\sum_{i=1}^{n} \mu_{1i} + \sum_{i=1}^{n} \mu_{2i}\right); \quad i = 1, 2, \ldots, n, \tag{7.17}$$

i.e., to assume that all the influence coefficients μ_{1i} and μ_{2i}, $i = 1, 2, \ldots, n$ all take the value of their average, as a constant. The feasibility of such simplification has been verified by the fitting and the test of a drilling model (see Chap. 8). Under this simplification, Eq. (7.8) becomes,

$$A = \sum_{i=1}^{n} I_{coi} V_i \Delta b_i \left[1 + \mu \frac{(\boldsymbol{T} + \boldsymbol{\Omega} \times \boldsymbol{\rho}_i - \boldsymbol{U}_{oi})^2}{\boldsymbol{U}_{oi}^2}\right], \tag{7.18}$$

And Eqs. (7.15) and (7.16) become

$$\sum_{i=1}^{n} \frac{I_{coi} V_i \Delta b_i}{U_{oi}^2} (\boldsymbol{T} + \boldsymbol{\Omega} \times \boldsymbol{\rho}_i) = \sum_{i=1}^{n} \frac{I_{coi} V \Delta b_i}{U_{oi}^2} \boldsymbol{U}_{oi} = \boldsymbol{A}, \tag{7.19}$$

And

$$\sum_{i=1}^{n} \frac{I_{coi} V_i \Delta b_i}{U_{oi}^2} [\boldsymbol{\rho}_i \times (\boldsymbol{T} + \boldsymbol{\Omega} \times \boldsymbol{\rho}_i)] = \sum_{i=1}^{n} \frac{I_{coi} V \Delta b_i}{U_{oi}^2} (\boldsymbol{\rho}_i \times \boldsymbol{U}_{oi}) = \boldsymbol{B}, \tag{7.20}$$

Respectively. Denote them by average sign $\bar{x}$, the above two equations can be written as

$$\boldsymbol{T} + \boldsymbol{\Omega} \times \overline{\boldsymbol{\rho}_i} = \boldsymbol{A}, \tag{7.21}$$

And

$$\boldsymbol{T} \times \overline{\boldsymbol{\rho}_i} + \overline{(\boldsymbol{\Omega} \times \boldsymbol{\rho}_i) \times \boldsymbol{\rho}_i} = -\boldsymbol{B}. \tag{7.22}$$

In the deduction the following rules were used $\overline{\boldsymbol{A}_i+\boldsymbol{B}_i}=\overline{\boldsymbol{A}_i}+\overline{\boldsymbol{B}_i}$, $\overline{\boldsymbol{A}_i\times\boldsymbol{B}_i}=-\overline{\boldsymbol{B}_i\times\boldsymbol{A}_i}$ and $\overline{\boldsymbol{C}\times\boldsymbol{A}_i}=\boldsymbol{C}\times\overline{\boldsymbol{A}_i}$ ($\boldsymbol{C}$ is a constant vector), etc. From (7.21) $\times\ \overline{\boldsymbol{\rho}_{\mathrm{i}}}$—(8.22), we have

$$\boldsymbol{T}\times\overline{\boldsymbol{\rho}_i}+(\boldsymbol{\Omega}\times\overline{\boldsymbol{\rho}_i})\times\overline{\boldsymbol{\rho}_i}-\boldsymbol{T}\times\overline{\boldsymbol{\rho}_i}-\overline{(\boldsymbol{\Omega}\times\boldsymbol{\rho}_i)\times\boldsymbol{\rho}_i}=\boldsymbol{A}+\boldsymbol{B},$$

$\boldsymbol{T}$ can be cancelled,

$$(\boldsymbol{\Omega}\times\overline{\boldsymbol{\rho}_i})\times\overline{\boldsymbol{\rho}_i}-\overline{(\boldsymbol{\Omega}\times\boldsymbol{\rho}_i)\times\boldsymbol{\rho}_i}=\boldsymbol{C},$$

where $\boldsymbol{C}=\boldsymbol{A}+\boldsymbol{B}$. From above equation we can work out $\boldsymbol{\Omega}$. The method is described as follows. Denote

$$\boldsymbol{\Omega}=\Omega_x\mathbf{i}+\Omega_y\mathbf{j}+\Omega_z\mathbf{k}, \tag{7.23}$$

$$\boldsymbol{\rho}_{\mathbf{i}}=x_i\mathbf{i}+y_i\mathbf{j}+z_i\mathbf{k}, \tag{7.24}$$

$$\boldsymbol{C}=C_x\mathbf{i}+C_y\mathbf{j}+C_z\mathbf{k}. \tag{7.25}$$

Then,

$$\begin{aligned} a_{11}\Omega_x+a_{12}\Omega_y+a_{13}\Omega_z&=C_x\\ a_{21}\Omega_x+a_{22}\Omega_y+a_{23}\Omega_z&=C_y\\ a_{31}\Omega_x+a_{32}\Omega_y+a_{33}\Omega_z&=C_x, \end{aligned} \tag{7.26}$$

where

$$\begin{aligned} a_{11}&=\overline{y_iy_i}-\overline{y_i}\,\overline{y_i}+\overline{z_iz_i}-\overline{z_i}\,\overline{z_i},\\ a_{22}&=\overline{x_ix_i}-\overline{x_i}\,\overline{x_i}+\overline{z_iz_i}-\overline{z_i}\,\overline{z_i},\\ a_{33}&=\overline{x_ix_i}-\overline{x_i}\,\overline{x_i}+\overline{y_iy_i}-\overline{y_i}\,\overline{y_i},\\ a_{12}&=a_{21}=\overline{x_i}\,\overline{y_i}-\overline{x_iy_i},\\ a_{13}&=a_{31}=\overline{x_i}\,\overline{z_i}-\overline{x_iz_i},\\ a_{23}&=a_{32}=\overline{y_i}\,\overline{z_i}-\overline{y_iz_i}, \end{aligned} \tag{7.27}$$

From linear system of Eqs. (7.26) we can work out Ω_x, Ω_y and Ω_z, and get $\boldsymbol{\Omega}_{\mathbf{m}}$ according to Eq. (7.23), substituting into Eq. (7.21), we obtain

$$\boldsymbol{T}_{\mathbf{m}}=\boldsymbol{A}-\boldsymbol{\Omega}_{\mathbf{m}}\times\overline{\boldsymbol{\rho}_{\mathbf{i}}}. \tag{7.28}$$

7.4 Non-free Cutting Example Analyses (I)

We have discussed the general solution of Eq. (7.9), and the common way of calculating $\boldsymbol{T}_{\mathbf{m}}$ and $\boldsymbol{\Omega}_{\mathbf{m}}$, hereinafter we will give some experimental examples to show the application of the general principle to various specific conditions and to check the correctness of the principle.

7.4.1 The Non-free Cutting with Double Edged Tool

The double edged cutting tool shown in Fig. 6.6c was used, the cutting edge angle θ is fixed as $\theta = 20°$, as shown in Fig. 7.3. The cutting conditions are also the same as described in Sect. 6.2.4.1, the total width of work piece is $b = 4$ mm, but the burdens b_l, b_r of the left and right edges are variable, not symmetrical ($b_l + b_r = b$). The widths of cut of the two edges are $b_l/\cos\theta$ and $a_{wr} = b_r/\cos\theta$.

From Eq. (6.16) and noticing that the width a_w was 4 mm, we have

$$F_c = 145.35(1 + 5.45 \times 10^{-4}\theta^2) \times a_w. \tag{7.29}$$

If the included angle between the overall chip-ejection direction $\boldsymbol{a}$ and longitudinal direction of the tool shank is η as shown in Fig. 7.3, then the deviations of $\boldsymbol{a}$ from natural chip-ejection directions of left and right edges are $\theta_l = \theta - \eta = 20° - \eta$ and $\theta_r = \theta + \eta = 20° + \eta$, respectively. And the total cutting force of the two edges is

$$F_c = \frac{145.35}{\cos 20°}\{b_l[1 + 5.45 \times 10^{-4}(20° - \eta)^2] + b_r[1 + 5.45 \times 10^{-4}(20° + \eta)^2]\}. \tag{7.30}$$

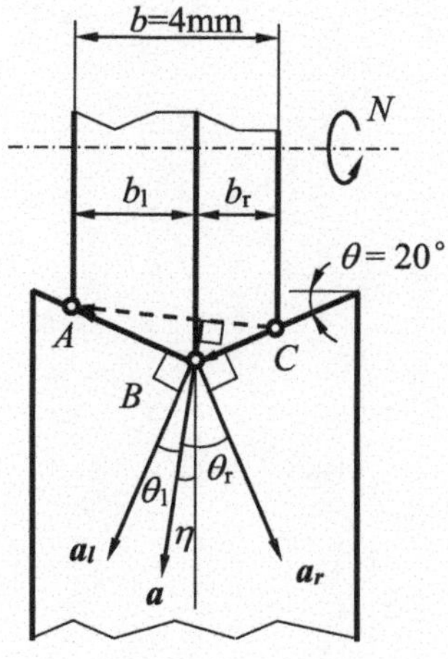

Fig. 7.3 Experiments of non-free cutting with a double edged tool

Considering $b_l + b_r = b = 4$, denoting $b_l - b_r = \Delta b$ and rewriting Eq. (7.30), we have

$$F_c = 753.59 - 8.433(\Delta b)^2 + 0.337(\eta - 5\Delta b)^2. \quad (7.31)$$

According to the principle of minimum energy dissipation the state parameter η must minimize cutting force F_c, so let $\partial F_c/\partial \eta = 0$, we have

$$\eta = 5\Delta b. \quad (7.32)$$

Substituting Eqs. (7.32) into (7.31) results in

$$F_c = 753.59 - 8.433(\Delta b)^2. \quad (7.33)$$

The measured in experiments and predicted with Eqs. (7.32) and (7.33) the overall chip-ejection angle η and cutting force F_c are given in Fig. 7.4. From the figure we can see that the really measured data and the predicted ones are agree with each other fairly well.

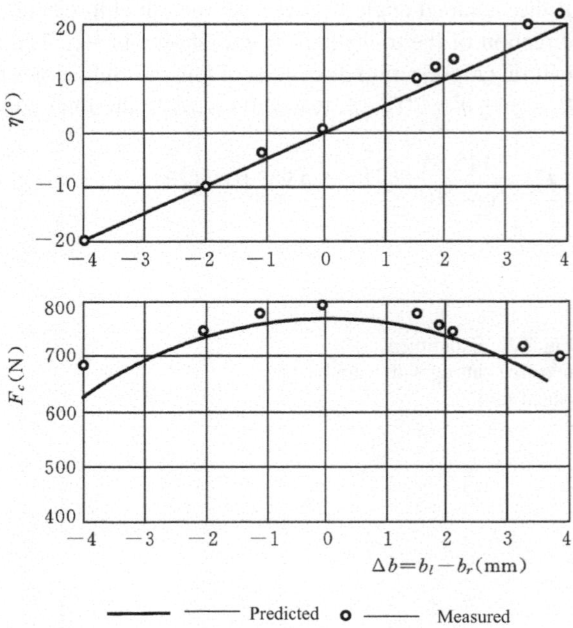

Fig. 7.4 The overall chip-ejection angle and the cutting force in non-free cutting with double edged tool

7.4.2 *Equivalent Cutting Edge*

7.4.2.1 An Example

We will introduce the conception of "equivalent cutting edge" through an example. As shown in Fig. 7.3, the two segments of cutting edges ***CB*** and ***BA*** taking part in cutting work may be regarded as two vectors, and their composite vector ***CA*** = ***CB*** + ***BA*** is called the equivalent cutting edge of the cutting process. Equivalent cutting edge comprehensively reflects the external effects of chip-ejection interference and chip-ejection compromise, or in other words, the chip-ejection vector of a multi-edged cutting tool performing non-free cutting just like the chip-ejection vector of their equivalent cutting edge performing free cutting. In the example (Fig. 7.3), the normal ***a*** to the "equivalent cutting edge" ***CA*** is just the real chip-ejection direction of the cutting process. From this point of view, the chip-ejection angle is

$$\tan \eta = \tan 20^\circ \Delta b/4. \tag{7.34}$$

The above result is very close to Eq. (7.32), as a matter of fact, substituting $\tan \eta \approx \eta$ and $\tan 20^\circ \approx 20^\circ$ into the above equation results in Eq. (7.32).

7.4.2.2 Generalized "Equivalent Cutting Edge"

In fact, the conception of "equivalent cutting edge" can be deduced from the principle of minimum energy dissipation and the average law, and some approximate methods have to be applied. In fact, during turning, planing, slotting and milling operations, the variation in cutting speed over all the working edge is rather small, and so $\boldsymbol{V_i}$ may be regarded as constant over the working edge. The difference in cutting ratio r_i over the working edge is also small, so chip-ejection speed $U_i = r_i V_i$ can be also regarded as a constant over the working edge. As a result of this, the rotation component $\boldsymbol{\Omega}$ in the chip-ejection motion can be ignored. Let $\boldsymbol{\Omega} = \mathbf{0}$, Eq. (7.19) becomes

$$\boldsymbol{T} = \overline{\boldsymbol{U_{oi}}(\mathbf{b_i}, \mathbf{a_i}, \gamma_{oi}, \lambda_{si}, \ldots)}, \tag{7.35}$$

i.e. the overall chip-ejection vector (translation motion ***T***) of the whole cutter is the weighted average of chip-ejection vectors $\boldsymbol{U_{oi}}$ of all its elementary cutting tools.

In Eq. (7.35), $\boldsymbol{U_{oi}}$ is the function of the independent variables $\mathbf{b_i}, \mathbf{a_i}, \gamma_{oi}, \lambda_{si}, \ldots$. The equation shows that we have to calculate the function $\boldsymbol{U}_{oi}(\mathbf{b_i}, \mathbf{a_i}, \gamma_{oi}, \lambda_{si}, \ldots)$; i = 1, 2… at first, and then take the average of $\overline{\boldsymbol{U_{oi}}}$ to get ***T***. Generally speaking, the order of functional calculation and taking average of independent variables cannot be inversed. However, if the function just has a weak non-linearity in the variation range of its independent variables, it may be regarded approximately as a linear function, and the function calculation and average calculation can be inversed, so, we have

$$\boldsymbol{T} = \boldsymbol{U}_{oi}(\overline{\mathbf{b_i}}, \overline{\mathbf{a_i}}, \overline{\gamma_{oi}}, \overline{\lambda_{oi}}, \overline{\cdots}). \tag{7.36}$$

where $\overline{\mathbf{b_i}}, \overline{\mathbf{a_i}}, \overline{\gamma_{oi}}, \overline{\lambda_{oi}}, \overline{\cdots}$ and etc. are the averages of the independent variables. Equation 7.36 expresses a cutting edge possessing average parameters. Such a cutting edge is called "equivalent cutting edge". Because this edge comprehensively reflects the influences of all the independent variables, it has a representativeness, and the chip-ejection motion of the edge, to some extent, might represents the real overall chip-ejection motion.

7.4.2.3 Special Equivalent Cutting Edge

In Sect. 7.4.2.2 we introduced a kind of "equivalent cutting edge", which has more generalized meaning, called generalized equivalent cutting edge. The example in Sect. 7.4.2.1 is, however, of special "equivalent cutting edge". To introduce special one, we should add more assumptions:

(i) Rake face is a plane, and the unit vector normal to it is denoted by **g**;
(ii) Cutting velocity $\boldsymbol{V}$ is the same for the whole cutting edge, and so the reference plane P_r is the same for the whole cutting edge;
(iii) The weight coefficients are only related to lengths of elementary cutting tools,

$$w_i = \Delta b_i \Big/ \left| \sum\nolimits_{i=1}^{n} \mathbf{b_i} \Delta b_i \right|, i = 1, 2, \ldots, n, \tag{7.37}$$

where $\mathbf{b_i}\Delta b_i$ is an edge's vector of ith elementary cutting tool.

According to Eq. (7.36), let's calculate the averages of various parameters:

$$\mathbf{b} = \overline{\mathbf{b_i}} = \sum_{i=1}^{n} \mathbf{b_i} w_i, \tag{7.38}$$

$$\mathbf{a} = \overline{\mathbf{a_i}} = \sum_{i=1}^{n} \mathbf{a_i} w_i, \tag{7.39}$$

$$\sin \lambda_s = \overline{\sin \lambda_{s\,i}} = \sum_{i=1}^{n} \sin \lambda_{s\,i} w_i, \tag{7.40}$$

$$\tan \gamma_n = \overline{\tan \gamma_{ni}} = \sum_{i=1}^{n} \tan \gamma_{ni} w_i, \tag{7.41}$$

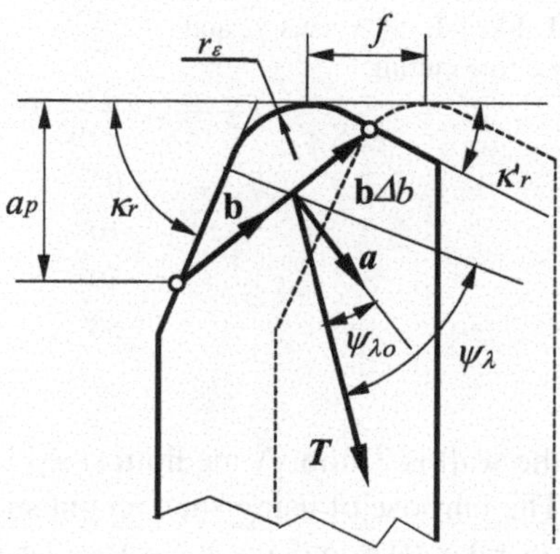

Fig. 7.5 The chip-ejection direction of equivalent cutting edge

where **b** is the unit vector in the direction of equivalent cutting edge. Substituting Eqs. (7.37) into (7.38) leads to,

$$\mathbf{b}\Delta b = \sum_{i=1}^{n} \mathbf{b}_\mathbf{i}\Delta b_\mathrm{i}, \tag{7.42}$$

It means that equivalent cutting edge, as a vector, is the vector sum of all the segment edges. In the above equation, $\Delta b = \left|\sum_{i=1}^{n} \mathbf{b}_\mathbf{i}\Delta b_i\right|$ is the length of vector sum, which is the length of special equivalent cutting edge. The edge defined in this way will certainly in the rake face.

Unit vector **a** is the weighted average of $\mathbf{a}_\mathbf{i}$ of all the segment edges, it is easy to verify that $\mathbf{a} = \mathbf{g} \times \mathbf{b}$, so, **a** is in rake face and perpendicular to **g** and **b**. Two unit vectors **a** and **b** regulate the orientation in the space of the equivalent cutting edge and the rake face. And calculated λ_s and γ_n according to Eqs. (7.40) and (7.41) are just the tool cutting edge inclination and the cutting edge normal rake of the equivalent cutting edge defined by Eq. (7.42).

The conceptions above described may generalized to curved edges. As a matter of fact, the connection line from the starting point to the end point of the working edge is just the equivalent cutting edge. And the chip-ejection direction determined according to edge inclination and the rake of the equivalent cutting edge is just the chip-ejection direction for the non-free cutting process after chip-ejection compromise (see Fig. 7.5).

7.4.2.4 Experimental Verification

In other to verify the method of equivalent cutting edge for predicting chip-ejection direction, turning experiments were conducted.

First, free cutting experiments, the end surface of a seamless steel tube (soft steel #20) was cut by a HSS cutter, work piece diameter is 100 mm, and the thickness of

Table 7.1 ψ_λ versus γ_n and λ_s, free cutting

γ_n	λ_s			
	0°	10°	20°	30°
−10°	0°	7.5°	13.3°	19.6°
0°	0°	6.4°	12.0°	18.3°
10°	0°	6.4°	13.3°	18.3°
20°	0°	8.1°	14.7°	21.8°
30°	0°	8.8°	16.4°	23.6°

the wall is 3 mm. A medium sized lathe was used. Cutting speed: $V = 76.5$ m/min. The purpose of using such an unusual high speed was to let chip makes clear marks on rake face and we have to measure the chip-ejection angle according to these marks. Because, however, the cutting time was short, so there was no remarkable wear on the cutter. Chip ejection angle ψ_λ was measured, the data of ψ_λ versus rake γ_n and λ_s are given in Table 7.1. The rake angle γ_o may through changing cutting temperature to change the friction coefficient between rake and the chip, and thus to change ψ_λ. The data in Table 7.1 deviate from Stabler law, this may be caused by the deviation of the cutting speed from ordinary cutting parameter.

Secondly, turning experiments with differently shaped tool nose. According to the way of "vector sum" determine equivalent edge, and determine the edge inclination λ_s and the normal rake γ_n of equivalent edge according to the orientation of the equivalent edge on the rake face. Based on the data in Table 7.1, through interpolation method determine the natural chip-ejection angle $\psi_{\lambda o}$ and chip-ejection direction $\boldsymbol{T}$ under free cutting of the equivalent edge. As shown in Fig. 7.5, $\psi_{\lambda o}$ is the angle enclose between the normal **a** to equivalent edge and $\boldsymbol{T}$, and ψ_λ is the angle enclose between the normal to the major edge and $\boldsymbol{T}$. The $\psi_{\lambda o}$ and ψ_λ labeled in Fig. 7.5 is their projection on the reference plane, and in fact they are defined and measured in rake plane.

Experimental conditions are listed in Table 7.2. Under various cutting tools and various cutting parameters, the equivalent edge predicted and really measured overall chip-ejection angle ψ_λ are both listed in Table 7.3. It is clear from the data that they are fairly closed to each other and the method of predicting chip-ejection direction by means of equivalent edge is feasible.

7.4.3 Re-Analysis of the Data in Tables 6.4 and 6.5

To analyze another example of non-free cutting, let's go back to Sect. 6.2.4.2, and Tables 6.4 and 6.5. According to Eq. (6.21), the cutting force of a triple blade cutter can be expressed as

Table 7.2 Chip-ejection angle experimental conditions, non-free cutting

No	$\kappa_r{}^\circ$	$\kappa_r'^\circ$	γ_n°	λ_s°	Chamfering			a_p mm	*f*mm/r
					$r_\varepsilon$mm	$\kappa_\circ^r$	*b*mm		
1	45	10	0	0	0	0	0	1.97	0.14
2	45	10	0	0	0	0	0	0.94	0.28
3	45	10	0	0	0	0	0	0.52	0.55
4	45	10	0	0	0	0	0	0.29	1.11
5	45	10	−10	0	0	0	0	0.53	0.55
6	45	10	10	0	0	0	0	0.53	0.55
7	45	10	20	0	0	0	0	0.54	0.55
8	45	10	30	0	0	0	0	0.57	0.55
9	45	10	0	5	0	0	0	0.54	0.55
10	45	10	0	−10	0	0	0	0.53	0.55
11	45	10	0	−20	0	0	0	0.48	0.55
12	45	10	0	−30	0	0	0	0.51	0.55
13	47.7	25.6	11.4	−14.4	0	0	0	0.54	0.55
14	45	0	0	0	0	0	0	0.97	0.28
15	45.6	10.2	12.8	3.1	0	19.7	1.98	1.58	0.14
16	45	10	0	0	1.09	0	0	0.88	0.28
17	45	8	0	0	2.20	0	0	1.08	0.14
18	45	10	0	0	3.36	0	0	1.04	0.14

Table 7.3 Chip-ejection angle, predicted by equivalent edge and really measured

No	Equivalent edge			ψ_λ° predicted	ψ_λ° measured
	γ_n°	λ_s°	$\psi_{\lambda o}^\circ$		
1	0	0	0	2	2.2
2	0	0	0	7.5	7.2
3	0	0	0	22	23.6
4	0	0	0	41	42.9
5	−9.3	3.6	2.7	24	28.0
6	9.3	−3.6	2.2	19.7	21.9
7	18.7	−7.5	5.9	16	15.3
8	28.4	−11.3	9.6	13.7	10.1
9	1.8	4.6	2.9	25	26.2
10	−3.3	−7.5	4.9	16.3	19.6
11	−8	−18.8	12	10.3	12.2
12	−12	−28.4	18	1	1.8
13	3.7	−17.9	11.2	15	15.8
14	0	0	0	7	5.1
15	13.2	−0.6	0.4	17	15.3
16	0	0	0	14	11.5
17	0	0	0	17	13.0
18	0	0	0	21.6	18.5

$$F_c = I_{co}\Delta b \sum_{i=1}^{3} [1 + \mu_2 (r - r_{oi})^2 / r_{oi}^2], \tag{7.43}$$

where, $r_{oi} = U_{oi}/V, i = 1, 2, 3$ is the natural cutting ratios of the blades, which are already given in Table 7.4, and $r = U/V$ is the cutting ratios of global chip-ejection. As a state parameter of the cutting process, r should be determined according to the principle of minimum energy dissipation, thus, let $\partial F_c/\partial r = 0$ and substituting Eq. (7.43) into it results in

$$r = \sum_{i=1}^{3} (1/r_{oi}) \Big/ \sum_{i=1}^{3} (1/r_{oi})^2. \tag{7.44}$$

Predicted values of r according to above Equation are listed in Table 6.5. A comparison of predicted r with the measured value shows that they are much closed.

7.5 Non-free Cutting Examples Analyses (II)

Taking forming turning with circular edged tools as an example, let's show how to analyze a nan-free cutting process on the bases of the principle of minimum energy dissipation. The tools used in experiments and analysis are shown in Fig. 7.6.

7.5.1 Non-linear Synthetic Method of Elementary Cutting Tools

As shown by Eq. (6.19), from the experimental data an empirical formula of cutting force generated by a unit length of cutting edge is fitted as

$$I = 1593a_c^{0.61}(1 + 0.76\theta^2), \tag{7.45}$$

which is the basis for further analysis, where θ is the forced deviation in chip-ejection angle (radian).

If each elementary cutting tool ejects chips in its natural chip-ejection direction, i.e. in the radius direction, then $\theta = 0°$. The elementary cutting force generated by edge Δl would have been

$$\Delta F_{co} = 1593a_c^{0.61}\Delta l, \tag{7.46}$$

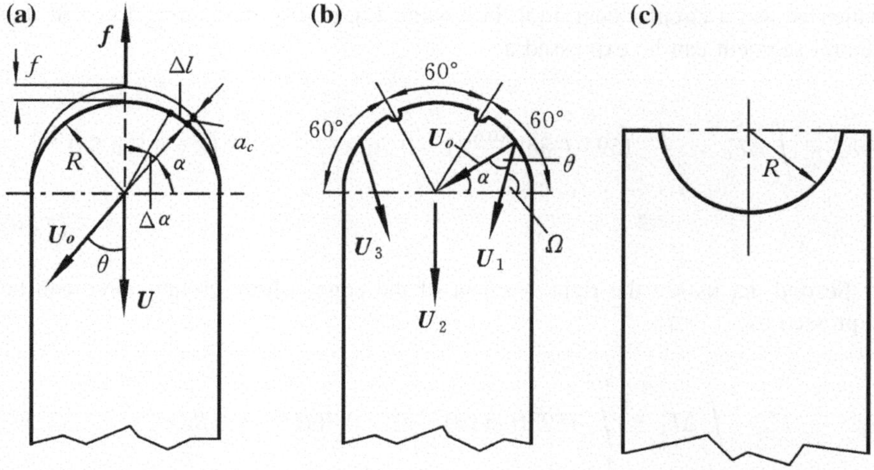

Fig. 7.6 The circular edge tools used in experiments and analysis. Plunge turning, work-piece: bronze, tool: HSS, $\gamma_{oi} = \lambda_{si} = 0°$. **a** Convex, without chip splitting slots. **b** Convex, with chip splitting slots. **c** Concave

And the cutting force of the whole tool (see Fig. 7.6a) would be calculated by simply superposing the elementary forces of all the elementary cutting tools

$$F_{co} = \int \Delta F_{co} = \int_0^{\pi} 1593(f \sin \alpha)^{0.61} R\mathrm{d}\alpha = 3647Rf^{0.61}, \quad (7.47)$$

where f is plunge feed rate, $f \sin \alpha = a_c$, α is the orientation angle of the elementary cutting tool and R = 2 mm is the radius of the circular edge. Some of the previous researchers just use this simply superposing method to calculate cutting force of complex shaped tools [70–72].

As a matter of fact, because of symmetry of the circular edge the common chip as a whole can only be ejected in the longitudinal direction on the tool shank. Because of this, the deviation angle of the real chip-ejection direction from the natural one is $\theta = \pi/2 - \alpha$. Substituting this into Eq. (7.45) and integrating lead to the real cutting force of the tool

$$F_c = \int \Delta F_c = \int_0^{\pi} 1593(f \sin \alpha)^{0.61}[1 + 0.76(\pi/2 - \alpha)^2]R\mathrm{d}\alpha = 5211Rf^{0.61}. \quad (7.48)$$

This shows an increase of 43% in the cutting force caused by chip-ejection interference under non-free cutting.

Let us see Fig. 7.6b, the two slots divide the circular edge into three segments, each of which ejects its chips independently. First, let us see the central segment, which is also a circular edge tool. Following Eq. (7.48), the cutting force of this central segment can be expressed as

$$F_c'' = \int \Delta F_c = \int_{\pi/3}^{2\pi/3} 1593(f\sin\alpha)^{0.61}[1 + 0.76(\pi/2 - \alpha)^2]R\mathrm{d}\alpha = 1413.5Rf^{0.61}, \tag{7.49}$$

Second, let us see the right segment of the edge, whose cutting force can be expressed as

$$F_c' = \int \Delta F_c = \int_{0}^{\pi/3} 1593(f\sin\alpha)^{0.61}[1 + 0.76(\Omega - \alpha)^2]R\mathrm{d}\alpha. \tag{7.50}$$

The parameter Ω in the equation is the overall chip-ejection angle of this segment (Fig. 7.6b), which is a state parameter and should be determined by the principle of minimum energy dissipation. Letting $\partial F_c'/\partial\Omega = 0$ yields

$$\Omega = \frac{1593f^{0.61}R\int_0^{\pi/3}\alpha\sin^{0.61}\alpha\mathrm{d}\alpha}{1593f^{0.61}R\int_0^{\pi/3}\sin^{0.61}\alpha\mathrm{d}\alpha} = 0.6403(\mathrm{rad}),$$

Substituting it into Eq. (7.50) and integrating lead to $F_c' = 1077Rf^{0.61}$.

Note that the elementary free-cutting force of the *i*th elementary cutting tool can be expressed as $\mathrm{d}F_{co} = 1593(f\sin\alpha)^{0.61}R\mathrm{d}\alpha$, and $\mathrm{d}\mu = dF_{co}/\int_0^{\pi/3}\mathrm{d}F_{co}$ is the normalized weighting coefficients, taking cutting force as the weight. Thus the above equation may be rewritten as

$$\Omega = \int_0^{\pi/3}\alpha\mathrm{d}\mu. \tag{7.51}$$

This is the embodiment of the weighted average relation expressed by Eq. (7.9) in this particular case, and the relation shows that the elementary cutting tool with larger cutting force has stronger influence on chip-ejection compromising.

Finally, let us see the left segment of the edge. Because of symmetry the cutting force F generated by this segment of the edge is equal to that of the right segment, i.e. $F_c''' = F_c'$, and the total cutting force of the entire tool is

$$F_{cs} = F_c' + F_c'' + F_c''' = 2F_c' + F_c'' = 3904Rf^{0.61}. \tag{7.52}$$

A comparison of above Equation with Eq. (7.48) shows that the chip splitting slots have reduced the cutting force by one fourth.

Table 7.4 The cutting force of circular edge tools, predicted and measured (N)

f mm/rev			0.019	0.0215	0.026	0.032	0.0325
Without chip splitting slots	Experimental	Convex	915	971	1137	1303	1331
		Concave	887	957	1109	1317	1331
	Predicted		929	1002	1125	1277	1429
With chip splitting slots	Experimental	Convex	702	735	832	930	988
		Concave	684	721	777	943	961
	Predicted		696	750	843	957	1010

Notice because of the stiffness limitation of the machining system the feed rate cannot be higher

7.5.2 A Comparison of Experiments with the Theory

The main cutting forces of the tools shown in Fig. 7.6 were theoretically predicted and experimentally verified. The results are listed in Table 7.4. The data in Table 7.4 has shown that the predicted results coincide with the experimentally measured data pretty good.

7.6 The Failure in Chip-Ejection Compromise and Natural Chip Splitting

Let's consider the case of double edged tool cutting (Fig. 6.6c). If the difference of natural chip-ejection vectors of the two edges are too large, the chip-ejection compromise may be failed, and in this case, to keep the integrity of the common chip is no longer feasible, therefore the chip-ejections of two cutting edges will depart at their own natural chip-ejection vectors. This phenomenon is called "natural chip splitting". Differing from artificial chip splitting in broaching, drilling or forming turning, where chip splitting is caused by the uncontinuity of the cutting layout, natural chip splitting, however, is caused by inherent law of the cutting process. Here is an analysis on such a phenomenon and an explanation for the interrupted part in Fig. 6.7.

Rewriting Eq. (6.16) to show the influence of depth of cut a_c yields

$$F_c = (4272.2 + 2.33\theta^2)a_c^{0.75}, \tag{7.53}$$

At a particular θ the natural chip splitting may, or may not appear. If it doesn't appear (chip is kept as a whole in the direction $\boldsymbol{U}$, see Fig. 6.6c), the cutting power is

$$A = F_c(\theta)V; \tag{7.54}$$

If it does (i.e. chip bifurcates into two branches), then two edges eject chips separately at their natural chip-ejection directions $\boldsymbol{U_o}$ and U'_o (Fig. 6.6c), and the cutting power is

$$A_o = F_c(0)V, \tag{7.55}$$

Besides, extra energy should be paid for tearing the chip into two, which may be estimated by the following equation

$$A_f = T(a_c/r)Vr = Ta_cV, \tag{7.56}$$

where, T is the force required for tearing the chip of the unit thickness, which is the surface energy of the unit area on the tear. Which possibility will be converted into a reality depends on which needs less energy, in other words, the principle of minimum energy dissipation is still supposed to be in effect. Because of this, the criterion for no chip splitting is

$$A < A_o + A_f, \tag{7.57}$$

And for chip splitting is

$$A > A_o + A_f, \tag{7.58}$$

Substituting Eqs. (7.53)–(7.56) into (7.57) and (7.58) yields

$$\theta < 0.655\sqrt{T}a_c^{0.125} \tag{7.59}$$

For no splitting and

$$\theta > 0.655\sqrt{T}a_c^{0.125} \tag{7.60}$$

For splitting.

From the above two equations we can see that for a certain material (T), the larger the θ and the thinner the a_c are the easier the chip splits. This has been verified in Fig. 6.7. In Fig. 6.7, $a_c = 0.07$ mm, when $\theta = 20°$, chip was kept as a whole, and when $\theta > 25°$, chip splitting appears as shown by interrupted line in Fig. 6.7. From Eqs. (7.59) and (7.60), the value of T in the condition of the experiment can be estimated as

$$1812 \text{ N/mm} < T < 2832 \text{ N/mm},$$

The value of T estimated is much greater than that given in ordinary data [69]. The physics of T is worth further investigation.

7.7 The Degree of Freedom Confinement and Free Cutting

7.7.1 *Coefficient of Non-free Cutting*

Coefficient of non-free cutting of a cutting process Φ is defined as

$$\Phi = A \Big/ \sum_{i=1}^{n} \Delta A_i, \tag{7.61}$$

where A, stands for the real cutting force (reckoning in chip-ejection interference), and $\sum_{i=1}^{n} \Delta A_i$ the sum of the incremental forces generated by all the elementary cutting tools each working separately (without interference).

Coefficient of non-free cutting is defined to indicate the degree of freedom confinement of a cutting process. $\Phi \geq 1$, and the bigger the coefficient is, the severer the chip-ejection interference is. If and only if $\Phi = 1$ then the cutting process is free.

7.7.2 *The Sufficient and Necessary Conditions for Realizing Free Cutting*

Theorem [2]A cutting process is free, i.e. $\Phi = 1$, if and only if the natural chip-ejection vectors $\boldsymbol{U}_{oi}$ (i = 1, 2,…, n) of all the elementary cutting tools form a rigid body motion, i.e.

$$\boldsymbol{U}_{oi} = \boldsymbol{T}_o + \boldsymbol{\Omega}_o \times \boldsymbol{\rho}_i, \quad i = 1, 2, \ldots, n. \tag{7.62}$$

where $\boldsymbol{T}_o$ and Ω_o are two constant vectors and in this case, the real chip motion is composed of translation $\boldsymbol{T}_o$ and rotation Ω_o (proof ignored).

Equation 7.62 has two special cases: $\boldsymbol{T}_o = \boldsymbol{0}$ or $\boldsymbol{\Omega}_o = \boldsymbol{0}$.

The theorem clarifies a prevalent misunderstanding: only when the cutting tool is a straight line edged one can it perform free cutting. This, unfortunately, is not true. As a matter of fact, the theorem says that a cutting process is free has no inevitable connection with whether or not the cutting edge is a straight line.

[2]The proof of the theorem is in the Appendix 7.2.

7.7.3 The Principle of Free Cutting

As revealed by the research, for arbitrarily edged cutting tool the principle of free cutting lies in properly designing the tool face with a purpose of dredging the chip-ejection so as to eliminate chip ejection interference from the segments of the working edge. As verified by experiments free cutting technique notably improves cutting condition, reduces the cutting force and the power consumed, perfects chip control and elongates the tool life (Chinese patent ZL 96 2 35153.9). These questions will be discussed in Chap. 10.

7.8 Remarks

7.8.1 Chip-Ejection Interference and Compromise

Modern cutting tools may be regarded as a combination of a series of elementary cutting tools and all of them are working concurrently. In the cutting process each elementary cutting tool tries to eject its chips according to its own natural chip-ejection vector, and this will certainly result in interference among chip-ejections. The interfering chip ejections have to compromise with each other so as to keep the integrity of the common chip. The compromised common chip motion must minimize the cutting power consumed by the whole cutter, which is the principle of minimum energy dissipation for cutting processes with complexly shaped tools. A great deal of experiments and analysis have verified or support this principle.

7.8.2 Non-free Cutting

Whenever chip ejection compromise is necessary due to existing of interference the cutting process is called non-free cutting. During a non-free cutting process elementary cutting tools normally cannot eject chips at their natural chip ejection vectors and this factor increases the cutting force and intensifies the chip deformation, causing many real problems in today's machining operations.

7.8.3 General Equation Governing Chip Ejection Motion

With an assumption of chip's rigid body motion and based on the principle of minimum energy dissipation, the general equation governing the translation and the rotation components of the chip ejection motion is deduced.

7.8.4 Average Law

From the equation a law of average is concluded: a state parameter, such as chip ejection angle, relating to overall chip motion is equal to the average of the same parameter relating to each elementary cutting tool weighted by its free cutting force. By means of the law, state parameters of a non-free cutting process are conveniently determined, and the influence of various factors on state parameters is promptly analyzed. The law is supported by experiments.

7.8.5 The Conditions for Realizing Free Cutting

A coefficient of non-free cutting is defined to indicate the degree of freedom confinement of a cutting process. The conditions for realizing free cutting are that the natural chip ejection vectors $\boldsymbol{U}_{oi}$ (i = 1, 2… n) of all the elementary cutting tools form a rigid body motion.

7.8.6 The Principle of Free Cutting

Properly designing the geometry of the tool rake may eliminate chip ejection interference and realize free cutting so as to increase the efficiency and to improve the quality of machining operations.

Appendix 7.1

The deduction of the general Eq. (7.9) governing the chip-ejection motion

As shown in Eq. (7.8), cutting power can be expressed

$$A = \sum_{i=1}^{n} I_{coi} V_i \Delta b_i \{1 + \mu_{1i} \frac{[(\boldsymbol{T} + \boldsymbol{\Omega} \times \boldsymbol{\rho}_i - \boldsymbol{U}_{oi}) \cdot \mathbf{b_i}]^2}{\boldsymbol{U}_{oi}^2} + \mu_{2i} \frac{[(\boldsymbol{T} + \boldsymbol{\Omega} \times \boldsymbol{\rho}_i - \boldsymbol{U}_{oi}) \cdot \mathbf{a_i}]^2}{\boldsymbol{U}_{oi}^2}\}, \tag{7.63}$$

where state parameters $\boldsymbol{T}$ and $\boldsymbol{\Omega}$ should be determined according to the principle of minimum energy dissipation. For this purpose, let $\partial A/\partial \boldsymbol{T} = \mathbf{0}$ and $\partial A/\partial \boldsymbol{\Omega} = \mathbf{0}$, and at last we get the general equation governing the chip-ejection motion

$$\begin{array}{l} \overline{\boldsymbol{U}'_i} = \overline{\boldsymbol{U}'_{oi}}; \\ \overline{\boldsymbol{\rho}_i \times \boldsymbol{U}'_i} = \overline{\boldsymbol{\rho}_i \times \boldsymbol{U}'_{oi}}, \end{array} . \tag{7.64}$$

This is Eq. (7.9). Here is the derivation process of the above equation. Both $\partial A/\partial \boldsymbol{T}$ and $\partial A/\partial \Omega$ are the derivative of a scalar with respect to a vector, and the result is vector

$$\partial A/\partial \boldsymbol{T} = (\partial A/\partial T_x)\mathbf{i} + (\partial A/\partial T_y)\mathbf{j} + (\partial A/\partial T_z)\mathbf{k}, \tag{7.65}$$

$$\partial A/\partial \Omega = (\partial A/\partial \Omega_x)\mathbf{i} + (\partial A/\partial \Omega_y)\mathbf{j} + (\partial A/\partial \Omega_z)\mathbf{k}. \tag{7.66}$$

The following equations of vector derivative were used in the derivation process

$$\partial(\boldsymbol{A} \cdot \boldsymbol{B})/\partial \boldsymbol{B} = \boldsymbol{A}, \tag{7.67}$$

$$\partial \boldsymbol{A}^2/\partial \boldsymbol{A} = \partial(\boldsymbol{A} \cdot \boldsymbol{A})/\partial \boldsymbol{A} = 2\boldsymbol{A}, \tag{7.68}$$

$$\partial(\boldsymbol{A} \times \boldsymbol{B}) \cdot \boldsymbol{C}/\partial \boldsymbol{A} = \boldsymbol{B} \times \boldsymbol{C}. \tag{7.69}$$

Which are similar to equations of scalar derivative, and may be deduced from basic definitions (where $\boldsymbol{C}$ is a constant vector). Now let's calculate $\partial A/\partial \boldsymbol{T}$, from Eq. (7.63),

$$\partial A/\partial \boldsymbol{T} = 2\sum_{i=1}^{n} I_{coi} V_i \Delta b_i \frac{1}{U_{oi}^2} \{\mu_{1i}[(\boldsymbol{T} + \boldsymbol{\Omega} \times \rho_{\mathrm{i}} - \boldsymbol{U_{oi}}) \cdot \mathbf{b_i}]\mathbf{b_i} \\ + \mu_{2i}[(\boldsymbol{T} + \boldsymbol{\Omega} \times \boldsymbol{\rho}_i - \boldsymbol{U_{oi}}) \cdot \mathbf{a_i}]\mathbf{a_i}\}$$

Considering Eq. (7.6), the above Equation may be written as

$$\partial A/\partial \boldsymbol{T} = 2\sum_{i=1}^{n} I_{coi} V_i \Delta b_i \frac{1}{U_{oi}^2} \{\mu_{1i}[(\boldsymbol{U_i} - \boldsymbol{U_{oi}}) \cdot \mathbf{b_i}]\mathbf{b_i} + \mu_{2i}[(\boldsymbol{U_i} - \boldsymbol{U_{oi}}) \cdot \mathbf{a_i}]\mathbf{a_i}\}$$

Rearranging leads to

$$\partial A/\partial \boldsymbol{T} = 2\sum_{i=1}^{n} \frac{I_{coi} V_i \Delta b_i}{U_{oi}^2} \{[\mu_{1i}(\boldsymbol{U_i} \cdot \mathbf{b_i})\mathbf{b_i} + \mu_{2i}(\boldsymbol{U_i} \cdot \mathbf{a_i})\mathbf{a_i}] \\ - [\mu_{1i}(\boldsymbol{U_{oi}} \cdot \mathbf{b_i})\mathbf{b_i} + \mu_{2i}(\boldsymbol{U_{oi}} \cdot \mathbf{a_i})\mathbf{a_i}]\}$$

From Eqs. (7.3) and (7.10), we have

$$\partial A/\partial \boldsymbol{T} = 2\sum_{i=1}^{n} \frac{I_{coi} V_i \Delta b_i}{U_{oi}^2} [\boldsymbol{U}_i' - \boldsymbol{U}_{oi}'] = \mathbf{0},$$

i.e.

$$\sum_{i=1}^{n} \frac{I_{coi} V_i \Delta b_i / U_{oi}^2}{\sum_{i=1}^{n} I_{coi} V_i \Delta b_i / U_{oi}^2} \boldsymbol{U}_i' = \sum_{i=1}^{n} \frac{I_{coi} V_i \Delta b_i / U_{oi}^2}{\sum_{i=1}^{n} I_{coi} V_i \Delta b_i / U_{oi}^2} \boldsymbol{U}_{oi}',$$

From Eqs. (7.11) and (7.12), we have $\overline{U_i'} = \overline{U_{oi}'}$, this is the first one of Eq. (7.9). Now let's prove the second equation. From Eq. (8.11),

$$\begin{aligned}\partial A/\partial \boldsymbol{\Omega} = 2\sum_{i=1}^{n} I_{coi}V_i\Delta b_i \frac{1}{U_{oi}^2}\{\mu_{1i}[(\boldsymbol{T}+\boldsymbol{\Omega}\times\boldsymbol{\rho}_i-\boldsymbol{U}_{oi})\cdot\mathbf{b_i}][\partial(\boldsymbol{T}+\boldsymbol{\Omega}\times\boldsymbol{\rho}_i-\boldsymbol{U}_{oi})\cdot\mathbf{b_i}/\partial\boldsymbol{\Omega}] \\ +\mu_{2i}[(\boldsymbol{T}+\boldsymbol{\Omega}\times\boldsymbol{\rho}_i-\boldsymbol{U}_{oi})\cdot\mathbf{a_i}][\partial(\boldsymbol{T}+\boldsymbol{\Omega}\times\boldsymbol{\rho}_i-\boldsymbol{U}_{oi})\cdot\mathbf{a_i}/\partial\boldsymbol{\Omega}\},\end{aligned}$$

where

$$\begin{aligned}\partial(\boldsymbol{T}+\boldsymbol{\Omega}\times\boldsymbol{\rho}_i-\boldsymbol{U}_{oi})\cdot\mathbf{b_i}/\partial\boldsymbol{\Omega} = \partial(\boldsymbol{\Omega}\times\boldsymbol{\rho}_i)\cdot\mathbf{b_i}/\partial\boldsymbol{\Omega} \\ \partial(\boldsymbol{T}+\boldsymbol{\Omega}\times\boldsymbol{\rho}_i-\boldsymbol{U}_{oi})\cdot\mathbf{a_i}/\partial\boldsymbol{\Omega} = \partial(\boldsymbol{\Omega}\times\boldsymbol{\rho}_i)\cdot\mathbf{a_i}/\partial\boldsymbol{\Omega}\end{aligned}$$

According Eq. (7.69), the above two equations equal to $\boldsymbol{\rho}_i\times\mathbf{b_i}$ and $\boldsymbol{\rho}_i\times\mathbf{a_i}$, respectively. And so

$$\begin{aligned}\partial A/\partial \boldsymbol{\Omega} = 2\sum_{i=1}^{n} I_{coi}V_i\Delta b_i \frac{1}{U_{oi}^2}\{\mu_{1i}[(\boldsymbol{T}+\boldsymbol{\Omega}\times\boldsymbol{\rho}_i-\boldsymbol{U}_{oi})\cdot\mathbf{b_i}]\boldsymbol{\rho}_i\times\mathbf{b_i} \\ +\mu_{2i}[(\boldsymbol{T}+\boldsymbol{\Omega}\times\boldsymbol{p}_i-\boldsymbol{U}_{oi})\cdot\mathbf{a_i}]\boldsymbol{\rho}_i\times\mathbf{a_i}\} = \mathbf{0},\end{aligned}$$

According to Eqs. (7.3) and (7.7), the parts of square brackets of above equation are equal to $(U_{ti} - U_{oti})$ and $(U_{ni} - U_{oni})$, respectively, both are scalar, and can move to the position right after the symbol $\times$, and considering Eq. (7.6) again, the above equation may be rewritten as

$$\begin{aligned}\partial A/\partial \boldsymbol{\Omega} = 2\sum_{i=1}^{n} I_{coi}V_i\Delta b_i \frac{1}{U_{oi}^2}\{\boldsymbol{\rho}_i\times[\mu_{1i}(\boldsymbol{U}_i-\boldsymbol{U}_{oi})\cdot\mathbf{b_i}]\mathbf{b_i}+\boldsymbol{\rho}_i \\ \times[\mu_{2i}(\boldsymbol{U}_{oi}-\boldsymbol{U}_{oi})\cdot\mathbf{a_i}]\mathbf{a_i}\} = \mathbf{0},\end{aligned}$$

i.e.

$$\begin{aligned}&\sum_{i=1}^{n}\frac{I_{coi}V_i\Delta b_i/U_{oi}^2}{\sum_{i=1}^{n} I_{coi}V_i\Delta b_i/U_{oi}^2}\boldsymbol{\rho}_i\times\{[\mu_{1i}(\boldsymbol{U}_i\cdot\mathbf{b_i})\mathbf{b_i}+\mu_{2i}(\boldsymbol{U}_i\cdot\mathbf{a_i})\mathbf{a_i}] \\ &=\sum_{i=1}^{n}\frac{I_{coi}V_i\Delta b_i/U_{oi}^2}{\sum_{i=1}^{n} I_{coi}V_i\Delta b_i/U_{oi}^2}\boldsymbol{\rho}_i\times[\mu_{1i}(\boldsymbol{U}_{oi}\cdot\mathbf{b_i})\mathbf{b_i}+\mu_{2i}(\boldsymbol{U}_{oi}\cdot\mathbf{a_i})\mathbf{a_i}]\} = \mathbf{0},\end{aligned}$$

Considering Eqs. (7.3), (7.10), (7.11) and (7.12), the above equation becomes $\overline{\boldsymbol{p}_i\times\boldsymbol{U}_i'} = \overline{\boldsymbol{p}_i\times\boldsymbol{U}_{oi}'}$, this is the second one of Eq. (7.9), which may be written as $\overline{\boldsymbol{U}_i'\times\boldsymbol{p}_i} = \overline{\boldsymbol{U}_{oi}'\times\boldsymbol{p}_i}$.

□

Appendix 7.2

The proof of a theorem about free cutting

Now let's prove that Eq. (7.62)

$$\boldsymbol{U}_{oi} = \boldsymbol{T}_o + \boldsymbol{\Omega}_o \times \boldsymbol{\rho}_i, \quad i = 1, 2, \ldots, n$$

is the sufficient and necessary condition for realizing a free cutting. First, to prove its sufficiency, i.e. to suppose that the above equation is true and to deduct that the cutting process is free. Substituting above equation into Eqs. (7.13) and (7.14) leads to

$$\begin{aligned}
&\sum_{i=1}^{n} \frac{I_{coi} V \Delta b_i}{U_{oi}^2} \{\mu_{1i}[(\boldsymbol{T} + \boldsymbol{\Omega} \times \boldsymbol{\rho}_i) \cdot \mathbf{b}_i]\mathbf{b}_i + \mu_{2i}[(\boldsymbol{T} + \boldsymbol{\Omega} \times \boldsymbol{\rho}_i) \cdot \mathbf{a}_i]\mathbf{a}_i\} \\
&= \sum_{i=1}^{n} \frac{I_{coi} V \Delta b_i}{U_{oi}^2} \{\mu_{1i}[(\boldsymbol{T}_o + \boldsymbol{\Omega}_o \times \boldsymbol{\rho}_i) \cdot \mathbf{b}_i]\mathbf{b}_i + \mu_{2i}[(\boldsymbol{T}_o + \boldsymbol{\Omega}_o \times \boldsymbol{\rho}_i) \cdot \mathbf{a}_i]\mathbf{a}_i\}; \\
&\sum_{i=1}^{n} \frac{I_{coi} V \Delta b_i}{U_{oi}^2} \boldsymbol{\rho}_i \times \{\mu_{1i}[(\boldsymbol{T} + \boldsymbol{\Omega} \times \boldsymbol{\rho}_i) \cdot \mathbf{b}_i]\mathbf{b}_i + \mu_{2i}[(\boldsymbol{T} + \boldsymbol{\Omega} \times \boldsymbol{\rho}_i) \cdot \mathbf{a}_i]\mathbf{a}_i\} \\
&= \sum_{i=1}^{n} \frac{I_{coi} V \Delta b_i}{U_{oi}^2} \boldsymbol{\rho}_i \times \{\mu_{1i}[(\boldsymbol{T}_o + \boldsymbol{\Omega}_o \times \boldsymbol{\rho}_i) \cdot \mathbf{b}_i]\mathbf{b}_i + \mu_{2i}[(\boldsymbol{T}_o + \boldsymbol{\Omega}_o \times \boldsymbol{\rho}_i) \cdot \mathbf{a}_i]\mathbf{a}_i\}.
\end{aligned}$$

Because we have many ways to divide elementary cutting tools, so $\mathbf{a}_i$, $\mathbf{b}_i$ and $\boldsymbol{\rho}_i$ are all variable and the only way to meet above equation is to let

$$\boldsymbol{T} = \boldsymbol{T}_o, \ \boldsymbol{\Omega} = \boldsymbol{\Omega}_o.$$

Substituting the above equation and Eq. (7.62) into Eq. (7.8) yields

$$A = A_{\min} = \sum_{i=1}^{n} I_{coi} V_i \Delta b_i = \sum_{i=1}^{n} A_i,$$

And from Eq. (7.61), we have $\Phi = 1$, so the cutting process is free (no chip-ejection interference). Because $\boldsymbol{T}_o$ and Ω_o minimize cutting power A, so $\boldsymbol{T}_o = \boldsymbol{T}_m$, $\boldsymbol{\Omega}_o = \boldsymbol{\Omega}_m$. So far we have proved the sufficiency. The following is the proof of the necessity.

If the process is a free cutting, then $\Phi = 1$, from Eq. (7.61), we have

$$A = A_{\min} = \sum_{i=1}^{n} I_{coi} V_i \Delta b_i = \sum_{i=1}^{n} A_i,$$

And from Eq. (7.8), we have

$$\boldsymbol{T}_{\mathrm{m}} + \boldsymbol{\Omega}_{\mathrm{m}} \times \boldsymbol{\rho}_i - \boldsymbol{U}_{oi} = \mathbf{0};\ i = 1,2,\ldots,n,$$

Therefore,

$$\boldsymbol{U}_{oi} = \boldsymbol{T}_{\mathrm{m}} + \boldsymbol{\Omega}_{\mathrm{m}} \times \boldsymbol{p}_i;\ i = 1,2,\ldots,n$$

This is to say that the natural chip-ejection vector form a rigid body motion, and its translation component and rotation component are $\boldsymbol{T}_{\mathrm{m}}$ and Ω_{m}, respectively. Comparison with Eq. (7.62) shows

$$\boldsymbol{T}_o = \boldsymbol{T}_{\mathrm{m}},\ \boldsymbol{\Omega}_o = \boldsymbol{\Omega}_{\mathrm{m}}.$$

□

Chapter 8
Non-linear Synthetic Method for Modeling Modern Cutting Tools

Based on the Non-linear Synthetic Method of Elementary Cutting Tools described in last chapter, and taking drilling force model as an example, this chapter will talk about the method of establishing a "functional model". The independent variable of a functional model is some kind of function, which is able to reflect the influence of geometric shape of a cutting tool on its cutting performance. Functional model is a bridge between the basic cutting theory and the cutting processes of modern complex shaped cutting tools, and therefore such kind of model has provide an approach for modeling, evaluating and optimizing advanced cutting tools. Traditional "parameter models" and the modeling method based on parameters are impossible or difficult to do so.

In establishing "functional models" of complex shaped cutting tools, the strategy and the technical approach adopted in this chapter are briefly described as following:

Local similarity of elementary cutting tools

The model of a category of cutting tools must be established on common features between all the tools belonging to the category. Without these common features, the model cannot be established. However, modern tools, even if they are belonging to the same category, such as twist drills, the geometry of their working part varies greatly from one to another, and on the whole, there is no any "global similarity". However, as previously stated, any tool can be regarded as a combination of a series of "elementary cutting tools", and among all the "elementary cutting tools" there exist certain "local similarities", a model of modern complex shaped cutting tools is based on nothing else than on this "local similarity". To find and to make use of this local similarity, a tool is divided along its edge into a series of elementary cutting tools.

The parameter model of elementary cutting tools

Regarding elementary cutting tools as "black boxes", through cutting experiments and data fitting so as to establish their parameter model.

H. Shi, *Metal Cutting Theory*, Springer Series in Advanced Manufacturing,
https://doi.org/10.1007/978-3-319-73561-0_8

Transformation of coordinates

Every elementary cutting tool has its own local coordinate system, it has to be converted into global coordinate system related to the whole cutter, so as to calculate the cutting force and cutting power.

Non-linear synthetic method of elementary cutting tools

The mechanical model of the whole cutter is formed by synthesizing all the local models of elementary cutting tools. The synthesis is not simple superposition, but a kind of "non-linear synthesis" in consideration of mutual actions among all the elementary cutting tools in their chip-ejections. The basic law governing this mutual actions is still the "principle minimum energy dissipation".

8.1 The Theoretical Framework and Modeling Procedures for Drilling Force Model

In this section, the main method and steps of modeling a drilling force will be introduced. The method and steps introduced here may be used to establish mechanical models for arbitrarily shaped drills; as a matter of fact, this method can be also used for modeling other hole machining tools, such as reamers.

8.1.1 Global Coordinates for a Twist Drill

According to the operating feature of a twist drill, we set a Cartesian coordinate system for the drill as shown in Fig. 8.1a, which is the global coordinate system.

8.1.2 The Geometrical Model and the Kinetic Model of Elementary Cutting Tools

8.1.2.1 The Location, the Shape and the Orientation of Elementary Cutting Tools

The position of an elementary cutting tool can be determined by its radius vector $\boldsymbol{\rho}_i$ (see Fig. 8.1),

$$\boldsymbol{\rho}_i = x_i\mathbf{i} + y_i\mathbf{j} + z_i\mathbf{k}, \tag{8.1}$$

where x_i, y_i, z_i; $i = 1, 2,\ldots, n$ are the coordinates of the points dividing elementary cutting tools (the subscript i is often omitted, if it doesn't cause confusions).

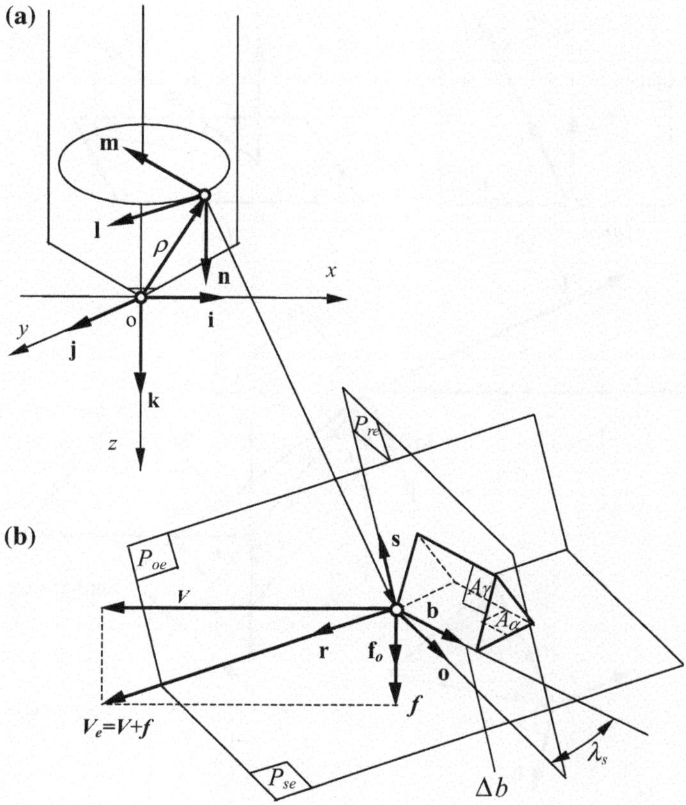

Fig. 8.1 The coordinate system. **a** Global coordinate system. **b** Local coordinate system

The shape and the orientation of an elementary cutting tool are uniquely determined by cutting edge length Δb and the two unit normal vectors **g**, **h** of the rake and the flank, respectively, as shown in Figs. 8.1b and 8.2c.

$$\Delta b_i = \sqrt{(x_{i+1} - x_i)^2 + (y_{i+1} - y_i)^2 + (z_{i+1} - z_i)^2}; \quad i = 1, 2, \ldots, n. \tag{8.2}$$

The unit vector along cutting edge is

$$\mathbf{b} = \mathbf{g} \times \mathbf{h} / |\mathbf{g} \times \mathbf{h}|, \tag{8.3}$$

The vector **b** also can be calculated according to the dividing points,

$$\mathbf{b_i} = \frac{(x_{i+1} - x_i)\mathbf{i} + (y_{i+1} - y_i)\mathbf{j} + (z_{i+1} - z)_i\mathbf{k}}{\sqrt{(x_{i+1} - x_i)^2 + (y_{i+1} - y_i)^2 + (z_{i+1} - z_i)^2}}; \quad i = 1, 2, \ldots, n. \tag{8.4}$$

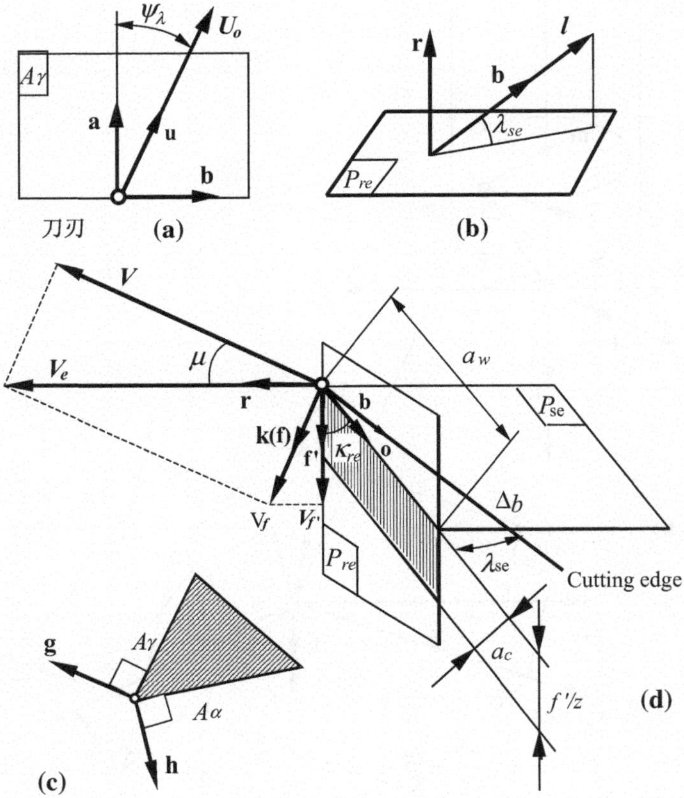

Fig. 8.2 Angles and cutting parameters of an elementary cutting tool. **a** Chip-ejection angle. **b** Cutting edge inclination. **c** Unit vectors normal to rake and flank, respectively. **d** The sizes of cut

The edge-line vector is $\Delta b_i \mathbf{b_i}$. And the unit vector in the rake face and normal to the edge (see Fig. 8.2a, c) is

$$\mathbf{a} = \mathbf{g} \times \mathbf{b}. \tag{8.5}$$

8.1.2.2 The Cutting Motion

Every elementary cutting tool has its own local coordinate system, which is based on the motion of the elementary cutting tool and its edge orientation. The angular speed of the drill rotation is

$$\boldsymbol{\Omega} = \frac{2\pi N}{60}\mathbf{k}(1/\mathrm{s}), \tag{8.6}$$

where, N is machine spindle revolutions (1/min). The cutting velocity of an elementary cutting tool, having a position radius ρ is

$$\boldsymbol{V} = 60\boldsymbol{\Omega} \times \boldsymbol{\rho}. \tag{8.7}$$

All the elementary cutting tools have a feed motion

$$\boldsymbol{V}_f = \frac{Nf}{1000}\mathbf{k}, \tag{8.8}$$

where f is the feed per revolution (mm/r). The resultant cutting speed vector is,

$$\boldsymbol{V}_e = \boldsymbol{V} + \boldsymbol{V}_f. \tag{8.9}$$

The reason why we have to calculate resultant cutting speed and later on we have to use "tool-in-use system" is that compared with feed speed, the cutting speed of those elementary cutting tools close to the drill center is not very large, and the feed speed is not negligible.

8.1.2.3 Local Coordinate Systems

The unit normal vector **r** of working reference plane P_{re} is

$$\mathbf{r} = \boldsymbol{V}_e/|\boldsymbol{V}_e| = \mathrm{r}_x\mathbf{i} + \mathrm{r}_y\mathbf{j} + \mathrm{r}_z\mathbf{k}. \tag{8.10}$$

The unit normal vector **s** of working cutting edge plane P_{se} is

$$\mathbf{s} = \boldsymbol{V}_e \times \boldsymbol{b}/|\boldsymbol{V}_e \times \boldsymbol{b}| = \mathrm{s}_x\mathbf{i} + \mathrm{s}_y\mathbf{j} + \mathrm{s}_z\mathbf{k}. \tag{8.11}$$

The unit normal vector **o** of working orthogonal **plane** P_{oe} is

$$\mathbf{o} = \mathbf{s} \times \mathbf{r} = \mathrm{o}_x\mathbf{i} + \mathrm{o}_y\mathbf{j} + \mathrm{o}_z\mathbf{k}. \tag{8.12}$$

Three coordinate planes P_{re}, P_{se}, P_{oe} and their normal vectors **r, s, o** form a rectangular coordinate system relating to an elementary cutting tool (see Fig. 8.1), and the cutting angles of every elementary cutting tool have to be defined and calculated in its own local coordinate system.

8.1.2.4 The Working Angles

According to definitions of cutting angles the formulae for calculating working angles are as follows.

Working cutting edge inclination the angle included between cutting edge and working reference plane (Fig. 8.2b),

$$\sin \lambda_{se} = \mathbf{b} \cdot \mathbf{r}. \tag{8.13}$$

Working orthogonal rake the angle included between rake face and working reference plane measured in working orthogonal plane. Making reference on Eq. (2.19), we have

$$\tan \gamma_{oe} = (\mathbf{g}, \mathbf{r}, \mathbf{o})/(\mathbf{g} \times \mathbf{o}) \cdot (\mathbf{r} \times \mathbf{o}), \tag{8.14}$$

where,

$$(\mathbf{g,r,o}) = \begin{vmatrix} \mathrm{g_x} & \mathrm{g_y} & \mathrm{g_z} \\ \mathrm{r_x} & \mathrm{r_y} & \mathrm{r_z} \\ \mathrm{o_x} & \mathrm{o_y} & \mathrm{o_z} \end{vmatrix}, \tag{8.15}$$

$$(\mathbf{g} \times \mathbf{o}) \cdot (\mathbf{r} \times \mathbf{o}) = \begin{vmatrix} \mathbf{o} \cdot \mathbf{o} & \mathbf{g} \cdot \mathbf{o} \\ \mathbf{r} \cdot \mathbf{o} & \mathbf{g} \cdot \mathbf{r} \end{vmatrix} = \begin{vmatrix} \mathrm{o_x o_x + o_y o_y + o_z o_z} & \mathrm{g_x o_x + g_y o_y + g_z o_z} \\ \mathrm{r_x o_x + r_y o_y + r_z o_z} & \mathrm{g_x r_x + g_y r_y + g_z r_z} \end{vmatrix}. \tag{8.16}$$

Working cutting edge angle the angle included between the projections of the cutting edge and feed vector onto working reference plane (see Fig. 8.2d). The unit vector **o** is in the direction of the projection of cutting edge $\Delta b\mathbf{b}$ onto working reference plane. The unit vector **f** in the feed direction is just **k** (Figs. 8.1a and 8.2d), the unit vector in the direction of projection of it onto reference plane is

$$\mathbf{f}' = \frac{\mathbf{k} - (\mathbf{k} \cdot \mathbf{r})\mathbf{r}}{|\mathbf{k} - (\mathbf{k} \cdot \mathbf{r})\mathbf{r}|}. \tag{8.17}$$

Working cutting edge angle is

$$\cos \kappa_{re} = \mathbf{o} \cdot \mathbf{f}'. \tag{8.18}$$

8.1.2.5 The Size of Cut

Undeformed chip section is measure in reference plane. Thickness of cut (Fig. 8.2d),

$$a_c = f' \sin\kappa_{re}/z, \tag{8.19}$$

where, z is number of teeth, for twist drill z = 2. And f' is the projection of vector $f\mathbf{f}$ onto reference plane,

$$f' = f(\mathbf{f} \cdot \mathbf{f}') = f \cos \mu, \tag{8.20}$$

where, μ is the angle included between $\boldsymbol{V}$ and $\boldsymbol{V_e}$,

$$\sin \mu = \mathbf{k} \cdot \mathbf{r}, \tag{8.21}$$

Width of cut

$$a_w = \Delta b \cos \lambda_{se}. \tag{8.22}$$

8.1.3 The Mechanical Model for Elementary Cutting Tools

In this section, a mechanical model of elementary cutting tools will be presented, as the basis for modelling the whole drill. The model consists of the following aspects:

8.1.3.1 Cutting Force

The cutting force of an elementary cutting tool performing free cutting is a vector function (measured in its local coordinate system),

$$\Delta \boldsymbol{I} = \Delta I_s \mathbf{s} + \Delta I_r \mathbf{r} + \Delta I_o \mathbf{o}. \tag{8.23}$$

where,

$$\left.\begin{aligned} \Delta I_s &= \Delta I_s(a_w, a_c, V_e, \gamma_{oe}, \lambda_{se}) \\ \Delta I_r &= \Delta I_r(a_w, a_c, V_e, \gamma_{oe}, \lambda_{se}) \\ \Delta I_o &= \Delta I_o(a_w, a_c, V_e, \gamma_{oe}, \lambda_{se}) \end{aligned}\right. . \tag{8.24}$$

In the above equations, only five variables $a_w, a_c, V_e, \gamma_{oe}, \lambda_{se}$ are taken into consideration. The influence of clearance is relatively small, and can be ignored. Besides, the work piece material and cutting tool material are the same for all the elementary cutting tools, the influence of them may be regarded as constants, thus they are not expressed in the above functions. The considerations also effective for the following functions.

8.1.3.2 Cutting Ratio

The cutting ratio of an elementary cutting tool is a scalar function,

$$r = U/V = r(a_w, a_c, V_e, \gamma_{oe}, \lambda_{se}). \tag{8.25}$$

8.1.3.3 Chip-Ejection Angle

Chip-ejection angle is also a scalar function,

$$\psi_\lambda = \psi_\lambda(a_w, a_c, V_e, \gamma_{oe}, \lambda_{se}). \quad (8.26)$$

From the above parameters we can calculate natural chip-ejection speed,

$$U_o = rV, \quad (8.27)$$

And chip-ejection vector,

$$\boldsymbol{U_o} = U_o(\mathbf{a}\cos\psi_\lambda + \mathbf{b}\sin\psi_\lambda). \quad (8.28)$$

Functions (8.24)–(8.28) may be obtained through theoretical analysis, here, however, we get those functions through cutting experiments and empirical formula fitting.

8.1.3.4 Perturbation of Chip-Ejection Vector

According to the principle of minimum energy dissipation, any perturbation of natural chip-ejection vector $\boldsymbol{U_o}$ will result in an increase in cutting power A or cutting force $\boldsymbol{F}$. After chip-ejection vector has been disturbed, the cutting force $\Delta\boldsymbol{F}$ of an elementary cutting tool can be expressed as

$$\Delta\boldsymbol{F} = \Delta F_s\mathbf{s} + \Delta F_r\mathbf{r} + \Delta F_o\mathbf{o} = \Delta\boldsymbol{I}K = (\Delta I_s\mathbf{s} + \Delta I_r\mathbf{r} + \Delta I_o\mathbf{o})K, \quad (8.29)$$

where,

$$K = 1 + \mu\frac{(\boldsymbol{U} - \boldsymbol{U_o})^2}{\boldsymbol{U_o^2}} \quad (8.30)$$

Reflects the influence of chip-ejection perturbation on cutting force. The coefficient μ has been simplified according to Eq. (7.17), and this coefficient should be calibrated experimentally.

8.1.4 *Coordinate Transformation and Cutting Power*

The cutting force and the chip-ejection vector are described in local coordinate system (s_i, r_i, o_i), which is related to the shape, orientation and cutting motion of the elementary cutting tool. To synthesize the cutting force of the whole drill (the torque and the thrust), we have to do coordinate transformation, so as to converse the description of various mechanical quantities from local coordinate systems to the global coordinate system, which is related to the global motion of the whole drill. The global coordinate system: the coordinate axes l_i, m_i, n_i are in the tangential, radial and axial directions, respectively (Fig. 8.1). The unit vectors in these directions $\mathbf{l_i}$, $\mathbf{m_i}$, $\mathbf{n_i}$ are expressed as

$$\begin{aligned} \mathbf{l_i} &= \mathrm{l_{xi}}\mathbf{i} + \mathrm{l_{yi}}\mathbf{j} + \mathrm{l_{zi}}\mathbf{k} = (-y_i\mathbf{i} + x_i\mathbf{j})\Big/\sqrt{x_i^2 + y_i^2} \\ \mathbf{m_i} &= \mathrm{m_{xi}}\mathbf{i} + \mathrm{m_{yi}}\mathbf{j} + \mathrm{m_{zi}}\mathbf{k} = -(x_i\mathbf{i} + y_i\mathbf{j})\Big/\sqrt{x_i^2 + y_i^2}, \\ \mathbf{n_i} &= \mathrm{n_{xi}}\mathbf{i} + \mathrm{n_{yi}}\mathbf{j} + \mathrm{n_{zi}}\mathbf{k} = \mathbf{k} \end{aligned} \tag{8.31}$$

where x_i, y_i, z_i are the coordinates of the position radius $\boldsymbol{\rho_i}$ in the coordinate system (x, y, z), see Eq. (8.1).

The cutting force $\Delta\boldsymbol{F_i}$ of i-th elementary cutting tool in systems (s_i, r_i, o_i) and (l_i, m_i, n_i) can be expressed as

$$\Delta\boldsymbol{F_i} = \Delta F_{si}\mathbf{s_i} + \Delta F_{ri}\mathbf{r_i} + \Delta F_{oi}\mathbf{o_i}, \tag{8.32}$$

And

$$\Delta\boldsymbol{F_i} = \Delta F_{li}\mathbf{l_i} + \Delta F_{mi}\mathbf{m_i} + \Delta F_{ni}\mathbf{n_i}, \tag{8.33}$$

Respectively. And the coordinate transformation formula from system (s_i, r_i, o_i) to system (l_i, m_i, n_i) is

$$\begin{bmatrix} \Delta F_{li} \\ \Delta F_{mi} \\ \Delta F_{ni} \end{bmatrix} = \begin{bmatrix} \mathbf{l_i}\cdot\mathbf{s_i} & \mathbf{l_i}\cdot\mathbf{r_i} & \mathbf{l_i}\cdot\mathbf{o_i} \\ \mathbf{m_i}\cdot\mathbf{s_i} & \mathbf{m_i}\cdot\mathbf{r_i} & \mathbf{m_i}\cdot\mathbf{o_i} \\ \mathbf{n_i}\cdot\mathbf{s_i} & \mathbf{n_i}\cdot\mathbf{r_i} & \mathbf{n_i}\cdot\mathbf{o_i} \end{bmatrix} \begin{bmatrix} \Delta F_{si} \\ \Delta F_{ri} \\ \Delta F_{oi} \end{bmatrix}. \tag{8.34}$$

The above formula should be calculated in (x, y, z) system. The torque generated by i-th elementary cutting tool is,

$$\Delta T_i = \Delta F_{li}\sqrt{x_i^2 + y_i^2} = \Delta I_{li}\sqrt{x_i^2 + y_i^2}K, \quad i = 1, 2, \ldots, n. \tag{8.35}$$

The thrust generated by i-th elementary cutting tool is,

$$\Delta R_i = \Delta F_{ni} = \Delta I_{ni}K, \quad i = 1, 2\ldots n. \tag{8.36}$$

where, ΔI_{li} and ΔI_{ni} in Eqs. (8.35) and (8.36) are the components in coordinate system (l_i, m_i, n_i) of the cutting force generated by i-th elementary cutting tool, the coordinate transformation formula is the same as Eq. (8.34). The coefficient K has been defined by Eq. (8.30). Vectors $\Delta\boldsymbol{T_i}$ and $\Delta\boldsymbol{R_i}$ can be expressed as

$$\Delta T_i = \Delta T_i\mathbf{k}, \Delta R_i = \Delta R_i\mathbf{k}, \quad i = 1, 2\ldots n. \tag{8.37}$$

Cutting power is

$$A = \sum_{i=1}^{n} \Delta A_i = 2\pi N \sum_{i=1}^{n} \Delta l_{li} \sqrt{x_i^2 + y_i^2} K. \tag{8.38}$$

Substituting Eq. (8.30) into above equation and considering Eq. (8.6) result in drilling power

$$A = 2\pi N \sum_{i=1}^{n} \Delta l_{li} \sqrt{x_i^2 + y_i^2} [1 + \mu \frac{(\boldsymbol{T} + \Omega \times \rho_{\mathrm{i}} - \boldsymbol{U}_{oi})^2}{\boldsymbol{U}_{oi}^2}]. \tag{8.39}$$

8.1.5 Chip-Ejection Interference and Its Compromise

According to the principle of minimum energy dissipation, derivation of Eq. (8.39) with respect to state parameters $\boldsymbol{T}$ and $\boldsymbol{\Omega}$ and letting them equal to zero, respectively, result to

$$\begin{aligned} \overline{\boldsymbol{U}_i} &= \overline{\boldsymbol{U}_{oi}}; \\ \overline{\boldsymbol{\rho}_i \times \boldsymbol{U}_i} &= \overline{\boldsymbol{\rho}_i \times \boldsymbol{U}_{oi}} \end{aligned} \tag{8.40}$$

The deduction procedure is very similar to that described in Appendix 7.1. And the upper transverse line means "weighted average" as defined in Eq. (7.11),

$$\overline{x} = \sum_{i=1}^{n} w_i x_i,$$

And

$$w_i = \frac{\Delta T_i}{U_{oi}^2} \Bigg/ \sum_{j=1}^{n} \frac{\Delta T_i}{U_{oi}^2}, \quad i = 1, 2, \ldots, n. \tag{8.41}$$

where,

$$\Delta T_i = \Delta l_{li} \sqrt{x_i^2 + y_i^2}. \tag{8.42}$$

According to the procedure described in 6.3.2 we can solve Eq. (8.40), obtained $\boldsymbol{T}_{\mathbf{m}}$ and $\boldsymbol{\Omega}_{\mathbf{m}}$, which minimize cutting power. Substituting $\boldsymbol{T}_{\mathbf{m}}$ and $\boldsymbol{\Omega}_{\mathbf{m}}$ into Eq. (8.39) results in real drilling power A_{min}. Substituting $\boldsymbol{T}_{\mathbf{m}}$ and $\boldsymbol{\Omega}_{\mathbf{m}}$ into Eqs. (8.35) and (8.36), respectively, and summing up lead to drilling torque and axial thrust.

8.1.6 A Brief Summary of Modelling Method and Procedures

To sum up, the drilling force model (Torque and thrust) of an arbitrarily shaped twist drill can be established according to the method and steps described in Fig. 8.3.

8.2 Geometrical Parameter Analysis and Mechanical Parameter Measurement of Elementary Cutting Tools[1]

Figure 8.4 shows standard twist drill and multi-facet drill used for modelling. To establish their models on the basis of local similarity of elementary cutting tools, each of their major cutting edges and chisels are divided into 1000 and 200 elementary cutting tools, respectively, and performed theoretical analyses and experimental measurements for their geometrical, kinetic and mechanical parameters.

8.2.1 Geometrical Parameter and Kinetic Parameter Analyses

Based on the surface equations of cutting parts of the drills to be modelled [65], through geometrical and kinetic analyses, ρ_i, $\mathbf{g_i}$, $\mathbf{h_i}$ of every elementary cutting tool were calculated, and then, according to Eqs. (8.1)–(8.22) the geometric parameters and the kinetic parameters $\mathbf{a_i}$, $\mathbf{b_i}$, Δb, Ω, V_i, f_i, V_{ei}, $\mathbf{r_i}$, $\mathbf{s_i}$, $\mathbf{o_i}$, λ_{se}, γ_{oe} ,κ_{re}, a_c,a_w; $i = 1, 2, \ldots, n$ are all worked out. As examples, Fig. 8.5 presents the distributions of working rakes of a standard twist drill and a multi-facet drill, and the distribution of working cutting edge inclination of a standard twist drill.

8.2.2 Mechanical Parameters

The mechanical property and the mechanical parameters of elementary cutting tools are obtained through cutting experiments. A CNC lathe was used in the

[1]This and the following sections some materials, including some experimental data and theoretical analysis are quoted from the Ph.D. dissertation [65] of Dr. Jane-Lin Wang, The University of Michigan. His Ph.D. work was partly under my direction when I was a visiting professor in the university (1992.9–1993.9).

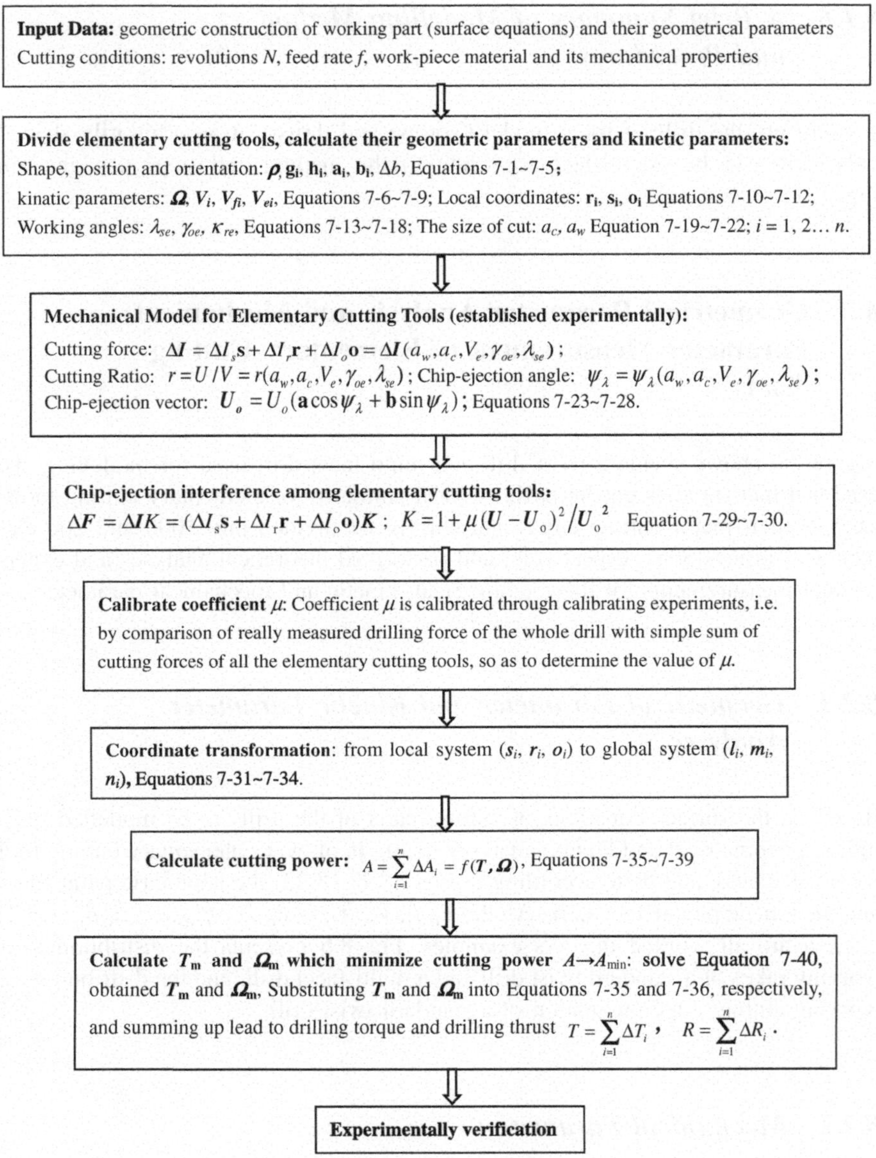

Fig. 8.3 The steps of modeling drill force

experiments. End face cutting of seamless steel tube was performed to simulate free cutting. The cutting force was measured by a Kistler Dynamometer, which was connected to an IBM-PC computer through a charge-amplifier and a A/D convertor. Tool material: Moly M-2 HSS. Work-piece material: cold draw steel S1018, HB 172–185. To keep the work-piece material in full accord with the drilling

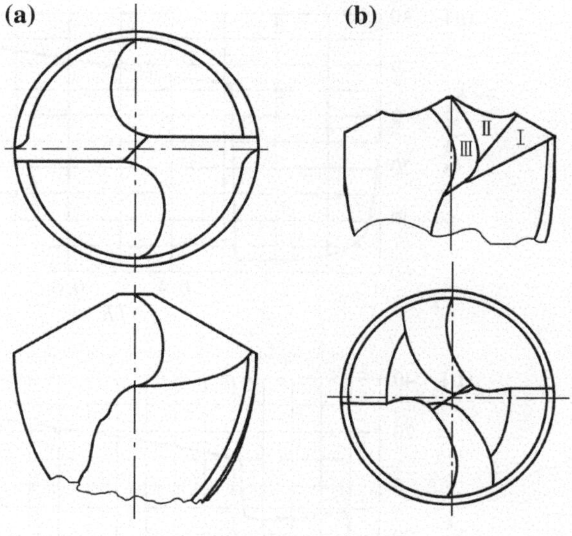

Fig. 8.4 Drills used for modeling. **a** Standard twist drill. **b** Multi-facet Drill (Qun drill)

experiments, a solid bar of diameter 101.6 mm and length 177.8 mm was piecewise dug into a tube of wall thickness 2.5–4.0 mm. The highest revolutions of the machine spindle was 3000 r/m.

8.2.2.1 Cutting Force of Elementary Cutting Tools

To obtain the cutting force of elementary cutting tools expressed by Eqs. (8.23)–(8.24), a 4^5 (5 factors, 4 levels) factorial design experiment was performed, $\gamma_{oe} = -20°, -10°, 20°, 30°$; $\lambda_{se} = 0°, 10°, 20°, 30°$; $a_c = 0.05, 0.10, 0.12, 0.15$ mm, $a_w = 2.5, 3.0, 3.5, 4$ mm; $V_e = 20, 25, 30, 35$ m/min. Clearance α_{oe} was fixed as 10°. In addition, some experiments with large edge inclination ($\lambda_{se} = 40°, 50°, 60°$) and large negative rake ($\gamma_{oe} = -30°, -35°, -40°$) were performed to simulate the cutting conditions of the chisel edge and near chisel edge. The variation of feed rate was as large as $f = 0.03$–0.3 mm/r, and the variation of cutting speed was as large as $V = 2$–45 m/min. Under different cutting conditions all together 96 experiments were conducted.

The empirical polynomial formulae fitted to cutting force on the basis of experimental data (for elementary cutting tools on the major edge):

$$F_r = Ca_w(r_1 + r_2a_c + r_3V_e + r_4\gamma_{oe}^2 + r_5\gamma_{oe}a_c + r_6\lambda_{se}^2 + r_7\lambda_{se}a_c + r_8\lambda_{se}V_e + r_9a_cV_e + r_{10}V_e^2 + r_{11}\lambda_{se}^3 + r_{12}V_e^3), \tag{8.43}$$

$$F_o = Ca_w(o_1 + o_2\gamma_{oe}\lambda_{se} + o_3\lambda_{se}^2 + o_4\lambda_{se}a_c + o_5\lambda_{se}V_e + o_6\gamma_{oe}^3 + o_7\lambda_{se}^3 + o_8a_c^3), \tag{8.44}$$

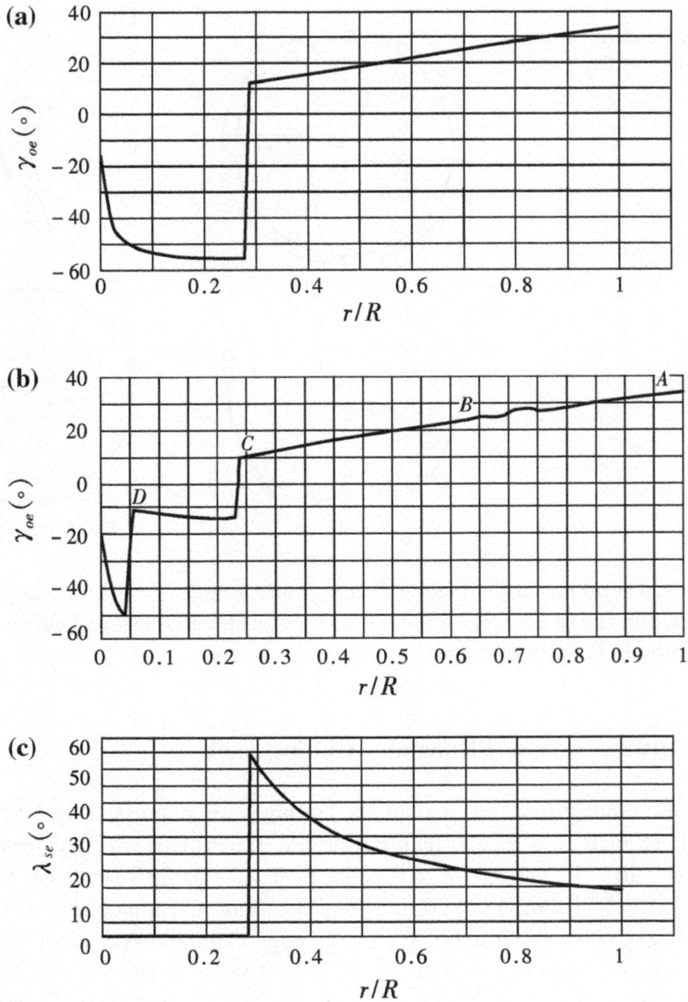

Fig. 8.5 Geometrical angle distributions of elementary cutting tools (examples). **a** Working rakes of a standard twist drill. **b** Working rakes of a multi-facet drill. **c** Working cutting edge inclination of a standard twist drill

$$F_s = Ca_w(s_1 + s_2\lambda_{se} + s_3a_c + s_4\gamma_{oe}^2 + s_5\gamma_{oe}\lambda_{se} + s_6\lambda\gamma_{oe}a_c + s_7\lambda_{se}^2 + s_8a_cV_e + s_9\gamma_{oe}^3). \tag{8.45}$$

where C = 0.00336 HB, HB represents the Brinell hardness of the work-piece material (between 172 and 185 HB). All other coefficients of the above force equations for main edge are shown in Table 8.1.

The empirical polynomial formulae fitted to cutting force on the basis of experimental data (for elementary cutting tools on the chisel edge):

Table 8.1 The coefficients in empirical formulae of cutting force of elementary cutting tools on major edge

i	r_i	o_i	s_i	i	r_i	o_i	s_i
1	316.4	1.788	41.32	7	9.694	0.001848	0.02349
2	1.526	−0.03301	−1.559	8	0.08723	−4129	19.85
3	−28.92	−0.09808	941.4	9	23.16		−0.003581
4	0.04577	29.35	0.08011	10	0.9452		
5	−31.82	0.02766	0.05225	11	0.002912		
6	−0.1714	0.0003193	−25.68	12	−0.0108		

$$F_{cr} = Ca_w(cr_1 + cr_2 a_c + cr_3 V_e + cr_4 \gamma_{oe}^2 + cr_5 \gamma_{oe} V_e + cr_6 V_e^2 + cr_7 a_c^3), \tag{8.46}$$

$$F_{cr} = 0, \tag{8.47}$$

$$F_{cs} = Ca_w(cs_1 + cs_2 \gamma_{oe} + cs_3 V_e + cs_4 \gamma_{oe} a_c + cs_5 \gamma_{oe} V_e + cs_6 a_c V_e + cs_7 V_e^2 + cs_8 a_c^3). \tag{8.48}$$

Because edge inclination at every point on chisel edge is zero, so F_{cr} constant be zero.

8.2.2.2 Cutting Ratio and Chip-Ejection Angle of Elementary Cutting Tools

The empirical formulae for cutting ratio and chip-ejection angle of elementary cutting tools were fitted on the bases of experimental data are

$$\begin{matrix} r = 0.54 - 0.0025\gamma_{oe} + 0.0033\lambda_{se} - 0.25a_c + 0.000033\gamma_{oe}^2 + \\ 0.013\gamma_{oe} a_c + 0.000034\gamma_{oe} V_e - 0.00014\lambda_{se} V_e \end{matrix}, \tag{8.49}$$

$$\psi_\lambda = 2.2 + 0.03\gamma_{oe} + 0.46\lambda_{se}^2 + 0.004\gamma_{oe}\lambda_{se} + 0.017\lambda_{se} - 0.00016\lambda_{se}^3. \tag{8.50}$$

The relations of chip-ejection angle $\psi_{\lambda o}$versus λ_{se} under a variety of γ_{oe} are given in Fig. 8.6. The data show that Stabler law is approximately correct. (Table 8.2)

8.2.2.3 Calibration of Coefficient μ

The coefficient μ in Eq. (8.30) was calibrated through a group of drilling experiments (including standard drills and multi-facet drills). By means of fitting the measured torque data and the thrust data, the values of μ for torque and for thrust are 355.6 and 624.2, respectively.

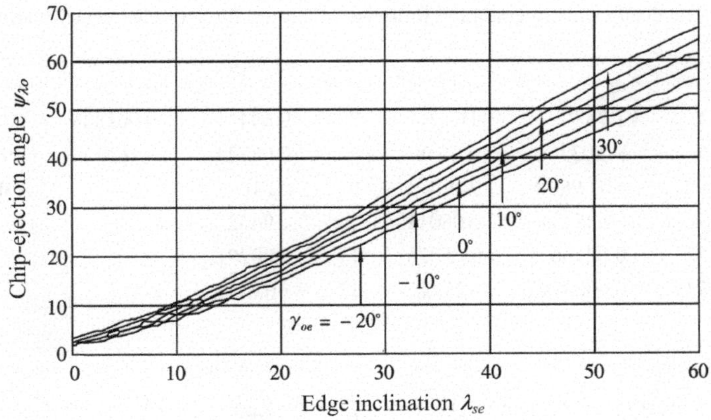

Fig. 8.6 Chip-ejection angle $\psi_{\lambda o}$ versus γ_{oe} and λ_{se}

Table 8.2 The coefficients in empirical formulae of cutting force of elementary cutting tools on chisel edge

i	cr_i	cs_i	i	cr_i	cs_i
1	45.12	543.8	5	−0.2299	−1.359
2	4593	6.465	6	0.2948	251.4
3	−17.53	−109.7	7	−83950	2.185
4	0.01324	−87.33	8		−100200

Based on above analyses and the experimental data, and according to the procedure described in Fig. 8.3, it is possible to predict the torque and the thrust of standard drills and multi-facet drills under a variety of cutting conditions.

8.3 The Experimental Verification of Mechanical Models of Drills

The predicted drilling torque and thrust by the above model have to be verified experimentally. The work-piece material was the same as in the experiments on elementary cutting tools previously discribed. The size of the work-piece is 76.2 × 50.8 × 12.7 mm. The diameters of the experimental drills were within 3.175–12.7 mm. The machine tool was a horizontal machining center, the highest revolution of its spindle is 1800 r/m. The drilling torque and the drilling thrust was measured by a Kistler Dynamometer, which was connected to an IBM-PC computer through a charge-amplifier, anti-aliasing filter and an A/D convertor. The data were recorded by the computer.

8.3.1 Drilling Experiments

The experiments of drilling solid matereal with standard twist drill and multi-faced drills were conducted, the mesured torque and thrust are listed in Tables 8.3 and 8.4.

The data listed in tables show that model predicted values are fairly close to the experimentally measured values, relative errors are ±15% and −19 to 12% for ΔT and ΔR, respectively, as shown in Fig. 8.7.

The error described above may be due to, first, the models were measured and fitted not accurate enough, secondly, the frictions between the drill margin and the hole wall, drill flank and the cutting surface, the chips and the hole wall did not be taken into consideration. Besides, the temperature in the real drilling area is higher than that in elementary cutting tool experiments, and this leads to difference in mechanical properties of the work-piece material, this also may be one of the reasons causing error. Anyway, the prediction accuracy of the model introduced is better than that of previous researchers [70–72].

8.3.2 Reaming Experiments

For the sake of eliminating the influence of chisel edge and specially investigating the model prediction accuracy for the major cutting edges, pre-drilled pilot holes, slightly bigger than the chisel diameter, were made on the work-piece and twist

Table 8.3 A comparison of predicted drilling force with measured ones, standard twist drills

Diameter D (mm)	Feed f (mm/r)	Rev. N (r/min)	Predicted		Measured		Relative error	
			Torque T (N cm)	Thrust R (N)	Torque T (N cm)	Thrust R (N)	ΔT (%)	ΔR (%)
3.175	0.0254	1500	21.4	137.8	18.1	129.3	18.2	6.6
3.572	0.0381	1400	34.8	275.8	30.9	250.5	12.7	10.1
3.969	0.0381	1400	44.0	351.5	40.7	318.4	8.1	4.1
4.366	0.0381	1400	45.4	392.9	49.3	354.2	−7.9	9.2
6.350	0.0432	1200	100.3	409.8	90.4	443.0	10.9	−7.5
7.938	0.0762	960	219.7	572.9	255.2	641.5	−13.9	−10.7
8.731	0.0762	873	298.4	623.0	305.1	753.3	−2.2	−17.3
9.128	0.0762	835	295.7	724.1	322.1	798.3	−8.2	−9.3
9.525	0.0762	800	315.1	887.7	355.6	843.8	−11.4	5.2
10.32	0.0762	739	431.3	815.6	412.3	923.7	4.6	−11.7
11.11	0.0762	620	454.6	1112.5	514.3	1214.5	−11.6	−8.4
12.70	0.0762	600	555.7	1187.5	588.7	1417.1	−5.4	−16.2
12.70	0.1270	600	882.1	1961.7	972.6	2116.2	−9.3	−7.3

Table 8.4 A comparison of predicted drilling force with measured ones, multi facet drills

Diameter D (mm)	Feed f (mm/r)	Rev. N (r/min)	Predicted		Measured		Relative error	
			Torque T (N cm)	Torque T (N cm)	Torque T (N cm)	Thrust R (N)	ΔT (%)	ΔR (%)
3.175	0.0254	1500	11.5	63.8	13.2	70.7	−13.4	−9.8
3.572	0.0381	1400	21.0	140.0	22.5	158.6	−6.7	−11.7
3.969	0.0381	1400	24.1	162.3	27.6	198.4	−12.1	−18.2
4.366	0.0381	1400	37.7	240.9	33.3	224.7	13.2	7.2
6.350	0.0432	1200	75.8	275.8	69.9	248.5	8.6	11.0
7.938	0.0762	960	159.8	363.8	178.1	398.0	−10.3	−8.6
8.731	0.0762	873	220.3	425.7	225.3	468.3	−2.2	-9.9
9.128	0.0762	835	267.0	577.1	242.5	506.7	10.1	13.9
9.525	0.0762	800	286.9	605.1	264.7	547.6	8.4	10.5
10.32	0.0762	739	371.1	751.3	334.6	646.6	10.9	16.2
11.11	0.0762	620	460.4	841.5	424.3	814.6	8.5	3.3
12.70	0.0762	600	519.1	982.1	484.2	902.7	7.2	8.8
12.70	0.1270	600	816.8	1525.3	739.9	1355.8	10.4	12.5

"Relative error" in above tables is defined as
$\Delta T\%$ = (predicted torque − measured torque)/measured torque × 100%
$\Delta R\%$ = (predicted thrust − measured thrust)/measured thrust × 100%

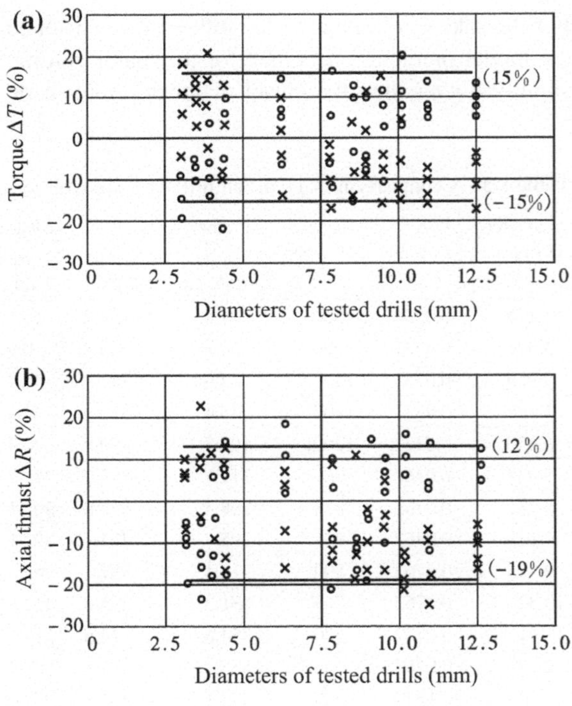

Fig. 8.7 The relative error of cutting force model predicted compared with that experiment measured. **a** Torque. **b** Axial thrust

drills were used to boring the pilot holes to a larger diameter. Cutting conditions: $f = 0.127$ mm/r, $N = 600$ r/min, drill diameter: $D = 12.7$ mm, the cutting diameter of the chisel edge is 2.464 mm for standard twist drill and 0.762 mm for multi-faced drills. The predicted and measured torque and thrust for standard twist drills and multi-faced drills are listed in Tables 8.5 and 8.6.

First, comparing the data in the columns of considering interference and without consideration of interference (just simple sum of elementary cutting tools) in above figures, we can see that cutting force could increase by 30–50% due to chip-ejection interference. Therefore such a factor is not often negligible.

Second, comparing measured data with predicted data (considering interference) and with the data of simple sum (ignore interference), respectively, we can see that only in the case of considering chip-injection interference the predicted cutting forces can be close to the really measured data, and ignoring this factor will result in very large prediction error. This might be main reason why the cutting force models suggested by previous researchers cannot predict accurately.

The third, it should be noticed that for standard twist drills, predicted values of cutting force are normally lower than experimentally measured ones, this is because some frictions did not consider in the model, as stated previously. As for multi-facet drills, when the pre-drilled pilot hole is smaller, i.e. most part of the major edge is cutting, the model predicted cutting force is little higher than that really measured, this is because the failure in chip-ejection compromise and natural chip splitting happened, as described in Sect. 7.6.

Table 8.5 A comparison of predicted drilling force with measured ones, with pre-drilled pilot holes, standard twist drills (drill diameter: $D = 12.7$ mm)

Pilot hole Diameter D (mm)	Predicted (considering interference)		Simple sum (ignore interference)		Measured	
	Torque T (N cm)	Thrust R (N)	Torque T (N cm)	Thrust R (N)	Torque T (N cm)	Thrust R (N)
3.175	795.5	769.5	579.5	526.8	872.4	741.2
3.572	775.0	664.2	532.3	432.0	832.0	672.3
3.969	771.6	553.1	508.6	393.2	821.4	632.7
4.366	735.2	516.0	505.2	303.4	800.7	606.1
5.556	680.1	445.0	445.3	193.9	734.9	479.2
6.350	648.3	400.4	435.4	185.6	642.7	417.8
7.938	490.2	274.3	308.2	122.1	546.4	304.9
8.731	411.8	243.4	267.0	101.5	451.6	253.0
9.128	376.5	190.0	246.4	82.9	392.4	223.2
9.525	335.4	160.1	205.1	65.4	363.5	171.0
10.320	266.3	118.0	148.7	51.2	294.0	145.6
11.110	154.3	79.2	83.0	40.3	163.7	97.3

Table 8.6 A comparison of predicted drilling force with measured ones, with pre-drilled pilot holes, multi-facet drills (drill diameter: $D = 12.7$ mm)

Pilot hole Diameter D (mm)	Predicted (considering interference)		Simple sum (ignore interference)		Measured	
	Torque T (N cm)	Thrust R (N)	Torque T (N cm)	Thrust R (N)	Torque T (N cm)	Thrust R (N)
3.175	693.5	793.8	456.2	551.5	661.2	746.5
3.572	673.2	722.3	413.0	507.9	651.4	701.3
3.969	607.4	611.2	347.5	412.3	602.3	576.1
4.366	604.6	557.5	336.8	341.4	585.2	538.2
5.556	487.6	406.8	276.2	222.9	461.6	369.2
6.350	476.9	358.6	267.1	184.9	459.5	348.0
7.938	365.0	262.3	205.0	133.7	386.0	291.4
8.731	340.1	221.6	195.2	108.6	366.7	253.6
9.128	325.5	203.9	149.3	100.3	341.3	219.2
9.525	285.3	164.1	136.4	87.0	305.6	176.3
10.320	231.0	110.5	115.0	70.3	254.0	126.9
11.110	143.6	85.0	86.7	48.2	172.1	93.5

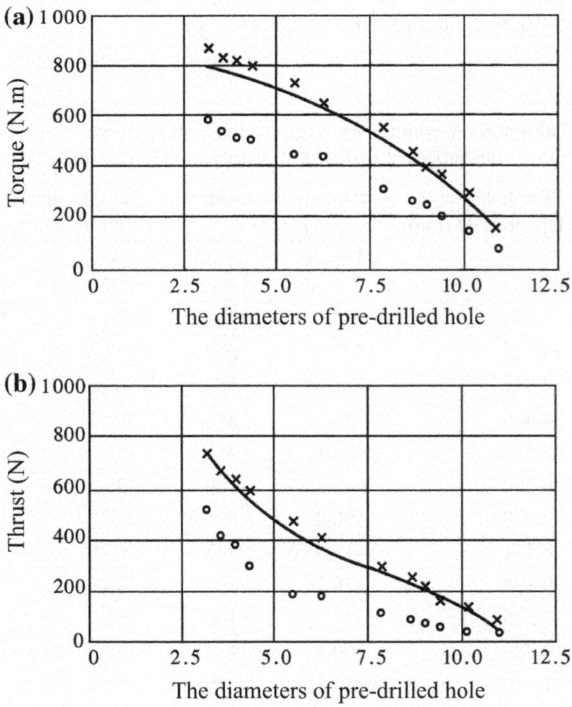

Fig. 8.8 Measured and predicted cutting force of boring with standard twist drills (considering and without consideration of chip-ejection interference). **a** Torque. **b** Axial thrust

Fig. 8.9 Measured and predicted cutting force of boring with multi-faced drills (considering and without consideration of chip-ejection interference). **a** Torque. **b** Axial thrust. Multiplication sign—Measured, open circle —The sum of elementary cutting tools (without consideration of chip-ejection interference), figure dash—Model predicted (considering of chip-ejection interference)

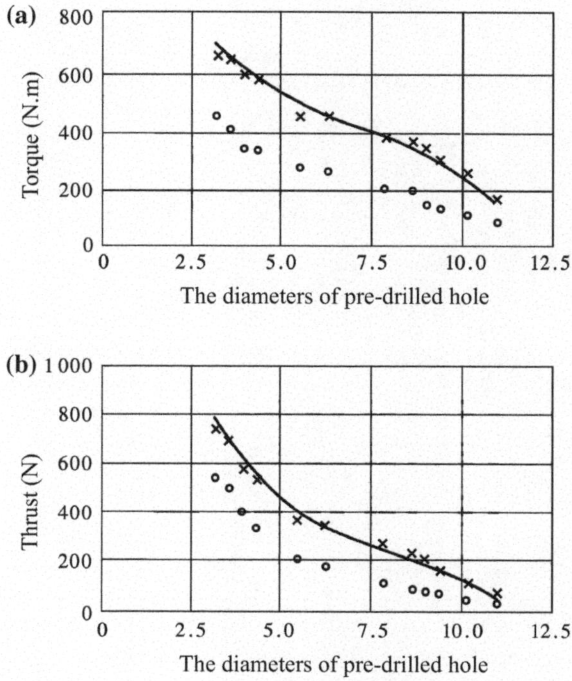

At last, the cutting force of multi-facet drills is obviously lower than that of standard twist drills, this is because the rake distribution of multi-facet drill is better than that of standard twist drills. Figures 8.5 and 3.11 are clearly show this feature. (Figs. 8.8 and 8.9)

Chapter 9
Bifurcation and Catastrophe in Metal Cutting Process

By means of theoretical analysis it is predicted that there must be Bifurcation and Catastrophe phenomena existing in cutting process, and we did find the phenomena in our cutting experiments. The root of the phenomena is the strong non-linearity of non-free cutting process. This finding is beyond people's traditional knowledge and understanding of metal cutting process, but has been fully verified by experimental facts. The research has revealed the complexity of a cutting process and the difficulty to control it, deepened people's understanding of the technological process of metal cutting, and required people to renew conception and strategy of machining process design and chip-control.

9.1 Introduction

As described in Chap. 6, there are two categories of parameters needed for describing a metal-cutting process: first, technological parameters, which are artificially controllable from outside the cutting process, therefore, they are also called control parameters; and secondly, physical parameters, which describe the physical state, so they are also called state parameters. Such parameters are restricted by some natural laws, and are determined from inside the cutting process, cannot be directly selected according to people's intention. The technological effects, people pursuant of a cutting process, such as accuracy and surface quality, are directly determined by physical parameters, which are in turn controlled by the technological parameters.

There are three hypotheses hidden in people's understanding of the technological process of metal cutting: first, the certainty of control, certain control parameters correspond to certain state parameters, and obtain certain technological effects;

H. Shi, *Metal Cutting Theory*, Springer Series in Advanced Manufacturing,
https://doi.org/10.1007/978-3-319-73561-0_9

second, the immediacy of control, the technological effects at a moment are completely determined by the control parameters at the moment only; third, the continuity of control, i.e., the control of technological effects are continuous and stable, that is, any tiny changes in control parameters only cause tiny changes in state parameter and technological effect.

If found in production or scientific experimentation of some phenomena, contrary to the above three hypotheses that people tend to be interpreted as something unknown interference. These three hypotheses have simplified the knowledge about technological process of metal cutting and the decision of process program. Unfortunately, studies in this chapter have shown that, because the presence of bifurcation and catastrophe, the three hypotheses mentioned above are not true. The actual situation of cutting process is much more complicated than people think about it. We will analyze and illustrate this issue through the following non-free cutting example.

9.2 A Theoretical Model of Non-free Cutting

Figure 9.1 shows a non-free cutting process, the major and the minor edges are cutting concurrently, and the interference exists between their chip-ejections. In the figure, $\boldsymbol{W}$ is the common shear velocity of shear planes I and II, $\boldsymbol{V}$ is cutting velocity vector and $\boldsymbol{U}$ is chip-ejection velocity vector.

According to the model of chip deformation, the total cutting power consumed can be expressed as,

$$A = 1.667 \times 10^{-5} V\tau \frac{a_c \cos\beta}{(\cos\beta - r\sin\beta)} \frac{\sqrt{1+r^2}}{r\cos\psi_\lambda}\Big[a_w\sqrt{1+r^2\cos^2\psi_\lambda} + \frac{1}{2}a_c\Big(\sqrt{1+r^2\sin^2\psi_\lambda} - \sqrt{\tan^2\psi_\lambda + r^2\sin^2\psi_\lambda}\Big)\Big],$$

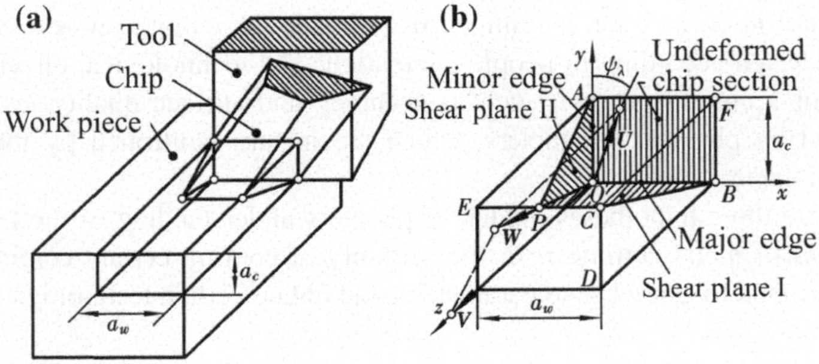

Fig. 9.1 A model of non-free cutting ($\gamma_o = 0°$, $\lambda_s = 0°$)

when $\tan\psi_\lambda < a_w/a_c$;

$$A = 1.667 \times 10^{-5} V\tau \frac{a_w \cos\beta}{(\cos\beta - r\sin\beta)} \frac{\sqrt{1+r^2}}{r\sin\psi_\lambda} [a_c\sqrt{1+r^2\sin^2\psi_\lambda} + \frac{1}{2}a_w(\sqrt{1+r^2\cos^2\psi_\lambda} - \sqrt{\cot^2\psi_\lambda + r^2\cos^2\psi_\lambda})],$$

when $\tan\psi_\lambda > a_w/a_c$;

$$A = 1.667 \times 10^{-5} V\tau \frac{\cos\beta}{\cos\beta - r\sin\beta} \frac{\sqrt{1+r^2}}{2r} [a_c\sqrt{a_w^2(1+r^2)+a_c^2} + a_w\sqrt{a_c^2(1+r^2)+a_w^2}],$$

when

$$\tan\psi_\lambda = a_w/a_c. \qquad (9.1)$$

where, a_w is width of cut,[1] a_c is thickness of cut, V is cutting speed,τ is stress at shear plane, β is friction angle between chip and rake face, r is cutting ratio, ψ_λ is chip-ejection angle. In this model, a_w, a_c, V, τ and β are control parameters while r and ψ_λ are state parameters, which determined chip-ejection velocity vector $\boldsymbol{U}$. Equation (9.1) shows that cutting power is related to both control parameters and state parameters,

$$A = f(r, \psi_\lambda; a_w, a_c, V, \tau, \beta). \qquad (9.2)$$

9.3 The Principle of Minimum Energy Dissipation and the Structure Stability

9.3.1 The Principle of Minimum Energy Dissipation

Control parameters have actually set some constraints for a cutting process, and there are many groups of state parameter evaluations all meet the constraint conditions, but, as shown by a good deal of experiment researches and theoretical analysis in last two chapters, only the group of state parameter values r_o, $\psi_{\lambda o}$, which minimize the cutting power, is the realizable one, i.e.,

$$A_{\min} = f(r_o, \psi_{\lambda o}; a_w, a_c, V, \tau, \beta). \qquad (9.3)$$

[1]The Deduction of this Formulae see the Appendix (9.1) of this chapter.

where, r_o, $\psi_{\lambda o}$ are the natural values of r, ψ_λ The values of r_o, $\psi_{\lambda o}$ determined the common chip-ejection vector U_o of the major and minor cutting edges. And by any means to force state parameters to departure from their natural values will inevitably lead to an increase in cutting power. This is so called the principle of minimum energy dissipation, and Eq. (9.3) defined an extremum problem.

9.3.2 *Digital Simulation of the Model*

For the sake of simplicity, control parameters V,τ, β were kept as constants and the relations of state parameters r, ψ_λ to the control parameters a_w, a_c are investigated in the way of digital simulation, i.e.

$$A = f(r, \psi_\lambda; a_\mathrm{w}, a_\mathrm{c})|V, \tau, \beta = \text{const.} \tag{9.4}$$

Under different a_w, a_c the value of r_o and $\psi_{\lambda o}$, which minimize A, were found numerically, and the following relations were established,

$$r_o = r_o(a_\mathrm{w}, a_c), \psi_{\lambda o} = \psi_{\lambda o}(a_\mathrm{w}, a_c). \tag{9.5}$$

9.3.3 *Structure Stability*

Under different a_w, a_c, the relations between cutting power A and chip-ejection angle ψ_λ are shown in Fig. 9.2. From the figure we can see that when a_w and a_c are in large disparity (Fig. 9.2a, e), the relation of cutting power A to chip-ejection angle ψ_λ has only one minimum, this is corresponding to a stable solution of the problem. When a_w and a_c are close to each other (Fig. 9.2c), however, the relation of A to ψ_λ has two minimums and one maximum, $\psi_{\lambda o1}$, $\psi_{\lambda o2}$, $\psi_{\lambda o3}$, and all of them satisfy the extremum condition, but the first solution and the third solution are stable and the second solution becomes unstable. This shows that the variation of control parameter causes the structure of the model changing, and the middle solution lost its stability [74].

9.3.4 *Bifurcation and Catastrophe*

When β is kept constant, the variation of r_o with a_w, a_c is very small, here let's first investigate the relation of chip-ejection angle $\psi_{\lambda o}$ (state parameter) to thickness of cut a_c (control parameter). As shown in Fig. 9.3, because the curve $\psi_{\lambda o} - a_c$ is folded, the one-to-one relation between state parameter and control parameter is no longer existing. For example, when a_c = 2.0 mm, there are three corresponding

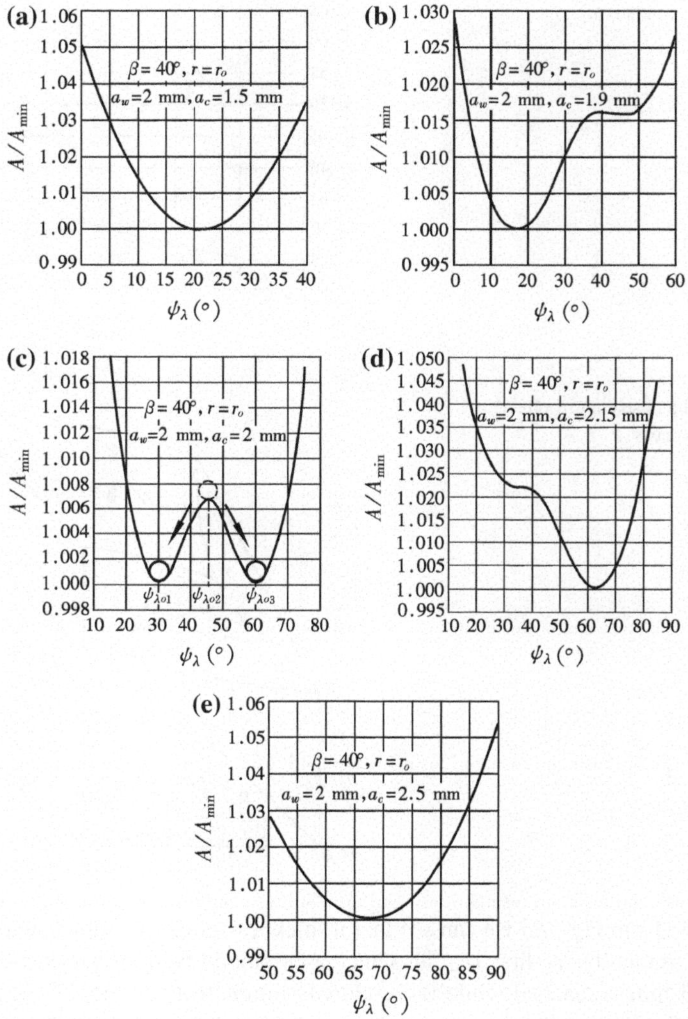

Fig. 9.2 Cutting power A versus chip-ejection angle ψ_λ ($\beta = 40°$, $r = r_o$)

values of $\psi_{\lambda o}$, i.e. 30°, 45° and 60°, they all let A get extremum. Among them, $\psi_{\lambda o} = 45°$ is unstable and cannot be realize; $\psi_{\lambda o} = 30°$ and $\psi_{\lambda o} = 60°$ which will realize, it depends on the historical process of a_c approaching 2.0 mm: if thickness of cut a_c from a smaller value and gradually increases to approach 2.0 mm, as indicated by arrow A, then, chip will ejected in the direction of $\psi_{\lambda o} = 30°$, if, however, thickness of cut a_c from a bigger value and gradually decreases to approach 2.0 mm, as indicated by arrow B, then, chip will ejected in the direction of $\psi_{\lambda o} = 60°$.

Fig. 9.3 $\psi_{\lambda o}$ versus a_c

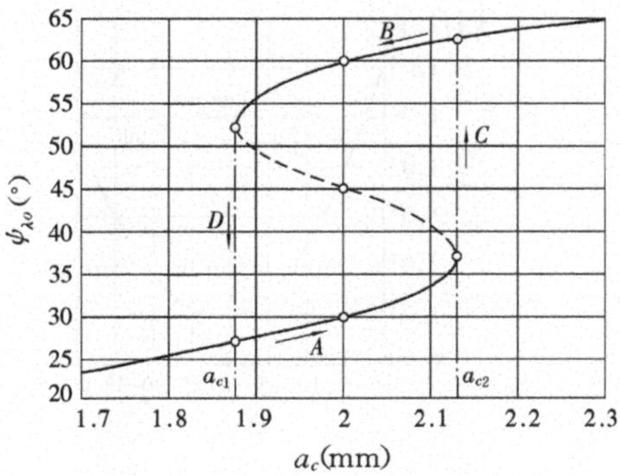

Fig. 9.4 The cusp catastrophe in cutting process

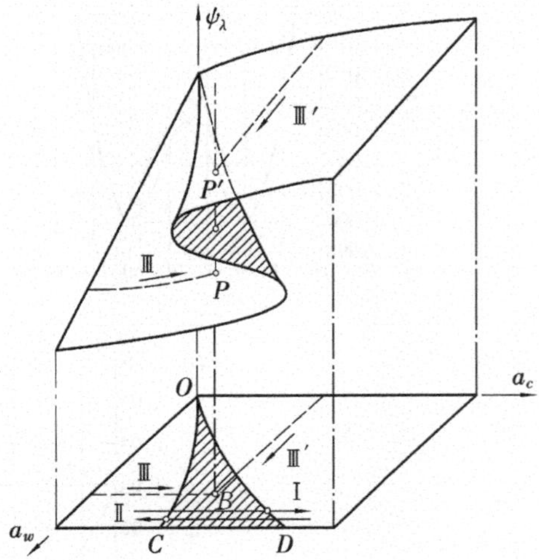

Besides, form Fig. 9.3 we can see that if thickness of cut a_c from a smaller value increases gradually, at first $\psi_{\lambda o}$ also increases smoothly until a critical point of $a_{c_2} = 2.13$ mm is reached while $\psi_{\lambda o}$ suddenly jumps from 37° to 63° as indicated by arrow *C*. If a_c varies in an opposite way, at the point of $a_{c_1} = 1.88$ mm $\psi_{\lambda o}$ will suddenly drop from 53° to 27°, as indicated by arrow *D*. It is clear that the jump point and the drop point differ from each other, and $a_{c_1} < a_{c_2}$. This is the so-called "lag" phenomenon, generally existing with catastrophes.

The relation of $\psi_{\lambda o}$ versus a_c, a_w is schematically expressed in Fig. 9.4 by a folded surface, which is called "balance surface". This is a typical cups catastrophe [74], the shaded area *OCD* in the control plane (a_c, a_w) is the projection of the folded part of the balance surface. When the control trajectory cross the boundary of the area, as indicated by arrows I and II, a big jump or drop in $\psi_{\lambda o}$ will happen, depending on the moving direction of the control trajectory. It would be also interesting to mention that the two control trajectories III and III′ have the same terminal point *B* but lead to quite different chip-ejection angle $\psi_{\lambda o}$ as indicated by points *P* and *P*′, respectively.

This shows that even though under the same control parameters, only the difference in control trajectories leads to the control point, i.e. the different in cut-in process, the state parameters (chip-ejection angle $\psi_{\lambda o}$) realized may be quite different. These characteristics are that the previous cutting theory did not expect, nor did give explanations.

9.4 The Experimental Verification of Bifurcation and Catastrophe

9.4.1 Cylindrical Turning Experiment

The purpose of this group of experiments is to find a bifurcation in the real cutting process. The experimental conditions: Lathe: J1-MAZAK-MATE 360 × 570; Work-piece: mild steel #45, $D = 98$ mm; Cutting tool, HSS, $\gamma_o = 0°$, $\lambda_s = 0°$, $\kappa_r = 90°$, $k_r' = 0°$; Cutting speed $V = 60$ m/min, a_p (depth of cut) = a_w (width of cut) = working length of the major edge, f (feed per revolution) = a_c (thickness of cut) = working length of the minor edge. The three components F_c, F_f, F_p of the cutting force are measured by a 3-D Kistler dynamometer and recorded by an *x*-*y* plotter. The friction angle in chip-rake interface is calculated from measured force components by $\tan\beta = (F_f^2 + F_p^2)^{1/2} \big/ F_c$, and the chip-ejection angle $\psi_{\lambda o}$ is determined by measuring the friction marks on the rake face with a microscope. Besides, the values of $\psi_{\lambda o}$ were also determined by on-line measured cutting force components: $\tan\psi_{\lambda o} = F_p / F_f$. The cutting ratio is determined by $r_o = f\ a_p/s_c$, where s_c is the area of the chip cross section. The experimental results are shown in Table 9.1. The theoretically predicted cutting ratio r_o versus experimentally measured ones are given in Fig. 9.5, and the theoretically predicted chip-ejection angle $\psi_{\lambda o}$ versus experimentally measured ones are given in Fig. 9.6.

Table 9.1 A comparison of predicted chip-ejection angle $\psi_{\lambda o}$ and cutting ratio r_o versus measured ones

a_w (mm)	a_c (mm)	$\beta(°)$	r_o		$\psi_{\lambda o}$		
			Measured	Predicted	Measured		Predicted
					From marks	From force	
4.0	0.2	48.0	0.362	0.380	4	8	2
3.0	0.3	43.8	0.411	0.430	7	13	3
2.0	0.5	41.1	0.465	0.460	22	20	8
1.0	1.0	48.5	0.367	0.390	58, 32	57, 45	60, 30
0.5	2.0	40.0	0.560	0.480	82	81	82
0.3	3.0	44.3	0.577	0.420	87	87	87
0.2	4.0	37.2	0.640	0.500	88	87	88

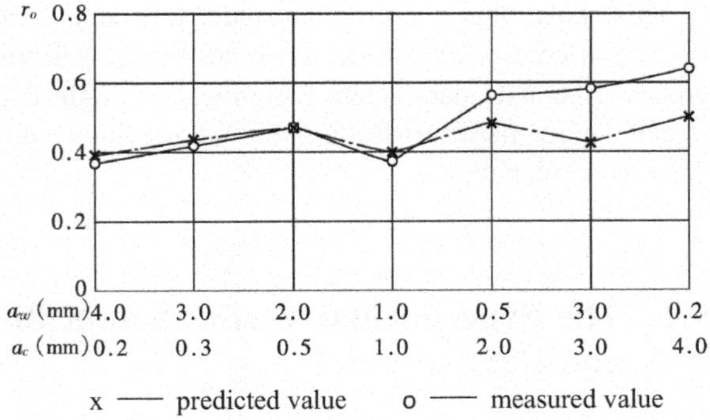

Fig. 9.5 Cutting ratio r_o, theoretical prediction versus experimental measurement

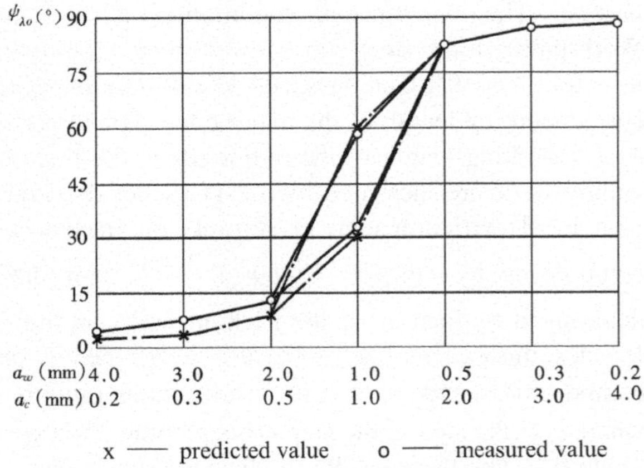

Fig. 9.6 Chip-ejection angle $\psi_{\lambda o}$, theoretical prediction versus experimental measurement. Multiplication sign—Predicted value, open circle—Measured value

The figures show a fairly good agreement between the theoretically predicted and experimentally measured values of $\psi_{\lambda o}$ and r_o under different cutting conditions. And at the point $a_w = a_c = 1$, chip-ejection angle $\psi_{\lambda o}$ does bifurcate itself into two values as predicted theoretically, they are corresponding to the two minimum points $\psi_{\lambda o_1}$ and $\psi_{\lambda o_3}$ shown in Fig. 9.2c.

When bifurcation appears, the friction marks on the rake face also bifurcate into two as shown in Fig. 9.7. This two directions are corresponding to the two minimum point $\psi_{\lambda o1}$ and $\psi_{\lambda o3}$ as shown in Fig. 9.2c. If the cutting process begin with a bifurcation condition ($a_w \approx a_c$), then, because the middle maximum point $\psi_{\lambda o2}$ is unstable, the chip-ejection angle (state parameter) may skip to $\psi_{\lambda o1}$, or to $\psi_{\lambda o3}$, depending on a very tiny random disturbance happened at the moment. This is a kind of "intrinsic randomness".

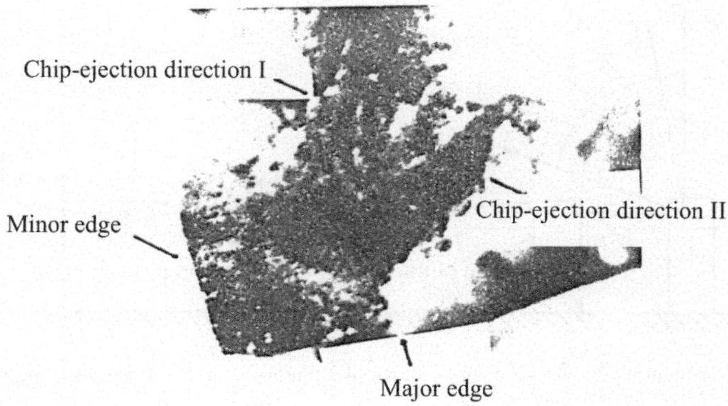

Fig. 9.7 Grazing bifurcation on the tool rake face when the bifurcation phenomenon appears

9.4.2 Low Speed Planing Experiment (I)

The purpose of this group of experiments is to identify the influence of cut-in process, i.e., the control history of a_c and a_w on chip-ejection angle $\psi_{\lambda o}$ (one of the state parameters). The experimental arrangement is the same as shown in Fig. 9.1. Machine tool: vertical milling machine, the feed motion of its table was taken as the main cutting motion, $V = 0.763$ m/min; work-piece: bronze plate, fixed on the table; cutting tool: HSS, $\gamma_o = 0°$, $\lambda_s = 0°$, attached to the milling head (without rotations). During the cutting process the width of cut and the depth of cut are kept constant $a_c = a_w = 0.1$, but the process begin with different initial cut-in history: first, a_c increases from 0 to 0.1 while a_w kept constant ($a_w = 0.1$); and secondly, a_w increases from 0 to 0.1 while a_c kept constant ($a_c = 0.1$). The work-piece was shaped as in Fig. 9.8a, b. The two cut-in process just correspond to the control trajectories III an III$'$ in Fig. 9.4. The chip-ejection angle $\psi_{\lambda o}$ under the same technological condition of $a_c = 0.1$ and $a_w = 0.1$ but with the two different cut-in processes are measured with microscope from the wear marks on the tool rake, and the results are: for the first case, $\psi_{\lambda o} = 20°–30°$, and for the second case, $\lambda_o = 50°–60°$.

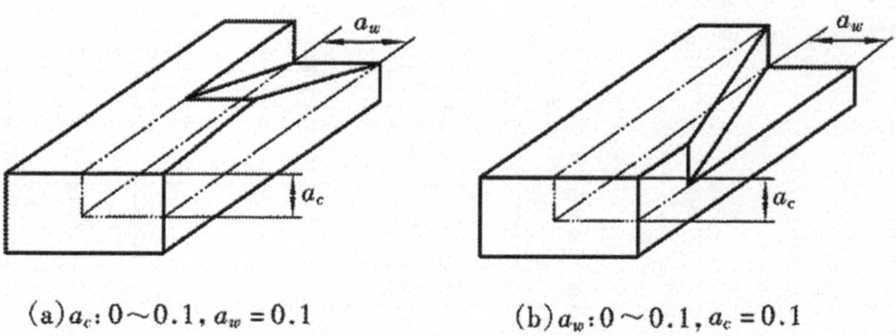

Fig. 9.8 The work-piece used for verification of bifurcation phenomenon experimentally, **a** a_c changes from 0 to 0.1, while $a_w = 0.1$, **b** a_w changes from 0 to 0.1, while $a_c = 0.1$

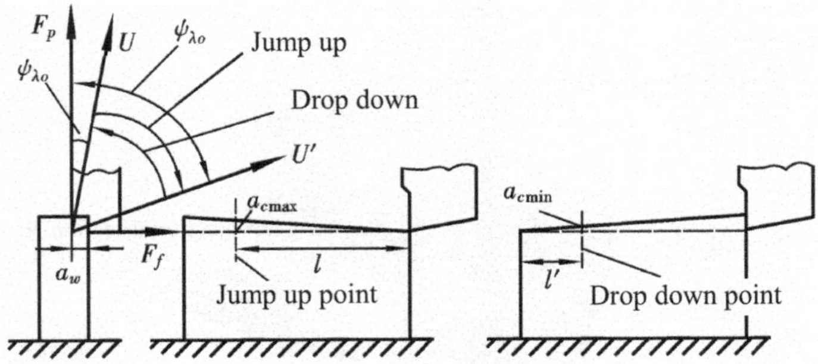

Fig. 9.9 Experimental scheme verifying catastrophe phenomena (jump up, drop down and lag)

9.4.3 Low Speed Planing Experiment (II)

The experimental conditions are the same as mentioned in Sect. 9.4.2. The experimental arrangement is shown in Fig. 9.9. The purpose of this experiment is to identify the "jump", "drop" and "lag" phenomena normally exist accompanying with bifurcation, as predicted by the theory. In the experiment, the width of cut is kept constant ($a_w = 0.1$), the thickness of cut a_c is gradually increasing or decreasing. The chip-ejection angle and its variations can be measured from photographs taken during cutting process (in the experiment, the sudden changes in chip-ejection direction were obviously observed). The experiments did show a jump of $\psi_{\lambda o}$ from 39° to 60° at a cutting length $l = 52$ ($a_{c,max} = 0.127$), while a_c increasing, and a drop of $\psi_{\lambda o}$ from 54° to 21° at a cutting length $l^{'} = 28$ ($a_{c,max} = 0.07$), while a_c decreasing. The jump and drop points do differ from each other as identified by the fact that $l > l^{'}$ ($a_{c,max} > a_{c,min}$), and this fact also indicate the "lag" phenomenon.

The cutting force F_f and F_p and chip-ejection angle $\psi_{\lambda o}$ recorded in cutting process are shown in Figs. 9.10 and 9.11. The former is the case of depth of cut

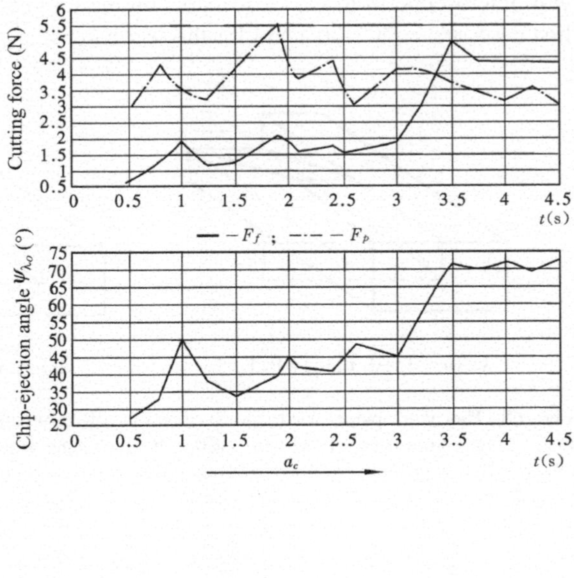

Fig. 9.10 The catastrophe phenomena in cutting force and chip-ejection angle at bifurcation point, when depth of cut gradually increasing

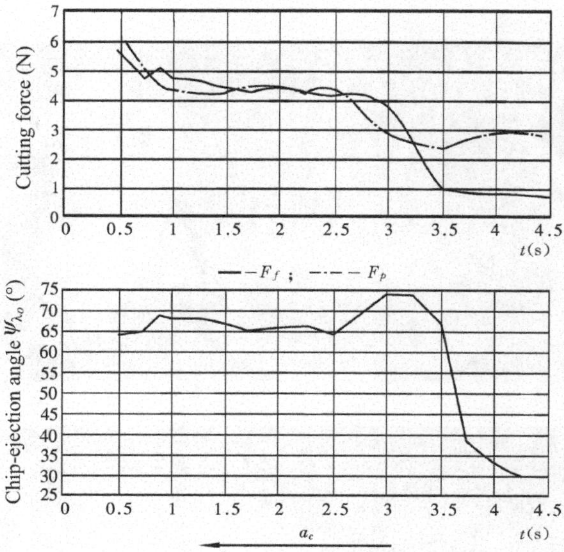

Fig. 9.11 The catastrophe phenomena in cutting force and chop-ejection angle at bifurcation point, when depth of cut gradually decreasing

increasing, and the latter is the case of depth of cut decreasing. The jump and the drop of chip-ejection angle $\psi_{\lambda o}$ can be clearly seen form Figs. 9.10 and 9.11, respectively.

At the moment of sudden change in chip-ejection direction, a piece of "s" shaped chip was produced, as shown in Fig. 9.12. Along with the jump in $\psi_{\lambda o}$ the shape of chip changes from right-hand helix to left-hand helix, as shown by Fig. 9.13a, b, and along with the drop in $\psi_{\lambda o}$ the shape of chip changes inversely.

Fig. 9.12 "S" shaped chip generated at the moment when catastrophe happened

(a)

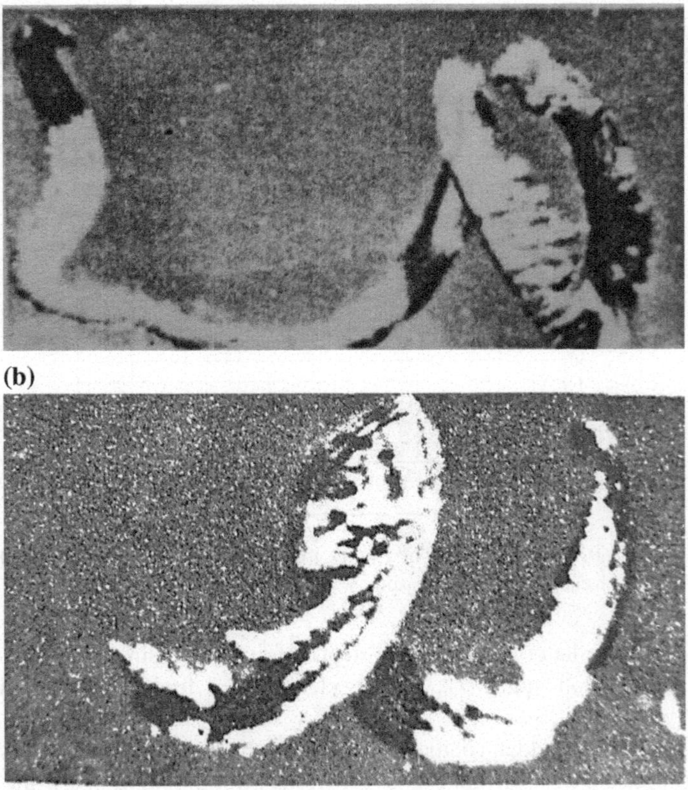

(b)

Fig. 9.13 The sudden change in chip shapes at the moment when catastrophe happened. **a** Right hand helix chip. **b** Left hand helix chip

9.5 Remarks

1. Based on the Principle of Minimum Energy Dissipation and the Non-Free Cutting Model proposed by the author, bifurcation and catastrophe in cutting processes are predicted and such predictions have been verified by experiments specially designed in this study.
2. From the view point of mechanics, a metal cutting process may be regarded as a problem of extremum: under the constraint condition set by the control parameters, the state parameters always minimize the cutting power consuming. Because of the strong non-linearity in the cutting process, the extremum points of cutting power may not be single, but multiple, and along with the variation of control parameters, the stability of each extremum point may change, i.e., appears the problem of structural stability. This is the physical root of bifurcation and catastrophe appearing in metal cutting processes.

3. Because of bifurcation and catastrophe, at a bifurcation point, the physical state and the technological effects cannot be fully determined by the control (technological) parameters, but "**intrinsic randomness**" exists, so, the **certainty of control** is no longer valid; Because of bifurcation and catastrophe, the physical state and the technological effects at a moment of a cutting process are not only related to the control parameters at the same moment, but also rely on the variation history of the control parameter, therefore, even if under the same control parameters, the physical state may be different, and this lead to different technological effects, the reason is only the different cut-in process, i.e., a cutting system has "**process dependence**", so, **immediacy of control** is no longer valid; Because of bifurcation and catastrophe, a tiny change in control parameters may result in very large sudden change in state parameters and thus, in technological effects, i.e., a cutting system has "**state discontinuity**", so, **continuity of control** is no longer valid. These factor have practical influence on chip control and cutting tool design, and should be fully taken into account.

Appendix 9.1

Deduction of Eq. (9.1).

A non-free cutting model with double edged cutting tool was presented in this chapter, as shown in Fig. 9.1, and its mathematical presentation is Eq. (9.1). Here, we will give a brief deduction of Eq. (9.1).

The unit vector in the direction of resultant force on rake face is

$$\mathbf{R}_1 = -\sin\beta\sin\psi_\lambda\,\mathbf{i} - \sin\beta\cos\psi_\lambda\mathbf{j} + \cos\beta\,\mathbf{k}, \tag{9.6}$$

where, β is the friction angle on the interface between the rake and the chips, and **i**, **j**, **k** are the unit vectors in the directions of axes x, y, z, respectively (they are not shown in Fig. 9.1). The resultant force imposed on chip by the rake is;

$$\boldsymbol{R} = R\mathbf{R}_1, \tag{9.7}$$

The vector of cutting velocity is,

$$\boldsymbol{V} = V\mathbf{k}, \tag{9.8}$$

The vector of chip-ejection velocity is,

$$\boldsymbol{U} = rV(\sin\psi_\lambda\mathbf{i} + \cos\psi_\lambda\mathbf{j}), \tag{9.9}$$

where, r is cutting ratio. The vector of shear velocity is,

$$\boldsymbol{W} = \boldsymbol{U} + \boldsymbol{V} = V(r\sin\psi_\lambda \mathbf{i} + r\cos\psi_\lambda \mathbf{j} + \mathbf{k}). \tag{9.10}$$

Shear speed, i.e., the module of vector $\boldsymbol{W}$ is,

$$|\boldsymbol{W}| = V\sqrt{1+r^2} \tag{9.11}$$

The unit vector in the shear direction is,

$$\mathbf{W_1} = \frac{r\sin\psi_\lambda \mathbf{i} + r\cos\psi_\lambda \mathbf{j} + \mathbf{k}}{\sqrt{1+r^2}} \tag{9.12}$$

The shear force, i.e., the projection of $\boldsymbol{R}$ on the shear direction $\mathbf{W_1}$,

$$\boldsymbol{R}\cdot\mathbf{W_1} = R(\mathbf{R_1}\cdot\mathbf{W_1}) = R\frac{(\cos\beta - r\sin\beta)}{\sqrt{1+r^2}} = \tau S, \tag{9.13}$$

where, τ is shear stress of the work-piece material, and S is the area of shear plane. So,

$$R = \frac{\tau S\sqrt{1+r^2}}{(\cos\beta - r\sin\beta)}. \tag{9.14}$$

The cutting power is,

$$A = R\cdot\cos\beta\cdot V = \frac{V\cdot\tau\cdot\cos\beta}{(\cos\beta - r\sin\beta)}\sqrt{1+r^2}\cdot S. \tag{9.15}$$

Shear plane is divided into two parts: the shear plane of major edge and the shear plane of minor edge, the areas of them are denoted as S_I and S_{II}, respectively. The calculation of S_I and S_{II} are divided into three cases:

The first case: $\tan\psi_\lambda < a_w/a_c$, in this case, vector W intersects with the top surface $AFCE$ of the work-piece, and the intersection point is P. What is shown in Fig. 9.1 is just this case. The coordinates of point P In the coordinate system xyz are $P_x = a_c\tan\psi_\lambda$, $P_y = a_c$ and $P_z = a_c/(r\cos\psi_\lambda)$. In this case, S_I and S_{II} are shown as Fig. 9.14a, b and the total area of the two shear plane

$$\begin{aligned} S &= S_{\mathrm{I}} + S_{\mathrm{II}} = \square BCEO - \Delta OPE + \Delta OAP \\ &= a_c\left(a_w - \frac{1}{2}a_c\tan\psi_\lambda\right)\sqrt{1+\frac{1}{r^2\cos^2\psi_\lambda}} + \frac{1}{2}\frac{a_c^2}{r\cos\psi_\lambda}\sqrt{1+r^2\sin^2\psi_\lambda}. \end{aligned} \tag{9.16}$$

The second case: $\tan\psi_\lambda > a_w/a_c$, in this case, vector W intersects with the side surface $FCDB$ of the work-piece, and the intersection point is P. The coordinates of

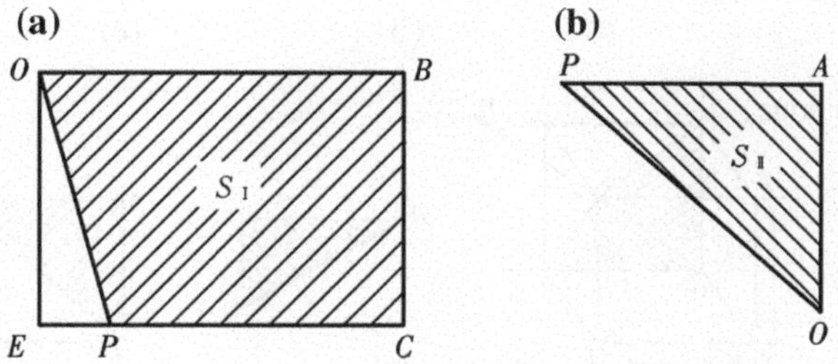

Fig. 9.14 The shear planes ($\tan\psi_\lambda < a_w/a_c$). **a** Major shear plane. **b** Minor shear plane

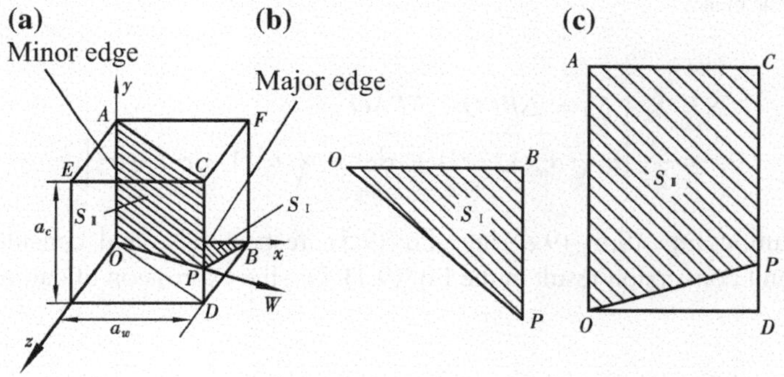

Fig. 9.15 The area of cut and shear planes ($\tan\psi_\lambda > a_w/a_c$). **a** Chip sections. **b** Major shear plane. **c** Minor shear plane

point P In the coordinate system xyz are $P_x = a_w$, $P_y = a_w/\tan\psi_\lambda$ and $P_z = a_w/(r\sin\psi_\lambda)$. In this case, S_I and S_{II} are shown as Fig. 9.15b and c, and the total area of the two shear plane is,

$$S = S_\mathrm{I} + S_\mathrm{II} = \Delta OBP + \square ACDO - \Delta OPD$$
$$= \frac{a_w}{r\sin\psi_\lambda}\left[a_c\sqrt{1 + r^2\sin^2\psi_\lambda} + \frac{1}{2}a_w\left(\sqrt{1 + r^2\cos^2\psi_\lambda} - \sqrt{\cot^2\psi_\lambda + r^2\cos^2\psi_\lambda}\right)\right]. \quad (9.17)$$

The third case, $\tan\psi_\lambda = a_w/a_c$, it is a critical case between the above two cases, the vector W intersects with the side ridge EC of the work-piece, and the intersection point is P, which coincides with the corner point C as shown in Fig. 9.16a. The coordinates of point P in the coordinate system xyz are $P_x = a_w$, $P_y = a_c$ and $P_z = a_w/(r\sin\psi_\lambda) = a_c/(r\sin\psi_\lambda)$. In this case, S_I and S_{II} are shown as Fig. 9.15b, c, and the total area of the two shear plane is,

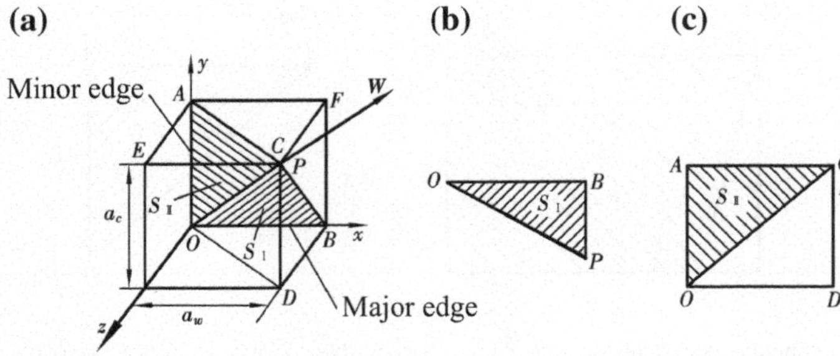

Fig. 9.16 The area of cut and shear planes ($\tan\psi_\lambda = a_w/a_c$). **a** Chip sections. **b** Major shear plane. **c** Minor shear plane

$$
\begin{aligned}
S &= S_{\mathrm{I}} + S_{\mathrm{II}} = \Delta BPO + \Delta PAO \\
&= \frac{1}{2r}\left[a_c\sqrt{a_w^2(1+r^2)+a_c^2} + a_w\sqrt{a_c^2(1+r^2)+a_w^2}\right],
\end{aligned} \tag{9.18}
$$

Substituting Eqs. (9.6)–(9.8) into Eq. (9.5), respectively, and considering the physical unit conversion result in the Eq. (9.1), i.e., the expression of cutting power.

Chapter 10
The Principle and the Law for Free-Cutting Tool Design

On the basis of in-depth study of the essence and the law of non-free cutting, in this chapter, a method of free cutting is proposed, and thus a feasible way is found for improving advanced cutting tool design and for optimizing machining process. The key point is properly designing the tool rake face, so as to dredge chip-ejections from different parts of edge line and to eliminate or to reduce the interference in chip-ejection, realizing free-cutting or quasi-free-cutting. The principle and the basic law of free-cutting tool design is proposed and experimental verifications are presented.

As described in Chaps. 6 and 7, today, almost all the actual cutting process is non-free cutting. Many problems of modern cutting processes are originated from the disadvantages arising from the non-free cutting.

About non-free cutting, there exists a misunderstanding: Only when the cutting edge is a single straight line the free-cutting can be achieved; and as long as there is more than one straight line edges at the same time participate in the cutting, or curved edge in the cutting, then, non-cutting is inevitable. This, unfortunately, is not true. As stated clearly in Chap. 7, A cutting process is free, if and only if the natural chip-ejection vectors $\boldsymbol{U}_{oi}$ (i = 1, 2 … n) of all the elementary cutting tools form a rigid body motion. Thus it can be seen that a cutting process is free or not has no inevitable connection with whether or not the cutting edge is a straight line. As a matter of fact, for arbitrarily shaped cutting edge, provided we properly design the rake face so as to dredge chip-ejections, can we eliminate or, to a large extent, to relieve chip-ejection interference and realize free-cutting or quasi-free-cutting.

Part of this chapter and next chapter is from the study of Mr. Yan Xing, a doctoral student of Huazhong University of Science and Technology and a member of my research group. His Ph.D. work was under my direction when I was a full professor in the university (1996–2000).

H. Shi, *Metal Cutting Theory*, Springer Series in Advanced Manufacturing,
https://doi.org/10.1007/978-3-319-73561-0_10

10.1 The Principle of Dredging Chip-Ejection Vectors to Eliminate Chip-Ejection Interference

Stabler [69] gave a simple empirical relation between chip-ejection angle $\psi_{\lambda o}$ and cutting edge inclination λ_s,

$$\psi_{\lambda o} \approx \lambda_s,$$

which is called Stabler Law. And in Chap. 6, it has been verified that within normal technological scope, the chip-ejection angle given by Stabler Law is approximately the natural chip-ejection direction which minimizes the cutting power. This has given the empirical relation of Stabler Law the theoretical explanation. Stabler Law has provided a method of properly design the edge inclination λ_s at each point of cutting edge to dredge chip-ejection and eliminate chip-ejection interference. In the follows we will explain this method through an example.

10.1.1 An Example

As shown in Fig. 10.1, the two cut-of tools are all double edged, and have the same shape in the projection on reference plane. The difference is that the tool in Fig. 10.1a has a flat rake face and the one in Fig. 10.1b has a roof-shaped rake face. In the former case, the chip-ejection vectors $\boldsymbol{U}_{o1}$ and $\boldsymbol{U}_{o2}$ interfere with each other, and the cutting process belongs to non-free cutting. In the latter case, because the both edges have edge inclination, their chip-ejections are get dredged and both are in the direction of tool shank axis, so, the chip-injection interference is eliminated, and its cutting process belongs to free-cutting.

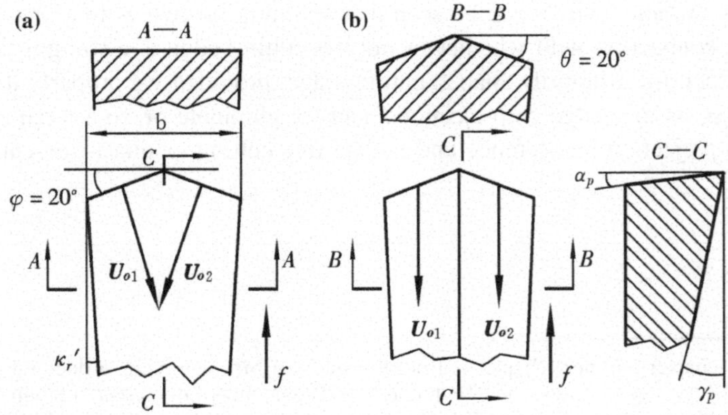

Fig. 10.1 The principle of free-cutting tool design. **a** Plane rake face. **b** Roof shaped rake face

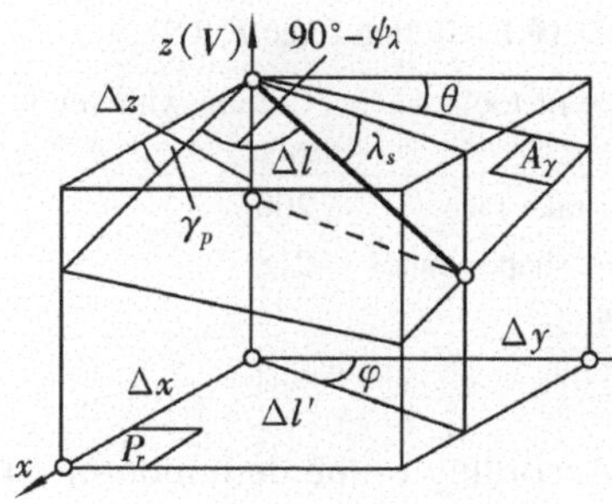

Fig. 10.2 The relations of angles related to free-cutting tool design

If Stabler law is true, as verified in Chap. 6, for the tool in Fig. 10.1, provided selecting roof angle θ according to edge shape angle $\varphi(\varphi = 90° - \kappa_r)$ and tool back rake γ_p (i.e., the rake in the chip-ejection direction), we can adjust the both the chip-ejection vectors $\boldsymbol{U}_{o1}$ and $\boldsymbol{U}_{o2}$ of left and right edges to the direction of the tool shank axis, as shown in Fig. 10.1b. The relation of θ, φ and γ_p is expressed by Fig. 10.2 and the following equation,

$$\tan\theta = \frac{\tan\varphi}{\cos\gamma_p}. \tag{10.1}$$

As a particular case, when $\gamma_p = 0°$, from Eq. (10.1), we have

$$\theta = \varphi. \tag{10.2}$$

10.1.2 Experimental Verification

The following experiments are for a comparison of the cutting properties of the two cutters in Fig. 10.1, and to reveal the effect of free-cutting.

Experiment I: Transverse Cutting Steel

Machine tool: J_1-MAZAK-MATE 360 × 570; Work-piece: mild steel #45, Φ 82 mm, L 300 mm, supported by chuck and back center; Cutting tool: cemented carbide YT14, Width of cutter b = 3 mm, $\gamma_p = 0°, \theta = \varphi = 20°, \kappa'_\gamma = 1°, \alpha_o = 4°$, as shown in Fig. 10.1; Cutting method: transverse cutting, $N = 400$ r/m, $f = 0.1$ mm/r. For the two kinds of cutters, flat rake face and roof shaped rake face, at first, cutting with only one of the two edges, and then cutting with double edges, respectively. For each cut, the cutting forces are measured with a KISTLER dynamometer (9257A) and a charge-amplifier (50006). Under each case several measures were taken, and then record their average. The results are listed in Table 10.1.

Table 10.1 Cutting force F_c (N)

Type of tools	Cutting with single edge	Cutting with double edge	Coefficient of non-free cutting
Flat rake face	200	498	1.245
Roof shaped rake face	218	443	1.016

According to the definition of coefficient of non-free cutting (see Sect. 7.7.1),

$$\text{coefficient of non-free cutting} = \frac{\text{cutting force with double edges cutting}}{2 \times \text{cutting force with single edges cutting}}. \tag{10.3}$$

The data in Table 10.1 show that compared with the cutting force of flat rake tool, the tool roof shaped rake reduce its cutting force by 11%. It should be noticed that the above two cutting tools have no any difference in their chip-ejections directions, in their rake angles at the chip-ejections directions and in their cutting parameters, the difference in cutting force is only caused by the effect of dredging the chip-ejection direction. As a matter of fact, the coefficient of non-free cutting for the cutter with roof shaped rake face is close to 1, this is the symbol of free cutting; as well as the coefficient of non-free cutting for the cutter with flat rake face is 1.245, this shows a relatively intense chip-ejection interference of non-free cutting.

Experiment II: Transverse Cutting Copper

In this group of experiments, the coefficients of non-free cutting were measured for the two kinds of cutters: flat rake face and roof shaped rake face, and each with different edge shape angle φ. The results are listed in Table 10.2. In the experiments, work-piece material was copper, the cutter material is HSS, $V = 21.87$ m/min, $f = 0.026$ mm/r, $\alpha_o = 12°$. The other conditions were the same as in Experiments I. The data in Table 10.2 show that for the cutting tools with flat rake face, while edge shape angle φ increasing from 0° to 35° the chip-ejection interference become more and more serious, the coefficients of non-free cutting rises to more than 1.5; for the cutting tools with roof shaped rake face, however, because the chip-ejection interference got dredged, the coefficients of non-free cutting kept in the vicinity of 1, this means that tools are performing free-cutting.

Table 10.2 The coefficients of non-free cutting

Edge shape angle φ	0°	10°	20°	30°	35°
Flat rake face	1.00	1.08	1.15	1.55	1.51
Roof shaped rake face	1.00	1.03	1.02	0.96	0.96

10.2 The Principle and the Basic Law of Free-Cutting Tool Design

10.2.1 General Principle

The fundamental way of design a free cutting tool is to adjust the shape of rake face (mainly to adjust edge inclinations), so as to let the natural chip-ejection vectors of all the elementary cutting tools form a rigid body motion, that is, to let the following equation true for every elementary cutting tool (see Fig. 10.3),

$$\boldsymbol{U}_{oi} = \boldsymbol{U}_i, \quad i = 1, 2, \ldots, n, \tag{10.4}$$

where, n is the number of elementary cutting tools joining cutting work, $\boldsymbol{U}_i$ is required chip-ejection vector of i-th elementary cutting tool, satisfying the condition that the overall chip-ejection motion is a rigid body motion. $\boldsymbol{U}_i$ is determined by following equation,

$$\boldsymbol{U}_i = \boldsymbol{T} + \boldsymbol{\Omega} \times \boldsymbol{\rho}_i, \quad \boldsymbol{i} = \boldsymbol{1}, \boldsymbol{2}, \ldots, \boldsymbol{n}, \tag{10.5}$$

where, $\boldsymbol{T}$ and $\boldsymbol{\Omega}$ are the translation and the rotation components of the overall chip-ejection motion, respectively, $\boldsymbol{\rho}_i$ is the position radius vector of i-th elementary cutting tool. $\boldsymbol{U}_{oi}$ is the natural chip-ejection vector of i-th elementary cutting tool, which is the function of rake angle, edge inclination, thickness of cut and width of cut, etc.,

$$\boldsymbol{U}_{oi} = \boldsymbol{U}_{oi}(\gamma_{oi}, \lambda_{si}, a_{ci}, a_{wi}), \quad i = 1, 2, \ldots, n, \tag{10.6}$$

Relevant independent variables may be obtained from geometrical and kinetic analyses, and the function may be work out by experiments and empirical formula fitting, or through a theoretical analysis (see Chap. 8).

The geometrical and cutting parameters γ_{oi}, λ_{si}, a_{ci} and a_{wi}, etc., are determined by two vectors $\mathbf{a_i}$ and $\mathbf{b_i}$, in the case that the overall chip-ejection motion has been fixed. Where $\mathbf{a_i}$ is the unit vector in the rake face and normal to the edge, and $\mathbf{b_i}$ is

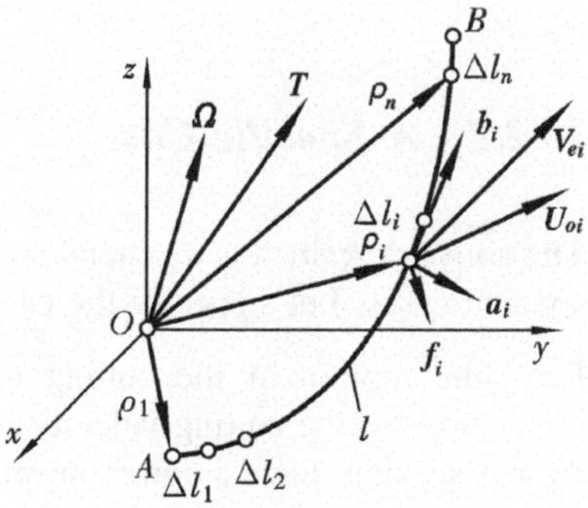

Fig. 10.3 A schematic diagram showing the key points of designing free-cutting tools

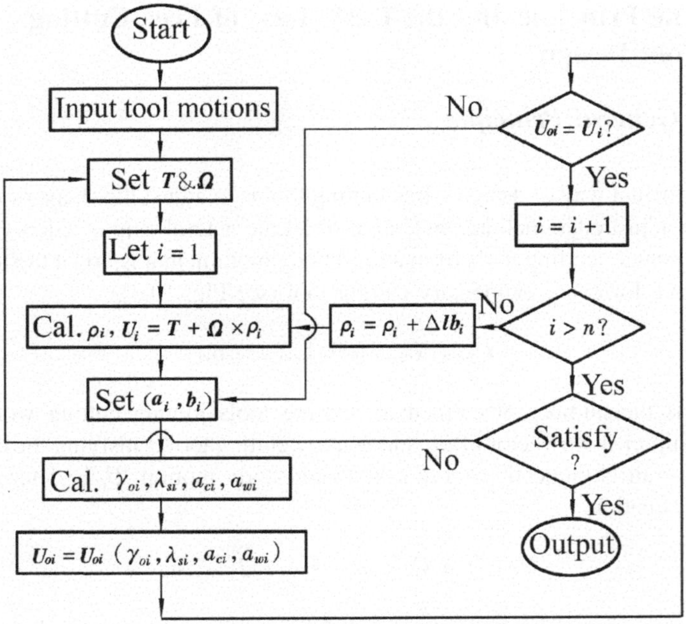

Fig. 10.4 The flow chart of free-cutting tool design

the unit vector in the direction of the edge. In the final analysis, the Eq. (10.6) can be expressed as,

$$U_{oi} = U_{oi}(\mathbf{a_i}, \mathbf{b_i}), \quad i = 1, 2, \ldots, n. \tag{10.7}$$

Substituting Eqs. (10.7) and (10.5) into (10.4) leads to,

$$U_{oi} = U_{oi}(\mathbf{a_i}, \mathbf{b_i}) = T + \Omega \times \rho_i, \quad i = 1, 2, \ldots, n, \tag{10.8}$$

The basic task of designing a free-cutting tool is to adjust vectors $\mathbf{a_i}$ and $\mathbf{b_i}$, to make the above equation satisfied, point by point from starting point A to the end point B (see Fig. 10.3). The flow chart is shown in Fig. 10.4.

10.2.2 A Specific Case

The above described is a general situation and the general methods, the follows is a specific cases. Let's specify the case:

First, the motion of the cutting tool with respect to the work-piece is only a translation, so the cutting velocity for all the elementary cutting tools are the same V, and so, they have a common reference plane P_r;

Second, the overall motion of the chip (when it just left the cutting edge) with respect to the cutting tool is also a translation, so, $\boldsymbol{\Omega} = 0$, $\boldsymbol{U}_i = \boldsymbol{T}$, $i = 1, 2, \ldots, n$.

For the most typical machining operations, such as planing, turning, end-milling, the above two assumptions are satisfied or approximately satisfied.

Let's see Fig. 10.5, the projection s' of edge line s on reference plane P_r is known. The rake face is supposed to be a cylinder with generatrix g. The translation motion $\boldsymbol{T}$ of chip-ejection is supposed in the direction of g, and the angle between g and the reference plane P_r is the rake in the direction of chip-ejection γ_ψ, which is the same for all the points of the edge line. According to Stabler Law, the natural chip-ejection direction is determined by cutting edge inclination λ_s. The task of tool design is to determine the edge line, which realizes free cutting. From the geometrical relations shown in Fig. 10.2 and supposing Stabler Law is true, we can deduce the formula for calculating the edge inclination λ_{si} of i-th elementary cutting tool, which guaranty the natural chip-ejection directions of all the elementary cutting tools in the direction of generatrix (i.e., no chip-ejection interference),

$$\tan \lambda_{si} = \sin \varphi_i \left(\frac{\cos \gamma_\psi}{1 - \sin \gamma_\psi}\right), \quad i = 1, 2, \ldots, n, \tag{10.9}$$

where, φ_i is the edge shape angle of i-th elementary cutting tool (the angle φ_i indicated in Fig. 10.5 is defined as negative). The orientation of i-th incremental edge Δl_i may be determined by its three components Δx_i, Δy_i and Δz_i. Once λ_{si} and φ_i are known, these components may be calculated as follows,

Fig. 10.5 Design of free-cutting tools

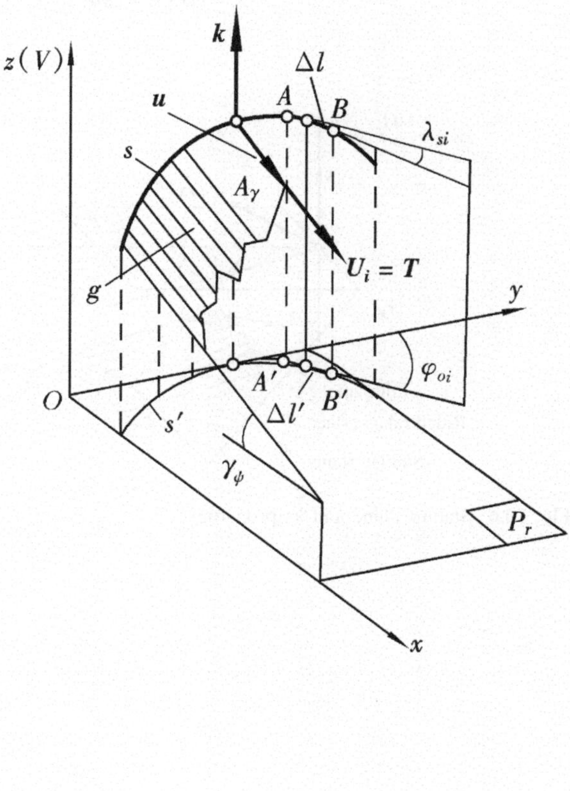

$$\begin{aligned} \Delta x_i &= \Delta l'_i \sin \varphi_i, \\ \Delta y_i &= \Delta l'_i \cos \varphi_i, \quad i = 1, 2, \ldots, n. \\ \Delta z_i &= \Delta l'_i \tan \lambda_{si}, \end{aligned} \tag{10.10}$$

where,

$$\Delta l'_i = \Delta l_i \cos \lambda_{si},$$

is the projection of Δl_i on reference plane. Connecting the successive incremental edges Δl_i results in the whole edge line.

10.2.3 A Theorem About the Edge Line of Free-Cutting [73]

See Fig. 10.6, set axis z in the direction of cutting velocity vector $\boldsymbol{V}$ (suppose that all the points of the edge line have the same cutting velocity), and unit vector in this direction is **k**, the reference plane P_r is perpendicular to $\boldsymbol{V}$. The unit vector in the direction of generatrix g of the rake face is **u** (suppose that the rake is a cylindrical surface).

The unit tangent vector to the edge line at a point A is **b**, and with respect to reference plane P_r: **b** going down, λ_s is positive, and going up, λ_s is negative, as shown in Fig. 10.6b. Therefore, we have

$$-\mathbf{b} \cdot \mathbf{k} = \sin \lambda_s; \tag{10.11}$$

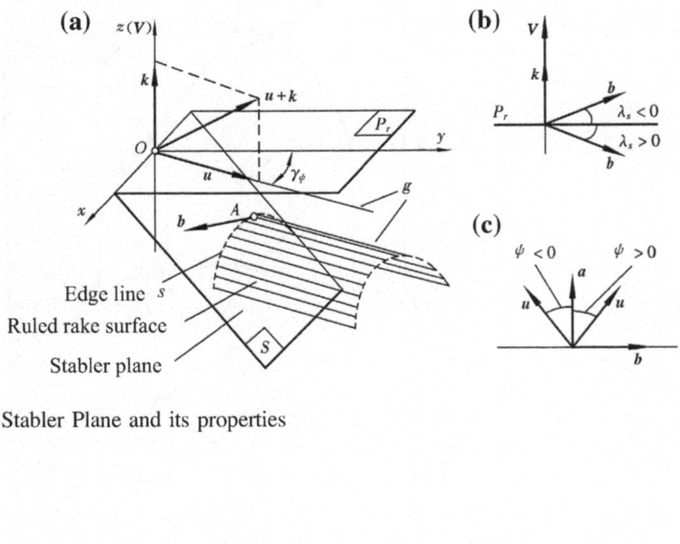

Fig. 10.6 Stabler Plane and its properties

With respect to **b**: if the unit vector **u** is in the same side of **a** as **b** is, then ψ is positive, otherwise, ψ is negative as shown in Fig. 10.6c. Therefore, we have

$$\mathbf{b} \cdot \mathbf{u} = \sin \psi. \tag{10.12}$$

The plane *S*, the normal of which is **u** + **k**, is call Stabler Plane. The following is a theorem about the plane.

Theorem *Any planar edge lines s on the Stabler Plane (say, the interrupted line in Fig. 10.6a) according to the Stabler Law, every part of the edge line ejects chips in the direction of* **u** *(that is, to realize free cutting).*

Proof Suppose there is a point *A* on the edge line *s*, the unit tangent **b** of the edge line at point *A* is on the Stabler Plane, and is perpendicular to **u** + **k**, so, $\mathbf{b} \cdot (\mathbf{u} + \mathbf{k}) = 0$, i.e., $\mathbf{b} \cdot \mathbf{u} = -\mathbf{b} \cdot \mathbf{k}$, substituting Eqs. (10.11) and (10.12) into the equation leads to $\lambda_s = \psi$. If Stabler law is satisfied at point *A*, i.e., $\lambda_s = \psi_\lambda$, this results in $\psi_\lambda = \psi$, and this means the segment of cutting edge at point *A* will ejects chips in the direction of **u**, however, point *A* is arbitrarily elected, and so, if every segment of the edge line *s* satisfys Stabler Law, then the whole edge will ejects chips in the direction of **u**, this is, the free cutting is realized.

□

The converse theorem is also true, i.e., If an edge line *s* is satisfying Stabler Law and ejection its chips in the direction **u** (that is, to realize free cutting), then *s* is surely a planar line on Stabler Plane.

10.2.4 Examples

The previous example (Fig. 10.1) is a particular case described by the theorem, readers may check it according to the theorem. Figure 10.5 described a more general case, but based on the theorem, it is not necessary to calculate the edge line piecewise. We can make Stabler Plane perpendicular to **u** + **k** at first, and then project the edge line *s*′ in the reference plane along the direction **k**, onto Stabler Plane, to get the real edge line *s*, and, at last, translating the generatrix *g* according to the edge line *s* (the guide line), so as to make the rack face, a ruled surface.

Let's see two more examples.

10.2.4.1 Example I

See Fig. 10.7, the generatrix *g* is parallel to the reference plane P_r (i.e., $\gamma_\psi = 0°$), Stabler Plane *S* tilted 45° with respect to the reference plane. Any edge lines *s* on the *S* plane, if Stabler Law is satisfied, all the elementary cutting tools on them will eject chips in the direction of *g* and perform free-cutting.

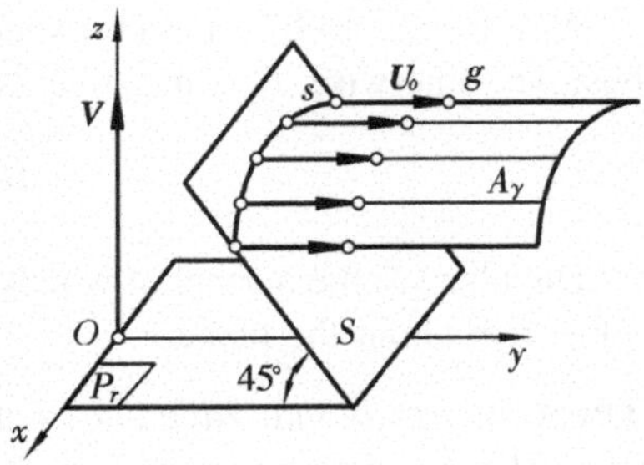

Fig. 10.7 An example of free-cutting tool design

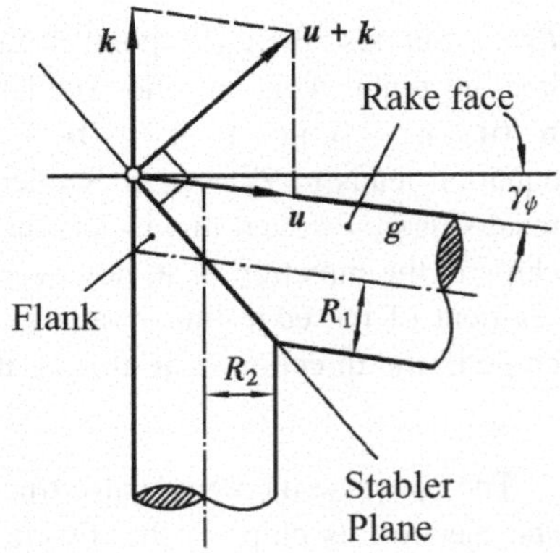

Fig. 10.8 Double cylinder shaped cutting tool

10.2.4.2 Example II

Double cylinder cutter, as shown in Fig. 10.8, has two cylindrical surfaces as its rake and clearance surfaces, respectively. The diameters of the two cylinders are the same: $R_1 = R_2$. Their intersection line is a planer line and on the Stabler Plane *S*. According to the theorem, if Stabler Law is satisfied, all the elementary cutting tools will eject their chips all in the direction of generatrix of the cylinder *g*, and thus the cutting process is free.

In the above example, the clearance is supposed to be 0°. In the case that both the rake and the clearance are neither zero, one of cylinders may be selected as cylindrical surface, and the intersection line of it with the Stabler Plane is taken as edge line, and the edge line translate along the axis of another cylinder surface so as to form another surface, which is no longer a cylindrical surface, but normally a elliptic cylinder.

The follows are two groups of experiments, verifying the effects of free cutting.

Experiments I: Transverse Feeding Turning

Cutting tool: HSS, double cylinder cutter, $R_1 = R_2 = 3.7$ mm, $\gamma_p = 0°$, $\alpha_p = 8°$, the rake cylinder and the clearance cylinder were formed by means of wire-electrode cutting processing. For comparison, flat rake cutters ($R_1 = \infty$, and the rest parameters no changes) were prepared. Work-piece: red copper bar, diameter 45 mm, rings with a width of 3.3 mm were precut on the periphery of the bar.

Table 10.3 Cutting force of double cylinder cutter with transverse feeding F_c(N)

Transverse feeding f (mm/r)	0.021	0.024	0.026	0.029	0.031
Flat rake cutting tool	679.5	693.4	776.6	845.9	859.8
Double cylinder cutting tool	499.9	513.1	610.2	693.4	776.6
Reduction in cutting force (%)	26.4	26.0	21.4	18.0	9.7

Table 10.4 Cutting force of double cylinder cutter with longitudinal turning F_c(N)

Longitudinal feeding f (mm/r)	0.1	0.2	0.3	0.4	0.5
Flat rake cutting tool	166	262	414	515	582
Double cylinder cutting tool	166	247	372	451	507
Reduction in cutting force (%)	0.0	5.7	10.1	12.4	12.9

$V = 27.6$ m/min. The cutting force will not be recorded until the circle edge totally cut in the surface of ring. The rest conditions were the same as previously described experiments. The results are listed in Table 10.3.

The data in the table show that without any changes in edge shape in the reference plane or in cutting conditions, only because of the change in rake design the cutting force reduced by 1/10 ~ 1/4, this is, of course, the effect of free-cutting.

Experiments II: Longitudinal Turning

Cutting tool: cemented carbide YT15, quadrilateral insert, grounded into a double cylinder cutting tool on a universal tool grinder in a special fixture. $R_1 = R_2 = 2\,\text{mm}, \gamma_p = 5°, \alpha_p = 8°, \kappa_r = \kappa_r' = 45°$. Besides, cutting tool with a flat rake face ($R_1 = \infty$) is also ground for comparison. Work-piece: mild steel #45, bar stork. $V = 83.7$ m/min, $a_p = 0.5$ mm (only the circle edge participates in cutting). The cutting force measured are listed in Table 10.4.

The data in the above table show that under the same conditions, the cutting forces of double cylinder cutter are lower than that of flat rake tools. And with the increase of longitudinal feed rate, the interference of chip-ejection for flat rake tool becoming more and more aggregative, and the cutting force reduction effect of double cylinder cutter becoming more obvious.

10.3 The Basic Law and the Constraint Conditions of Chip-Ejection Dredging Design for Free-Cutting Tools

10.3.1 The Basic Law

To summarize, the basic law of free-cutting tool design is, generally speaking, adjusting the normal vector $\mathbf{a_i}$ and the tangent vector $\mathbf{b_i}$ point by point of the edge line, to make the natural chip-ejection vectors from all the segments of the edge line form a rigid body motion; or, specifically, when the cutting velocities and the chip-ejection velocities of all the points of the cutting edge may regarded as constants, adjusting the edge inclination angles to let the natural chip-ejection from all the segments of the edge line be in the same direction.

10.3.2 The Development and Test of Fully Free-Cutting Carbide Inserts

According to the basic law described above, a fully free-cutting carbide inserts was designed and produced, cooperated with manufacturer as shown in Fig. 10.9. The material is YW3.

In cutting experiments, the chip-ejection directions were measured by means of a reading microscope, which checks the included angle ε between the friction mark and the generatrix of the rake cylinder. Every test repeated 3 times, and the average of the three measures were taken as the data. The results are listed in Table 10.5.

The data in above table shows that under various cutting conditions, the deviations of chip-ejection directions from the direction of generatrix are relatively small. This means that chips ejections are approximately in the same direction, i.e., the direction of generatrix, being independent of cutting parameters. Therefore, the developed inserts have really dredged chip-ejections and, to a large extent, eliminated chip-ejection interference. Besides, this result should be also significant for chip control.

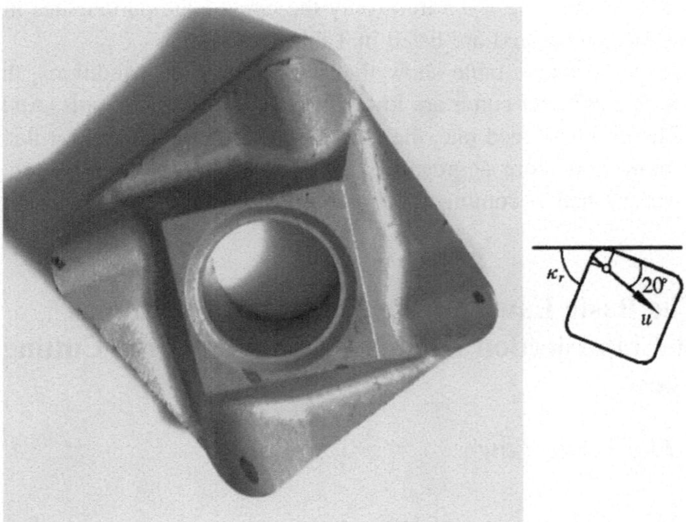

Fig. 10.9 First developed carbide insert (fully free-cutting)

Table 10.5 Chip-ejection directions measured on a fully free-cutting tool

a_p (mm)	f (mm/r)	V (m/min)	ε
1	0.1	80	5°20′
1	0.2	80	5°10′
1	0.3	80	5°30′
2	0.1	80	2°
2	0.2	80	1°50′
2	0.3	80	1°54′
3	0.1	80	56′
3	0.2	80	1°
3	0.3	80	48′
1	0.2	120	4°40′
2	0.2	120	2°
3	0.2	120	1°

However, the testing of the first developed inserts were not ideal: the cutting force of the testing inserts is about 10% higher than that of other contrastive inserts, and prone to vibration during its cutting process.

The further research shows that in dredging chip-ejections we have to abide by some constraint conditions, and trespassing the constraints, pursuing fully dredging chip-ejection interference is not the best scheme.

10.3.3 The Key Factor Controlling Cutting Ability of an Elementary Cutting Tool

To deeply analyze the reason why the tested insert did not realize the expected effects, let's see the distributions of geometric angles of the tested inserts. Figure 10.10a–c show the distributions of cutting edge inclination λ_s (absolute value), cutting edge normal rake γ_n and the rake angle in direction of chip-ejection γ_ψ, respectively. The abscissa in the figures is the central angle of the transition circular arc from major to minor cutting edges. Major and minor cutting edges are the straight lines tangential to the transition circular arc at the starting point and the end point, respectively. The angles on major and minor cutting edges are kept constants and take the value at central angle (abscissa) equal to 0° and 90°, respectively.

The figures shows that the rake angle in direction of chip-ejection γ_ψ on major, transition and minor cutting edges keeps unchanged (12°). As shown by Eq. (10.9), because the requirement of adjusting chip-ejection direction, and the farther the point from the highest point (central angle = 20°) is, the larger the correcting amount of chip-ejection angle φ is, and the larger the cutting edge inclination λ_s will be required. At the place of minor edge or close to minor edge, the edge inclination λ_s becomes very large, reaches 50° (negative).

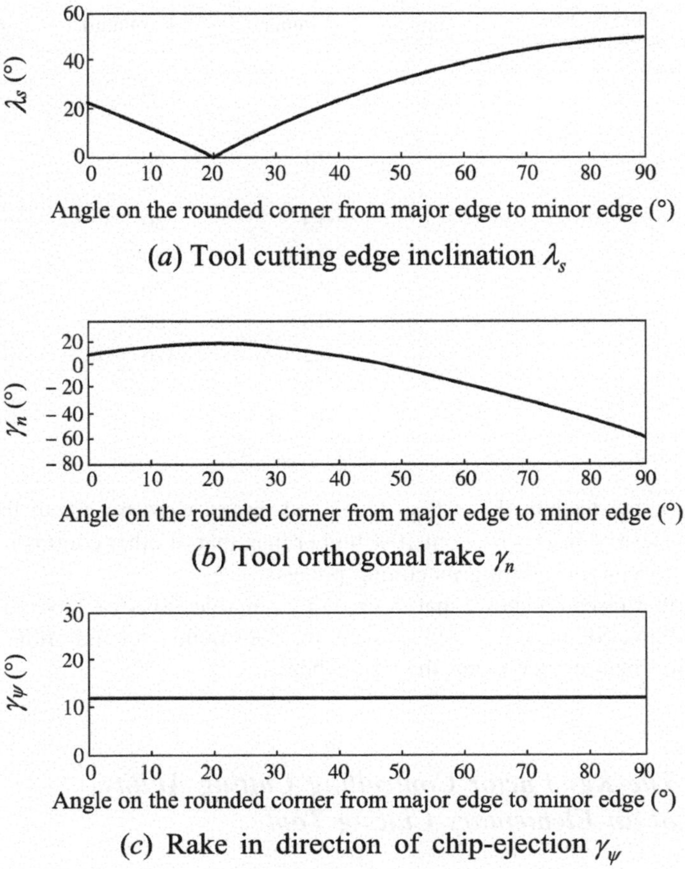

Fig. 10.10 Angle distribution of tested insert. **a** Edge inclination. **b** Orthogonal rake. **c** Rake in direction of chip-ejection

While the rake in the direction of chip-ejection γ_ψ keeps constant, the normal rake γ_n rapidly decreases as edge inclination λ_s (absolute value) increasing (Fig. 10.10b). At the place of minor edge or close to minor edge, γ_n reaches $-60°$, becoming very large negative rake.

Equation (10.13) and Fig. 10.11 show the relation of γ_n with λ_s (γ_ψ = const.). It can be seen that the normal rake γ_n rapidly decreases as edge inclination λ_s increasing.

$$\sin\gamma_n = \frac{\sin\gamma_\psi}{\cos^2\lambda_s} - \tan\lambda_s. \tag{10.13}$$

Fig. 10.11 The relation of γ_n with λ_s (γ_ψ = const.)

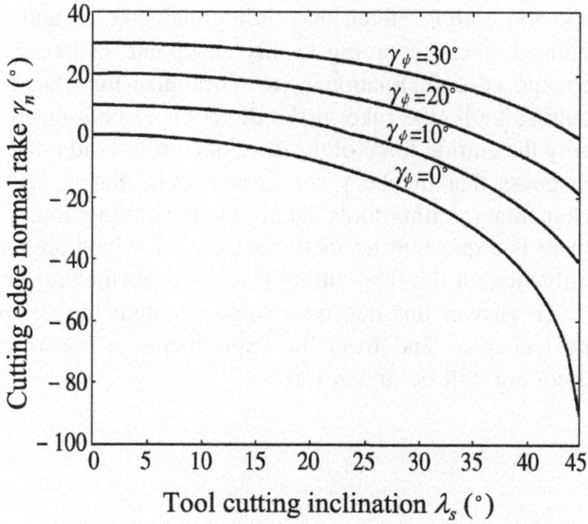

It is known that the relation of tool orthogonal rake γ_o and normal rake γ_n is,

$$\tan \gamma_o = \tan \gamma_n / \cos \lambda_s. \tag{10.14}$$

From the above two equations we have the relation of γ_o and λ_s as shown in Fig. 10.12.

The above two figures show that while the rake in the direction of chip-ejection γ_ψ keeps constant, the normal rake γ_n and the orthogonal rake γ_o rapidly decreases as edge inclination λ_s (absolute value) increasing. At first, they decrease slowly, and then they decrease rapidly.

It is generally acknowledged, that under oblique cutting, the rake in the direction of chip-ejection γ_ψ is the basic parameter which directly determine the cutting state

Fig. 10.12 The relation of γ_o with λ_s (γ_ψ = const.)

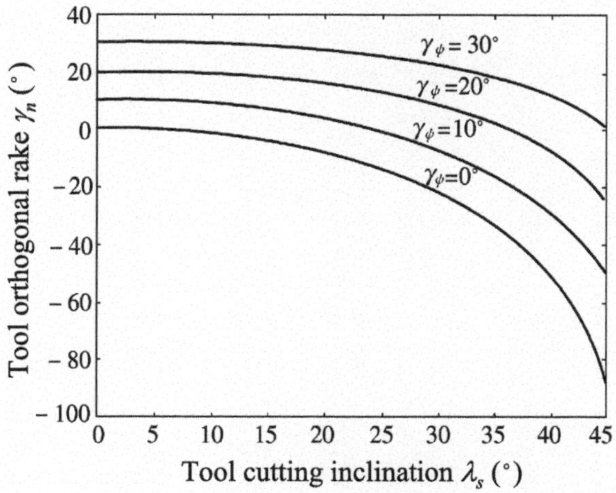

and the cutting effect, and orthogonal rake γ_o and normal rake γ_n only play an indirect role. According to this viewpoint, in free-cutting insert design, while we change edge inclination λ_s to harmonize the chip-ejection of all the elementary cutting tools, the rake in the direction of chip-ejection γ_ψ was kept constant, but why the cutting force of the developed insert did not decrease but increase? It's easy to guess that maybe γ_o or γ_n in action, that is, the negative γ_o or γ_n of part of elementary cutting tools deteriorate the cutting ability of the whole tool. Therefore, there is a question needs to be clarified: which angle, γ_ψ, γ_n or γ_o, has the highest influence on the free-cutting process of elementary cutting tools.

To answer this question, some distinguishing experiments will be done in the next chapter, and from the experiments, a constraint condition to chip-ejection dredging will be drawn forth.

Chapter 11
The Constraint Conditions for Free-Cutting Tool Design

In this Chapter, distinguishing experiments were performed, so as to verify which of the rakes: the rake in the direction of chip-ejection γ_{ψ}*, edge normal rake* γ_n *or orthogonal rake* γ_o *is the basic parameter determining cutting state and machining effect. From the experimental results, a constraint condition to the designing free-cutting tools will be drawn forth. Together with the basic law of free-cutting tool design presented in last chapter, they will lay foundation for the design and optimization of free-cutting tools. On the basis of these achievements, a new round development of free-cutting cemented carbide inserts was performed and their superiority will be experimentally verified.*

Under oblique cutting, edge normal plane, tool orthogonal plane and the plane of chip-ejection are on longer coincide with each other, then, the rake in which plane will have the most important influence on cutting performance of the cutter? This is still a very controversial issue [73, 76–79]. In designing free-cutting tools, for the sake of keeping the cutting performance of all the elementary cutting tools in a reasonable condition, while dredging chip-ejection vectors, we must have a clear understanding on such issue. In this chapter, we changed some rakes, but keeping cutting edge normal rake, tool orthogonal rakeor rake in the direction of chip-ejection unchanged, respectively, and measuring the cutting ratio and cutting force, as evaluation indicator, we studied which rack has key influence on the cutting performance of elementary cutting tools.

H. Shi, *Metal Cutting Theory*, Springer Series in Advanced Manufacturing,
https://doi.org/10.1007/978-3-319-73561-0_11

11.1 The Cutting Experiments Distinguishing Which Rake Plays Most Important Role in Influencing Cutting Tool's Performance

11.1.1 Experimental Conditions

Cutting method: Pipe end cutting, without fluid, tool cutting edge angle $\kappa_r = 90°$.

Work-piece: Hot-rolling seamless pipe, mild steel #20, diameter $D = 99.2$ mm, wall thickness $t = 2$ mm. For the measurement of cutting ratio a longitudinal groove of width $w = 2$ mm was precut on the work-piece.

Cutting Tools: material is HSS, the geometric parameters are listed in Table 11.1. The cutters No. 1–4 have the same edge normal rake γ_n, and edge inclination λ_s increases successively; The cutters No. 1, 5–8 have the same rake γ_ψ in direction of chip-ejection, but edge inclination λ_s increases successively (the edge normal rake γ_n and the tool orthogonal rake γ_o decrease rapidly and turned negative); The cutters No. 1, 9–11 have the same tool orthogonal rake γ_o, but edge inclination λ_s increases successively (rake in direction of chip-ejection γ_ψ increases rapidly, but the edge normal rake γ_n only has little change). Tool clearance in the direction of chip-ejection is 6° for all the experimental tools. In calculation, chip-ejection angles ψ_λ are evaluated according to Stabler Law.

Free-cutting: The largest edge inclination $\lambda_s = 38°$. Because of edge inclination, the cutting speed and the direction of cutting velocity differ from point to point on the cutting edge. But the difference is very limited: the largest variation in cutting speed along the working edge is only 4%, and largest variation in the direction of cutting velocity is only 1.78°. So, the cutting conditions can be approximately regarded as free-cutting (see Fig. 11.1).

Cutting speed: To cut mild steel with HSS tool, the normal cutting speed must be 20–40 m/min. This range of speed, however, is very prone to built-up edge, and

Table 11.1 The cutting tools used in experiments

Angles / No.	λ_s (°)	γ_n (°)	γ_o (°)	γ_ψ (°)	α_n (°)
1	0	14.0	14.0	14.0	6.0
2	10	14.0	14.2	15.4	6.1
3	20	14.0	14.9	19.3	6.4
4	30	14.0	16.1	25.6	6.9
5	10	12.6	11.8	14.0	6.1
6	20	8.1	8.6	14.0	6.4
7	30	−0.6	−0.7	14.0	6.9
8	38	−11.8	−16.0	14.0	7.6
9	10	13.8	14.0	15.2	6.1
10	20	13.2	14.0	18.6	6.4
11	30	12.2	14.0	24.1	6.9

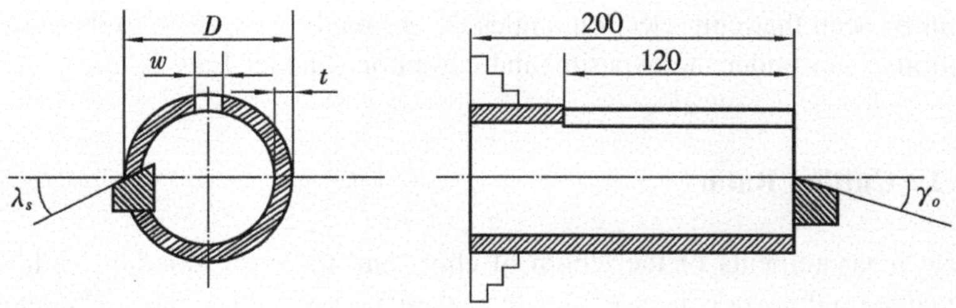

Fig. 11.1 The work-piece and the cutting method in the experiments of cutting performance of elementary cutting tools

to avoid the influence of built-up edge, in the experiments we increased the speed to 68.6 m/min (work-piece revolutions 220 r/min). High speed may shorten tool life, considering that the cutting time during experiments is relatively short, we still used such a high speed. Besides, using such a high speed it is convenient to measure chip-ejection direction from the friction marks on rake face.

Equipment and instrumentation: Universal tool and cutter grinder MQ6025A, fine turning lathe CA6140, equipped with continuously adjustable-speed motor, reading microscope 45×, tachometer HMZ10000, three-way strain dynamometer, calibrated with proving rings, dynamic resistance strain instrument YD-15, and strip chart recorder [73].

11.1.2 Experimental Results

11.1.2.1 Chip-Ejection Direction

Considering that for the cutters of No. 5–8 (having the same rake in direction of chip-ejection γ_ψ) the chip-ejection directions are determined according to Stabler Law, so, it is necessary to check the error caused by the law beforehand. For this purpose, after each cut with the cutter No. 5–8, the angles between edge and the friction marks (their projections on reference plane) were measured by a reading microscope, and, then, the real chip ejection angles on rake face were calculated, and these results are listed in Table 11.2.

Table 11.2 Chip-ejection angle ψ_λ, verification of Stabler Law

No.	*f* mm/r			
	0.028 (°)	0.05 (°)	0.15 (°)	0.2 (°)
5 (λ_s = 10°)	9.8	10.3	10.6	10.5
6 (λ_s = 20°)	20.8	20.6	20.9	20.5
7 (λ_s = 30°)	30.9	30.7	30.6	30.9
8 (λ_s = 38°)	38.3	38.5	38.6	38.5

It can be seen that chip-ejection angles ψ_λ are really very close to the tool edge inclination λ_s, so under the experimental condition, Stabler Law is true.

11.1.2.2 Cutting Ratio

Accurate measurements of the length of chips and the corresponding path of the cutting edge on the work-piece enable the calculation of cutting ratio. The length of a chip is measured with a soft thread along the middle of the chip, and the path of the cutting edge is also measured along the middle diameter of the work-piece, which is 303.36 mm according to Fig. 11.1. Under a stable cutting condition, 10 chips were collected and their lengths were measured, the average of the measures was taken as the chip length. Each cut repeated for three times, the average of the three results was taken as the final result. The results of cutting ratio are listed in Appendix 11.1.

Figures 11.2, 11.3 and 11.4 give the variations of cutting ratio versus the variation of edge inclination, while the tool orthogonal rake, cutting edge normal rake and the rake in direction of chip-ejection is kept unchanged, respectively.

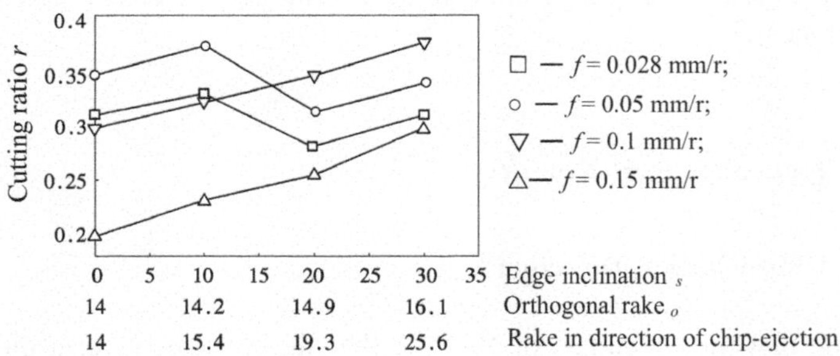

Fig. 11.2 r versus λ_s ($\gamma_n = 14°$)

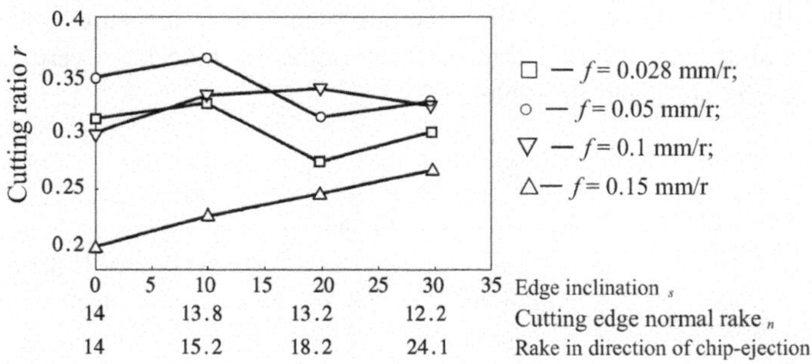

Fig. 11.3 r versus λ_s ($\gamma_o = 14°$)

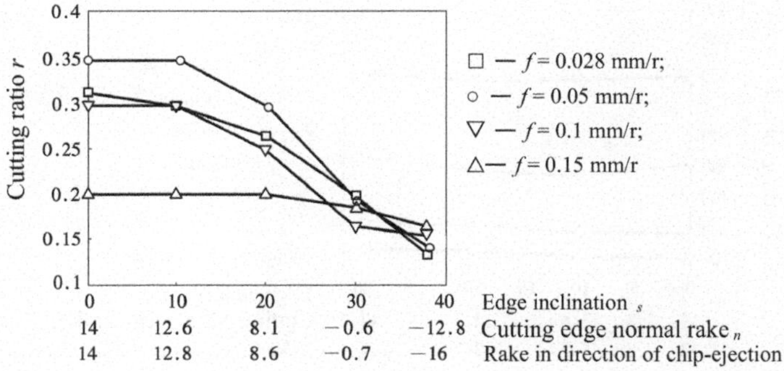

Fig. 11.4 r versus λ_s ($\gamma_\psi = 14°$)

11.1.2.3 Cutting Force

The three cutting force components, cutting force F_c, feed force F_f and back force F_p, were all measured. Three measurements were made for each cut, and their average was taken as the final result. All the results of cutting force measurements are given in Appendix 11.2.

Figures 11.5, 11.6 and 11.7 give the variations of cutting force versus the variation in edge inclination, while the tool orthogonal rake, cutting edge normal rake and the rake in the direction of chip-ejection is kept unchanged, respectively.

In Figs. 11.2, 11.3, 11.4, 11.5, 11.6 and 11.7 under the abscissa axis besides the value of edge inclination, another two rows of rake values, which are varying with edge inclination, also listed, for the convenience of analyzing the influence of the rakes on cutting performance of the cutters.

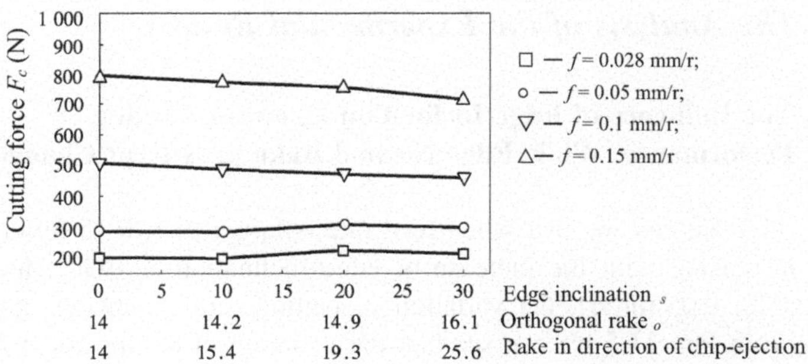

Fig. 11.5 F_c versus λ_s ($\gamma_n = 14°$)

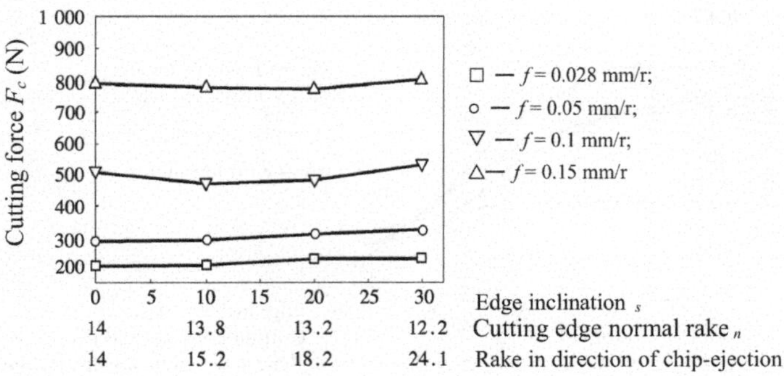

Fig. 11.6 F_c versus λ_s ($\gamma_o = 14°$)

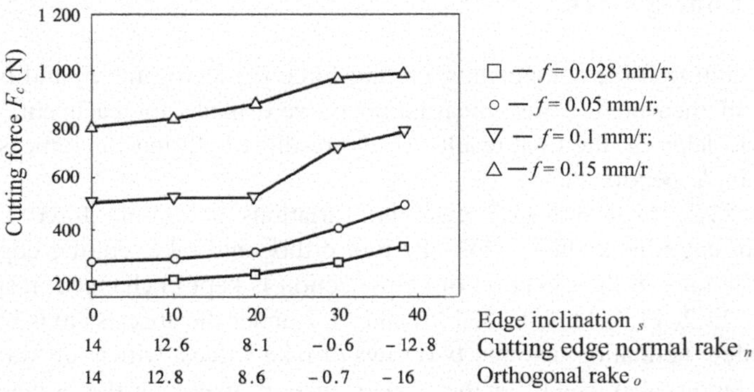

Fig. 11.7 F_c versus λ_s ($\gamma_\psi = 14°$)

11.1.3 The Analysis of the Experimental Results

11.1.3.1 The Influence of Edge Inclination λ_s on the Cutting Performance, While Edge Normal Rake γ_n Is Kept Changed

From Fig. 11.2 we can see that when feed rate is larger (0.1, 0.15 mm/r), cutting ratio r is increasing with the increase of edge inclination λ_s; when feed rate is smaller (0.028, 0.05 mm/r), the variation of cutting ratio fluctuates, no obvious tendency. From Fig. 11.5 we can see that the variation of cutting force F_c is very small while edge inclination λ_s is increasing.

The experimental results about cutting ratio r and cutting force F_c show that while edge normal rake γ_n is kept unchanging the cutting performance has not changed much and has improved slightly.

11.1.3.2 The Influence of Edge Inclination λ_s on the Cutting Performance, While Tool Orthogonal Rake γ_o Is Kept Unchanged

From Fig. 11.3 we can see that the variation of cutting ratio r has no obvious tendency, similar to the situation of keeping edge normal rake γ_n unchanged. This is mainly because in the test range, while tool orthogonal rake γ_o is kept unchanged, the variation in edge normal rake γ_n not more than 2°. From Fig. 11.6 we can see that the variation of cutting force F_c is very small while edge inclination λ_s is increasing.

Thus it can be seen, while tool orthogonal rake γ_o is kept unchanged, the increase in edge inclination λ_s has a little influence on the cutting performance of the elementary cutting tools.

11.1.3.3 The Influence of Edge Inclination λ_s on the Cutting Performance, While the Rake γ_ψ in Direction of Chip-Ejection Is Kept Unchanged

While the rake in direction of chip-ejection γ_ψ is kept unchanged, the increase of edge inclination λ_s causes monotonic decrease in cutting ratio r (Fig. 11.4) and the monotonic increase in cutting force F_c (Fig. 11.7). This shows that the cutting process is deteriorating. The season is that while the rake in direction of chip-ejection γ_ψ is kept unchanged, the increase of edge inclination λ_s causes decreases in the edge normal rake γ_n and tool orthogonal rake γ_o, and increase in the intensity of chip deformation. At first, the deterioration is slow, and when edge inclination λ_s is larger than 20°, the deterioration becomes sharp, cutting ratio r falls fast (Fig. 11.4) and cutting force F_c increase rapidly (Fig. 11.7).

Therefore, under the situation of keeping the rake γ_ψ in direction of chip-ejection unchanged, increase of edge inclination λ_s within limits will has no serious influence on cutting performance of the cutters, and this lefts space and possibilities for dredging chip-ejection by means of adjusting edge inclination λ_s; but edge inclination λ_s cannot be too large, otherwise, the cutting performance of the cutter will deteriorate rapidly, this has set a constraint condition for chip-ejection dredging.

The experimental results not only answered the question: why the cutting performance of the cemented carbide inserts we developed for a full chip-ejection harmonization in the first round did not satisfactory, and also clarified the doubtful point: which rake has the key influence on the cutting performance of the cutting tool, the rake in direction of chip-ejection γ_ψ, or the edge normal rake γ_n (or, tool orthogonal rake γ_o)? The answer is the latter.

To sum up, the two points the free-cutting tool design should abide by:

Basic law: by means of selecting proper distribution of cutting edge inclination λ_s, so as to coordinate the chip-ejection vectors from all the elementary cutting tools and to reduce chip-ejection interference as far as possible.

Constraint condition: all the edge normal rake γ_n, or, the tool orthogonal rake γ_o must within reasonable scope, so as to prevent cutting performance from remarkable deterioration.

The above two points are contradictory. The task of free-cutting tool design is to dredge chip-ejections furthest within the scope tolerated by the above constraint condition.

11.2 The Development of Free-Cutting Cemented Carbide Insert in the Second Round

According to the basic law and the constraint condition above described, the second round of development of free-cutting inserts was conducted.

11.2.1 The Design of Free-Cutting Cemented Carbide Insert

The newly developed cemented carbide insert is presented in Fig. 11.8.

Fig. 11.8 Free-cutting insert (second round developed)

As stated before, to fully coordinate chip-ejections and to make them being in the same direction will leads to very small edge normal rake γ_n, or even very large negative edge normal rake γ_n, or, tool orthogonal rake γ_o, as shown in Figs. 10.10, 10.11 and 10.12. And the experimental results of elemental cutting tools in this chapter have shown that this situation will certainly leads to deterioration of the cutter's cutting performance. That is, full coordination of the chip-ejection has to be at the cost of cutting performance, and, thus, it is not an ideal result.

In the second round development, we changed our strategy, no longer simply pursue a full chip-ejection coordination but try to make a balance between chip-ejection coordination and cutting performance, on the premise of all the elementary cutting tools having a reasonable cutting performance, to eliminate chip-ejection interference as far as possible. The distributions of cutting angles along major-chamfered corner-minor edges are shown in Fig. 11.9. This figure may be compared with Fig. 10.10. Besides, we reduced the radius r of chamfered corner edge, from 2.4 to 1.2 mm.

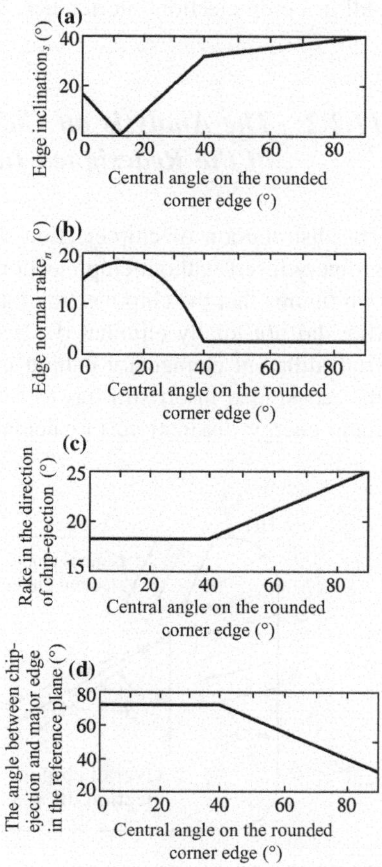

Fig. 11.9 The angle distributions of the insert (second round developed)

The orientation of the rake face can be determined by edge inclination λ_s (Fig. 11.9a) and edge normal rake γ_n (Fig. 11.9b), and according to Stabler Law, natural chip-ejection direction and the rake angle in that direction can be determined (Fig. 11.9c).

In design, we consciously controlled the distribution of edge normal rake γ_n, as shown in Fig. 11.9b. From major edge, to chamfered corner edge, to minor edge, when edge normal rake γ_n goes down to a certain degree (corresponding to the point of central angle = 40°), began to let γ_n keep constant, without further decrease. For this purpose, first, we control the momentum of the growth in edge inclination λ_s (Fig. 11.9a), and second, we raise the rake in direction of chip-ejection γ_ψ as shown in Fig. 11.9c. This means that we relaxed restrictions on chip-ejection direction, as shown in Fig. 11.9d. The ordinate of the figure, i.e., the angle η between chip-ejection and major edge in the reference plane may be approximately regarded as the complement angle of chip ejection angle $\psi_{\lambda o}$. And in design the set value of η is 77° (see Fig. 11.9d). As shown in the figure, the angle η on the part of the chamfered corner edge and the minor edge, the central angle of which is larger than 40° will differ from the set value. This means that the redesigned free-cutting insert still has chip-ejection interference, but has been greatly reduced.

11.2.2 The Analysis on Chip-Ejection Compromise of the Redesigned Insert

The distrubution of chip-ejection direction is shown in Fig. 11.10a, b is that of ordinary insert without chip-ejection dredging. It is clear from a comperison of the two figures that the chip-ejection interference or conflict have been greatly relieved (but did not totally eliminated). Because of the scatter of chip-ejection directions from different elementary cutting toots, the real overall chip-ejection direction of the redesigned insert still has to be calculated according to the principle of minimum energy dissipation. In calculation we ignoring the chip-ejection interference

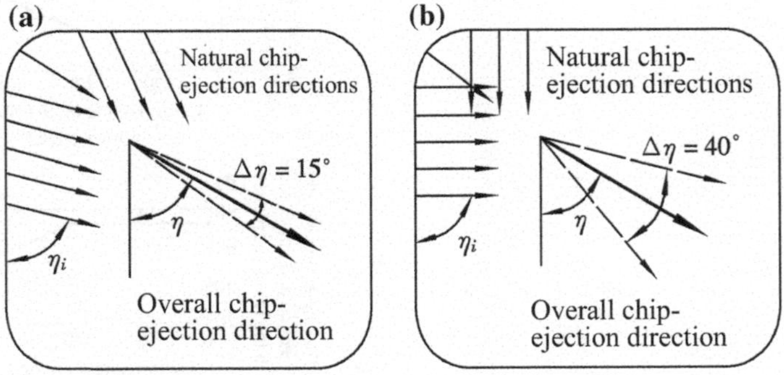

Fig. 11.10 The distributions of chip-ejection directions. **a** Free-cutting insert. **b** Ordinary insert

caused by the difference in chip-ejection speed and only consider the chip-ejection interference caused by the difference in chip-ejection direction.

Based on Eq. (7.29), we have the formula for calculating cutting force of i-th elementary cutting tools,

$$F_{ci} = (145.35 + 0.07925\,\theta^2)l_i$$

where, θ is the angle of deviation of the real chip-ejection direction from natural chip-ejection direction, in Fig. 11.10, is the angle ($\eta - \eta_i$), l_i is the edge length of the i-th elementary cutting tool.

The exponent of the thickness of cut a_c in the formula of cutting force is evaluated as 0.61 (it is only used for comparison, the error of this exponent would not change the tendency of the calculating results). So, the formula for calculating cutting force of i-th elementary cutting tools can be expressed as,

$$F_{ci} = \left[145.35 + 0.07925(\eta - \eta_i)^2\right]\left(\frac{a_{ci}}{0.07}\right)^{0.61} l_i,$$

Or,

$$F_{ci} = 736 + 0.4(\eta - \eta_i)^2 a_{ci}^{0.61} l_i, \tag{11.1}$$

And the total cutting force is,

$$F_c = \sum_{i=1}^{n} [736 + 0.4(\eta - \eta_i)^2] a_{ci}^{0.61} l_i. \tag{11.2}$$

where, n is the number of elementary cutting tools, η is the angle between the projections of overall chip-ejection direction and the major edge onto reference plane, and η_i is the angle between the projections of the natural chip-ejection direction of i-th elementary cutting tool and the major edge onto reference plane (as shown in Fig. 11.10). Angle η describes the overall chip-ejection direction, it is a state parameter, and should be determined by the principle of minimum energy dissipation. For this purpose, let

$$\frac{\partial F_c}{\partial \eta} = 0, \tag{11.3}$$

Substituting Eq. 11.2 into above Equation results in

$$\eta = \frac{\sum a_{ci}^{0.61} l_i \eta_i}{\sum a_{ci}^{0.61} l_i}. \tag{11.4}$$

Denoting

$$\frac{a_{ci}^{0.61} l_i}{\sum a_{ci}^{0.61} l_i} = \mu_i, \tag{11.5}$$

the weighted coefficients, leads to

$$\eta = \sum_{i=1}^{n} \mu_i \eta_i. \tag{11.6}$$

It means that the angle η related to overall chip-ejection is equal to the weighted average of the same angles η_i related to all the elementary cutting tools, and the weights relate to the edge length l_i and the thickness of cut a_{ci} of all the elementary cutting tools.

The suitable ranges of depth of cut and the range of feed rate for the designed free-cutting insert are 0.5–4 mm and 0.1–0.51 mm/r, respectively. In the ranges according to Eq. 11.6 calculated overall chip-ejection angles η are listed in Table 11.3, and the same angles η for the ordinary insert without chip-ejection dredging are also listed in the same table for a comparison.

From the above table we can see that, for the designed insert (with chip-ejection dredging) the variation of overall chip-ejection direction $\Delta\eta$ under different cutting conditions is 16.4°, and for the ordinary insert (without chip-ejection dredging), the variation is 43.6°, this means that after chip-ejection dredging design the influence of cutting condition on the chip-ejection is much smaller than that without chip-ejection dredging design, and this is certainly in favor of chip control.

The coefficient of non-free cutting Φ (see Sect. 6.7.1) can be calculated according to

$$\Phi = \frac{\sum_{i=1}^{n} [736 + 0.4(\eta - \eta_i)^2] a_{ci}^{0.61} l_i}{\sum_{i=1}^{n} 736 a_{ci}^{0.61} l_i}. \tag{11.7}$$

The values of Φ for the inserts with and without chip-ejection dredging, calculated according to above equation, are also listed in Table 11.3. It is clear that

Table 11.3 Overall chip-ejection angle (in reference plane) η and the coefficient of non-free cutting Φ

a_p (mm)	f (mm/r)	η (°)		Φ	
		Redesigned	Ordinary	Redesigned	Ordinary
4.0	0.10	74.4	82.3	1.03	1.15
3.0	0.20	73.0	79.2	1.05	1.21
2.0	0.30	70.8	73.2	1.08	1.30
1.0	0.41	65.8	57.2	1.13	1.38
1.0	0.51	64.3	55.3	1.15	1.42
0.5	0.51	58.0	38.7	1.17	1.27

after chip-ejection dredging design, the coefficient of non-free cutting Φ is obviously smaller than that of the ordinary insert (without chip-ejection dredging design). Furthermore, when feed rate become larger and the chip-ejection interference becomes more serious, the effect of chip-ejection dredging design is getting more obvious.

11.2.3 The Cutting Experiments of Redesigned Cemented Carbide Inserts

Cutting experiments were conducted with both of the redesigned free-cutting insert and a comparative insert provided by manufacture, so as to investigate the advantage of cutting performance of the redesigned insert over the comparative insert in the aspects of chip-ejection interference, cutting force, chip breaking performance and tool life. The material quality and manufacturing technique are the same for both inserts, only the groove profiles and the edge line shapes are different. The groove profile of the comparative insert is shows in Fig. 11.11.

The experimental conditions and the equipment, instrument are as follows:

Fine turning lathe CA6140, equipped with continuously adjustable-speed motor, bar stock of steel #45 for work-piece material, mechanical clamping tool shank, three-way strain dynamometer, dynamic resistance strain instrument YD-15, and strip chart recorder, reading microscope 45×, tachometer HMZ10000. The experiments are conducted under cutting speed 80 m/min.

From the experiments we can see that the redesigned free-cutting inserts can perform normal cutting under depth of cut a_p = 4 mm and feed rate f = 0.3, 0.41, 0.51, and they have the potential of elongating cutting edge and increasing depth of cut.

11.2.3.1 The Experiments of Cutting Force

The results of the experiments are listed in Table 11.4. Every experiment repeated twice, the data in the table are the averages of the twice experiments. Because of the power limitation, the experiments with combination of big depth of cut and large feed rate did not conduct.

From the data in the above table, the average reduction of the cutting force generated by free-cutting insert is 12.3% lower than that of comparative insert. This

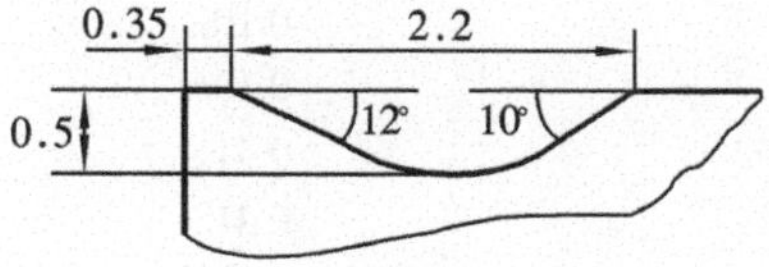

Fig. 11.11 The groove profile of the insert for comparison

Table 11.4 Cutting force F_c, the results of experiments

Feed rate f (mm/r)			0.1	0.15	0.2	0.3	0.41
Depth of cut a_p (mm)	1	Comparative insert	322.6	403.1	515.7	724.9	881.8
		Free-cutting insert	262.3	362.9	415.4	636.4	797.4
		Variation (%)	−19	−10	−13	−14	−10
	2	Comparative insert	596.2	757.1	966.3	1 288.3	1 630.9
		Free-cutting insert	507.7	700.8	877.8	1 114.7	1 409.3
		Variation (%)	−15	−7	−9	−14	−14
	3	Comparative insert	861.7	1 174.1	1 448.1		
		Free-cutting insert	761.2	1 014.2	1 288.3		
		Variation (%)	−12	−14	−11		

is, obviously, due to the rational allocation of the cutting angles along the edge on the redesigned free-cutting edge, and the chip-ejection interference get dredged, the cutting work becomes light-footed and the energy consumption is thus reduced.

11.2.3.2 The Experiments of Chip-Ejection Direction

The overall chip-ejection angle (in reference plane) is measured by a reading microscope, which reads the friction marks on the rake and measure the angle between the projections of the mark and the major edge. The set value of η is 77°. The experimental results are listed in Table 11.5.

From the variation of the angle η shown in the table we can see that in the range of cutting parameters used in the experiments, the variation of η for free-cutting insert is only 12°; and that for the comparative insert is 36°. This means that the free-cutting design not only have the function of restricting the scatter of chip-ejection directions of all the elementary cutting tools and reduce the

Table 11.5 The experimental results of chip-ejection angle

Depth of cut (mm)	Feed rate (mm/r)	Chip-ejection angle (in reference plane) η (°)	
		Comparative insert	Free-cutting insert
3	0.10	78	76
2	0.10	70	76
2	0.41	57	72
1	0.41	42	64

interference across them, but also have the effect of restricting the scatter of chip-ejection directions under different cutting conditions, reducing the influence of cutting parameters on the overall chip-ejection direction.

11.2.3.3 The Experiments of Chip Breaking

Based on the chips we collected, the chip breaking area chart is drawn as shown in Fig. 11.12. The full line *a* is for free-cutting insert, and the interrupted line *b* is for comparative insert (the right and upper borders in the chart are the limits of the parameters used in cutting experiments).

From Fig. 11.12, we can see that chip breaking range of free-cutting insert is wider than that of comparative insert. And as we observed in experimental process, that the free-cutting insert can steadily keep chip breaking effect for quite a long time; the cutting process of comparative insert, however, was frequently forced to be stopped due to the "snarl" chips winding round the work-piece.

Possibly, the reason why free-cutting insert has a better chip breaking effect is that the cross section of its chips is upward convex and such chips are easier to be broken.

Manufacture conducted more elaborate chip breaking test for our designed inserts, the results is giving in Fig. 11.13.

The digits in the figure are form ISO standard of chip types. The first digit shows the chip's forms, the second digit shows the length, the density and the continuity of the chip type. For example, 62 means discontinuous curved chips, 52 means short cone shaped helical chips, 42 means short washer shaped helical chips and 22 means short pipe shaped helical chip, etc.

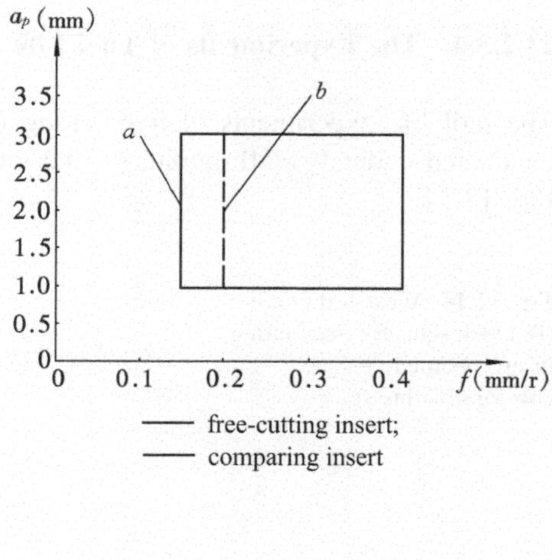

Fig. 11.12 Chip breaking area. **a** Free-cutting insert. **b** Comparative insert

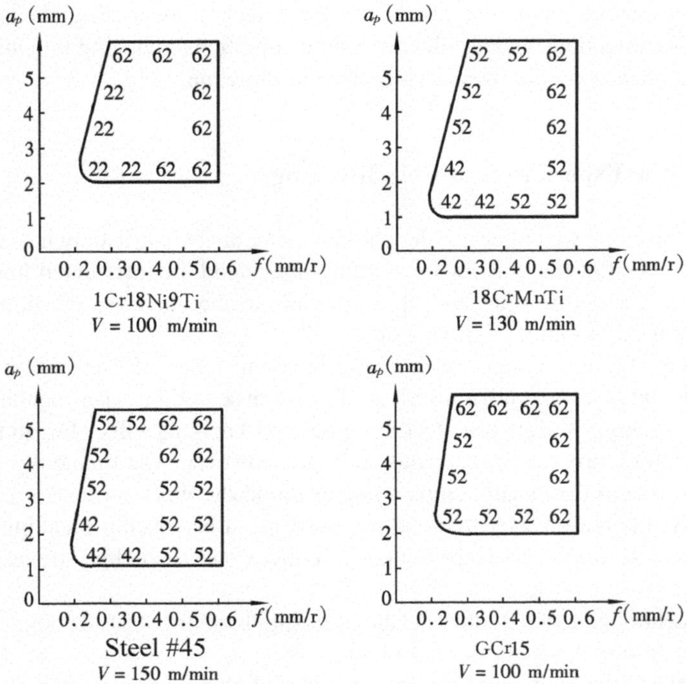

Fig. 11.13 The test results on chip control conducted by manufacture

Figure 11.13 shows that developed free-cutting insert is able to perform favorable chip control in a large scope of work-piece materials and cutting parameters.

11.2.3.4 The Experiments of Tool Life

The tool life experiments of free-cutting insert and the comparative insert were conducted under $V = 80$ m/min, $f = 0.2$ mm/r, $a_p = 2.5$ mm, results are shown in Fig. 11.14.

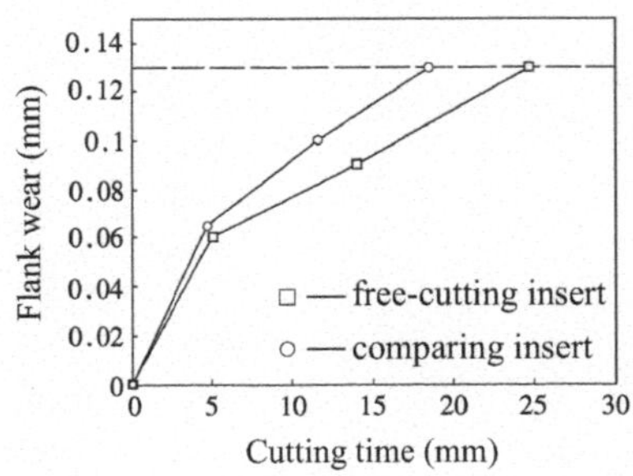

Fig. 11.14 Wear test curve (I). Open square—free-cutting insert, open circle—comparative insert

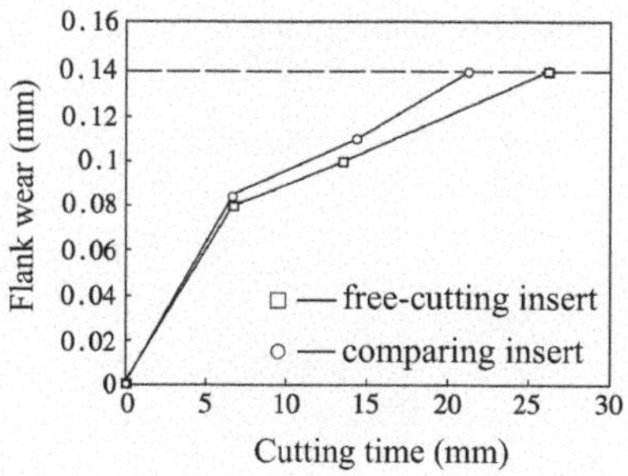

Fig. 11.15 Wear test curve (II). Open square—free-cutting insert, open circle—comparative insert

When flank wear is 0.13 mm, the tool life of designed free-cutting insert is 24.54 min, and that of comparative insert is 18.58 min, the former is 32% higher than the latter.

Tool life experiments of free-cutting insert and the comparative insert were conducted under $V = 120$ m/min, $f = 0.2$ mm/r, $a_p = 2$ mm. The results are in Fig. 11.15.

When flank wear is 0.14 mm, the tool life of designed free-cutting insert is 26.46 min, and that of comparative insert is 21.55 min, the former is 22.8% higher than the latter.

The reason why the redesigned free-cutting insert has a higher tool life is, first, the free-cutting design has reduced the chip deformation and the energy dissipation, and the heat generation is thus reduced, and cutting temperature decreased; second, the corner of free-cutting insert is of convex shape, unlike ordinary tools, the corner of them are concave shaped groove. Therefore, the point of free-cutting insert is stronger, and heat dissipation is quicker, so, wear is slower; and third, free-cutting insert has a longer cutting edge, and thus distributed cutting load.

To sum up, first, the designed free-cutting insert reduced cutting force, and has a smaller cutting power consumption; second, the free-cutting insert has a stable chip-ejection direction, and less affected by the cutting parameters, so, it has a wider and more stable range of chip breaking, and the third, the free-cutting insert has a longer tool life.

The research and development on the free-cutting inserts is at the elementary stage. However, the achievements, as stated above, reveals that the research on "free-cutting method" is an issue worth paying attention.

Appendix 11.1 The Results of Cutting Ratio Measurement

The purpose, method and conditions of the experiments are stated in Sect. 11.1.2. The values of measured cutting ratio are listed in Table 11.6.

Table 11.6 Cutting ratio r

No.	Cutting angle		Feed rate f (mm/r)			
	λ_s (°)	γ_n (°)	0.028	0.05	0.1	0.15
1	0	14	0.31	0.346	0.297	0.198
2	10	14	0.33	0.373	0.322	0.231
3	20	14	0.28	0.313	0.346	0.254
4	30	14	0.31	0.34	0.376	0.297
5	10	12.6	0.297	0.346	0.297	0.198
6	20	8.1	0.264	0.297	0.247	0.198
7	30	−0.6	0.198	0.194	0.165	0.185
8	38	−12.8	0.135	0.142	0.155	0.165
9	10	13.8	0.323	0.363	0.33	0.224
10	20	13.2	0.27	0.31	0.336	0.244
11	30	12.2	0.297	0.323	0.32	0.264

Appendix 11.2 The Results of Cutting Force Measurement

The purpose, method and conditions of the experiments are stated in Sect. 11.1.2.

The results of cutting force experiments, feed force F_f, back force F_p and cutting force F_c are listed in Tables 11.7, 11.8 and 11.9, respectively.

Table 11.7 Feed force F_f(N)

No.	Cutting angle		Feed rate f (mm/r)			
	λ_s (°)	γ_n (°)	0.028	0.05	0.1	0.15
1	0	14	149.30	212.62	313.18	559.00
2	10	14	138.13	186.55	305.73	536.65
3	20	14	145.58	194.00	298.28	506.85
4	30	14	130.68	179.10	261.04	417.35
5	10	12.6	149.30	201.45	335.53	581.34
6	20	8.1	179.10	234.97	462.16	626.04
7	30	−0.6	208.89	410.02	611.14	797.37
8	38	−12.8	387.67	491.96	722.88	906.41
9	10	13.8	144.09	199.21	320.63	562.72
10	20	13.2	160.48	223.79	320.63	521.75
11	30	12.2	167.92	223.79	320.63	555.27

Table 11.8 Back force F_p(N)

No.	Cutting angle		Feed rate f (mm/r)			
	λ_s (°)	γ_n (°)	0.028	0.05	0.10	0.15
1	0	14	0.00	0.00	0.00	0.00
2	10	14	0.00	−2.44	−7.43	−12.41
3	20	14	−7.43	−17.40	−32.36	−40.34
4	30	14	−17.40	−37.45	−62.28	−96.19
5	10	12.6	0.00	−2.44	−7.43	−12.41
6	20	8.1	−7.43	−17.40	−42.32	−62.28
7	30	−0.6	−27.37	−37.45	−77.24	−127.11
8	38	−12.8	−44.33	−65.27	−127.11	−147.05
9	10	13.8	0.0	−2.44	−7.43	−12.41
10	20	13.2	−7.43	−17.40	−34.36	−40.34
11	30	12.2	−27.37	−37.45	−67.27	−102.17

Table 11.9 Cutting force F_c(N)

No.	Cutting angle		Feed rate f (mm/r)			
	λ_s (°)	γ_n (°)	0.028	0.05	0.1	0.15
1	0	14	201.96	282.42	507.70	789.31
2	10	14	197.93	278.39	483.65	765.17
3	20	14	218.05	298.50	467.47	749.08
4	30	14	210.00	290.46	451.38	708.85
5	10	12.6	214.02	294.48	523.79	813.44
6	20	8.1	232.53	314.60	564.02	869.77
7	30	−0.6	280.40	403.10	708.85	966.32
8	38	−12.8	338.74	491.61	765.17	979.91
9	10	13.8	201.96	282.42	467.47	773.21
10	20	13.2	222.07	302.53	475.52	765.17
11	30	12.2	222.07	314.60	523.79	797.35

Part III
Non-Linear Machine Tool Dynamics

Chapter 12
A Summary of the Theory on Machine Tool Chatter

In this Chapter, the basic conceptions about dynamic systems and the outline of the linear theory of machine tool chatter are briefly introduced as the foundation necessary for readers to understand the author's research achievements of non-linear chatter theory, which will be explained in next chapters.

During machining process, vibration may be generated, that is, besides nominal cutting motion, there is also a periodical motion superposed between cutter and work-piece. Cutting vibration may lead to a series of undesirable effects, and sometimes, may cause serious results. First of all, the relative motion between the cutter and the work-piece will deteriorate the surface quality of machined parts, affect the performance of the parts, and shorten their life. Secondly, while vibrating, the real instantaneous thickness of cut may fluctuate about the set nominal thickness of cut. When vibration getting severe, the cutting edge may, sometimes, jump out of the work-piece material, the thickness of cut reduces to zero; and sometimes the cutting edge may deeply prick into the work-piece, the instantaneous thickness of cut may increase multifold. Besides, vibration also cause periodical change in the working rake and working clearance. All these factors leads to a periodical alternating dynamic cutting force, the amplitude of it is even bigger than the static cutting force. This alternating dynamic cutting force alternating makes the working portion of the tool fatigued, causes tool tip breaking, and speeds up the wear of the machine tool parts, makes the connections between parts loose, and the machine tool thus loses its precision. In addition, severe vibration also generates big noise, pollutes the environment and hazards operator's health. The vibration has frequently become one of the main barriers, which limits the productivity.

In recent years, because of the development of machinery industry, cutting vibration control problems become more acute and urgent. On one hand, due to the development of precision machining and ultra-precision machining technology, precision machining achieved micron level, or nanoscale, even minimal vibration during cutting process will make the size or surface roughness of the processed parts out of tolerance, becoming waste products. On the other hand, new cutting

H. Shi, *Metal Cutting Theory*, Springer Series in Advanced Manufacturing,
https://doi.org/10.1007/978-3-319-73561-0_12

tool materials, such as ceramic, diamond and CBN, in increasingly widely applied in machining, these materials with high hardness, heat resistance is good, capable of cutting with high velocity, improving productivity. However, they tend to be brittle, easily chipping, especially sensitive to vibration. Therefore, when using this type of tool material, vibrations must be strictly controlled. Meanwhile, the metal cutting and grinding are developing towards high velocity and heavy-cut direction, makes the possibility of vibration increases. Difficult to machine materials, such as titanium alloys, have been increasingly widely used in industry, they are easy to provoke vibrations during machining. In addition, the development of mechanical structure and machine parts towards light-weight trend and thin-wall type is prone to vibration in cutting process.

Therefore, it is necessary to study the mechanism of vibration and the methods of preventing vibration in cutting process, so as to strictly control the vibrations.

Vibration in metal-cutting process is not just about harm, if suitably applied, we may turn bane into a boon, achieving an unexpected effect, and benefit. In recent years there are a lot of research in this area, such as vibration cutting for machining hard-to cut materials; cutting vibrations caused surface ripples are used as oil storage space for precise matching surface to slow the wear and tear of parts, and in metallic fibers production in order to meet the needs of manufacturing composite materials, vibration cutting also used, and so on. For these purposes, also requires deep understanding and strict control of vibration in cutting processes.

The vibration generated in metal cutting process may be categorized as two types: forced vibration and self-excited vibration. Forced vibration is a kind of vibration excited by periodic vibration source, inner machine or afferent from outside. The generating mechanism, the phenomena and the law of forced vibration are elaborated in ordinary books about mechanical vibration. This book, however, is mainly devoted to the self-excited vibration in metal cutting process, i.e., chatter.

As early as in 1940s, the chatter theory of metal cutting had already come out, thereafter it made substantial development. Traditional chatter theory, however, has been limited to linear theory. Linear theory successfully predicted and explained the critical condition for chatter to happen, i.e., the stability threshold, but cannot predict and explain the various phenomena after the chatter happens, such as the root of chatter amplitude stability, the relation of stabilized amplitude and cutting parameters, jump phenomenon and the influence of external disturbance on the stability of cutting process. These phenomena are, however, of great significance for industrial production and the development of disciplines.

The development of contemporary era in automatic machining system and artificial intelligence manufacturing presents an urgent requirement for the on-line monitoring and real-time control of the machine tool chatter, and provides a new possibility for the realization of this goal. On-line monitoring and real-time control of the machine tool chatter is the most dynamic development direction. The study links surveillance, analysis, forecasting and control to an organic whole, changing traditional offline testing, modeling to online monitoring, and online monitoring to real-time control. This research route and direction help to make traditional machine tool chatter theory and knowledge better serve industrial production, thus adding new vitality to the discipline.

12.1 The Features and the Mechanism of Self-excited Vibration

Self-excited vibration (see [79], Chaps. 11, 13) is mostly caused by parasitic feedback in the system. It is much more difficult to reveal its mechanism and to control this kind of vibration than of forced vibration. The reason is that forced vibration should has one or more “excitation sources”, as long as we trace back to the sources, and delete these sources, or blocking the path from the sources to the cutting area, it is not difficult to suppress the vibration.

However, the self-excited vibration has no external alternating excitation source, and the vibration is caused by internal causes, namely internal feedback. It is not easy to identify and separate these internal feedbacks, which parasitize in the normal operation of a machine or device.

For readers to correctly understand the follow-up contents, here, it is necessary to briefly introduce some basic conceptions related to self-excited vibration (i.e., chatter) [79].

12.1.1 Phase Plane, Equilibrium Points, Limit Cycles and Their Stability

12.1.1.1 Phase Plane

Self-excited vibration may be intuitively expressed in phase space. For single-degree-of-freedom systems, phase space is the phase plane containing two axes: vibration displacement $x = x_1$ and vibration velocity $\dot{x} = x_2$, as shown in

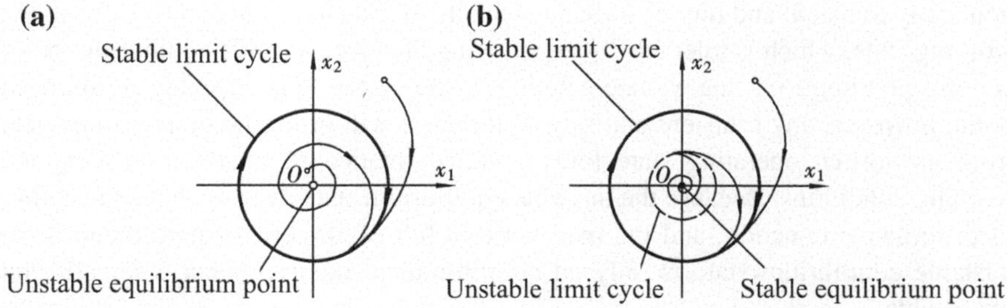

Fig. 12.1 Trajectory, trajectory portrait, equilibrium point and limit cycle. **a** Unstable equilibrium point + stable limit cycle. **b** Stable equilibrium point + unstable limit cycle + stable limit cycle

Fig. 12.1. Every point on phase plane is called "state point". Each state point represents a combination of particular vibration displacement and particular vibration velocity. While the system vibrating, both the displacement and the velocity are all continuously changing, and the state point makes a curve on the phase plane, called trajectory; the whole of trajectories forms trajectory portrait. The trajectory portrait reflects the overall dynamic behavior of the system.

The equilibrium point and the limit cycle on the phase plane represents a stationary state and a periodic motion, respectively, which have practical significance, and are worth studying deeply.

12.1.1.2 Equilibrium Point and Its Stability

When a dynamic system is in equilibrium state, its velocity ($\dot{x} = \dot{x}_1 = x_2$) and acceleration ($\ddot{x} = \dot{x}_2$) are both equal to zero. The points on the phase plane satisfying this condition are called equilibrium points. Obviously, a system located at one of such points cannot move by itself, because it has neither initial velocity nor imposed force (acceleration is zero).

The equilibrium points are divided into two categories: stable and unstable. The difference is not in the states themselves of the equilibrium, but is the movement trend that the system slightly deviates from its equilibrium point: is the tendency to return to the equilibrium point, stay near the equilibrium point or tend to drift away from the equilibrium point, accordingly, the equilibrium point is called asymptotically stable, merely stable or unstable, respectively.

We must understand the connections and differences of the two conceptions, the "equilibrium" and the "stability". An equilibrium point is not necessarily stable, however, stability, must refer to the behavior around a certain equilibrium point. These conceptions are of great concern in understanding the stability of a metal cutting system. Generally speaking, if a cutting system can working stably, its working point should be an equilibrium point, but this is not enough, we should say, it must be working at a stable equilibrium point. This is because a real system will be certainly disturbed by various perturbances, and only a stable equilibrium point has the ability of disturbance rejection, thus is able to keep the system steadily working around the equilibrium point. For a system working at a stable equilibrium point, any transient and tiny disturbance will only cause transient and tiny change in working state, which is tolerable by engineering practice. A stable equilibrium point is corresponding to a steady-state solution of the system. For unstable equilibrium point, however, any transient and tiny disturbance will bring the system to deviate from its normal operating state forever and dramatically, totally destroying the working conditions. Because the unstable equilibrium state does not have the ability of disturbance rejection, and the real world is full of all sorts of perturbations, so unstable equilibrium state is only an equilibrium in theory, and it's actually not observable.

The second type of equilibrium point, i.e., merely stable equilibrium point is an intermediate state, which is corresponding a stable vibration.

12.1.1.3 Limit Cycle and Its Stability

As we know [79], because of conservation of mechanical energy, the vibration trajectory of a conservative system must be a closed curve. Because of the energy exchange with the outside world, metal cutting system is a non-conservative system. In the phase portrait of a non-conservative system there also has a kind of closed trajectory, as shown by full thick line and interrupted line in Fig. 12.1a, b. Differing from the conservative system, the closed trajectory of such non-conservative systems possess the following features:

First, when operating around a closed trajectory, the total mechanical energy of a system is not conserved, it absorbs energy and dissipates energy, and the total mechanical energy is constantly changing. But after a cycle, the energy absorption and energy dissipation must be balanced, the system return to the original state and a new cycle of operation begins.

Second, closed trajectory of a non-conservative system possesses a kind of "attraction force", which can attract the trajectories starting from the other points (not on the closed trajectory) to the closed trajectory. The closed trajectories expressed by thick line in Fig. 12.1a, b possess such property. Therefore, the closed trajectory of a non-conservative system represents a periodical motion independent of initial conditions. Such a periodical motion is determined by feature of the system itself, and initial condition change (Fig. 12.1a), or initial condition change in a certain range (Fig. 12.1b) do not change the periodic motion the system has finally approached. As is generally known, closed trajectory of a conservative system is determined by initial conditions, and the trajectories starting from different initial conditions do not intersect each other.

Closed trajectories of a non-conservative system, possessing above described characteristics, are called "limit cycle". A limit cycle corresponds to a periodic solution of the system. Besides steady-state solution (equilibrium point), periodic solution (limit cycle) is also an important mode of motion. So, the study on limit cycles is of important engineering significance.

Limit cycles may be categorized as two types: stable and unstable. The limit cycles expressed by thick line in Fig. 12.1a, b are stable, and the limit cycles expressed by interrupted line in Fig. 12.1b are unstable. Generally speaking, for a closed trajectory c, if any of its nearby trajectories are kept nearby, it is said that c is orbitally stable; if the trajectories in the vicinity of c, go to c as $t \rightarrow \infty$, then c is asymptotically stable; if the nearby trajectories tend to get away from c, then it says c is orbitally unstable.

Sable limit cycles have orbital stability, it possesses the ability of disturbance rejection. While the system running along a stable limit cycle, even if an outside disturbance makes it temporally departure from the limit cycle, it will eventually goes back to the cycle, or kept nearby, and not to produce too much deviation. The unstable limit cycle, however, is only a theoretical possibility, due to the inevitable disturbance in the actual environment, therefore, it is impossible to observe the movement along an unstable limit cycle in practice.

In Fig. 12.1, two typical configurations of limit cycles and equilibrium points are given. In figure *a*, the origin is an unstable equilibrium point, and the limit cycle that surrounds the origin is stable; in figure *b*, the origin is an stable equilibrium point, and the limit loop that surrounds the origin is unstable (interrupted line), the outer layer is a stable limit loop (solid line). In the case of figure *b*, the origin for the small disturbance (within the dotted ring) is stable, while the disturbance is large (over the dotted ring) it is unstable. This origin is called "finite amplitude instability".

The combination of equilibrium point and limit cycles can be used to describe the process and characteristics of self-excited vibration. In the case shown in Fig. 12.1a, the system could not rest in the center of the equilibrium point, any tiny initial disturbance will provoke the rising vibrations, and finally the vibration is constrained on the stable limit cycle. In the case shown in Fig. 12.1b, if the disturbance does not exceed the threshold specified by the dotted ring, the system can be stabilized at its central equilibrium point; Once the disturbance crosses this threshold, it immediately stirs up the increasing self-excited vibration, and finally, vibration is stabilized in the outer (stable) limit cycle expressed by solid line.

12.1.2 The Features of Self-excited Vibration

Free vibration is excited by certain initial conditions [79], forced vibration is caused by some continuous process excitation. So, does "self-excited vibration" do not need to be excited by the outside world and is able to start automatically? As an example, let's see Fig. 12.1a, if the state of the system stays at the origin, it is impossible to move away from that point, because the origin is equilibrium point, and it has been already stated previously, a system in equilibrium point is impossible to leave the point by itself. However, if any accidental external disturbance makes the system deviate slightly from the equilibrium point (even a little bit), due to the instability nature of the near vicinity of the equilibrium point, system state will be immediately forced along a planar spiral quickly deviate from that point, which produces a sharp rise in vibration.

Therefore, the occurrence of self-excited vibration requires two conditions: first, the instability of the system near the equilibrium point, and second, the external disturbance which forces the operating point of the system to slightly deviate from the equilibrium point. Thus, self-excited vibration is impossible without external excitation. In fact, it should be said that the self-excitation is a "perturbation-excited vibration" because it always initiates in a random external disturbance. The second condition is always satisfied simply because the size and form of the disturbance do not matter, and the real world is always full of all sorts of perturbations. So, as long as the first condition once formed, namely the instability of the system itself around the equilibrium point once formed, is completely without waiting for, there will be some random disturbance come immediately, causing a sharp rising vibration.

Because of this, the study of self-excited vibration is different from the study of free vibration and forced vibration: don't need to study the form and amplitude of the occasional disturbances which triggering the vibration, also need not discuss the relationship between the disturbance and the vibration it aroused by, but focus on the mechanism and the laws of forming the instability of system itself.

Figure 12.1b presents also a self-excited vibration system, however, unlike the above talked one, it requires that the disturbance must have a certain size to arouse the self-rising vibration, namely that the amplitude must exceed the dashed circle in the figure. This situation is called "hard self-excited vibration", where the radius of the dotted circle is called the "threshold" of the excitation, and the situation shown in Fig. 12.1a is called "soft self-excited vibration".

Once excited, the amplitude of the self-excited vibration increases rapidly, which is the opposite of free vibration. In the latter case, the amplitude is attenuated due to the presence of the damping. This indicates that there is a factor which is opposite to damping in the self-excited vibration system, which is known as "negative damping". Negative damping is actually a kind of "help", which not only does not hinder the vibration, but also promotes the vibration. In addition to negative damping, the coupling between various modes of the multi-degree-of-freedom system can also be a contributing factor to the vibration [79], and we will talk about this later on.

The maintenance and expansion of self-excited vibration does not require constant and alternating excitation. This distinguishes it from the forced vibration. The self-excited vibration is maintained and expanded by the interaction between the various parts of the system. Therefore, there must be some kind of feedback relationship in the self-excited vibration system to realize the interaction between the various parts of the system.

After the self-excited vibration being inspired, the amplitude increase is of course impossible to be infinite. In fact, when the amplitude rises to a certain extent, it stabilizes itself and forms a stable periodic oscillation. It is important to note that the form and period of such periodic vibration are determined by the characteristics of the system itself, it has nothing to do with the initial value that excited the vibration, as shown in Fig. 12.1. This once again shows that the exact form of the disturbance that stirs the self-excitation vibration is not important.

The trajectory and the period of the steady-state self-excited vibration have nothing to do with the initial conditions of the motion, this is an important difference between self-excited vibration and undammed free vibration. We know that in the latter case, the trajectory of the movement or its period (if the system is non-linear) is determined by the initial conditions. The amplitude auto-stability of self-excited vibration is due to some nonlinear factors in the system. To describe and explain this phenomenon, non-linear model and non-linear theory are required. However, if only the condition of starting to have self-excited vibration, namely the "stability threshold" need to be studied, a linear model can be used to solve the problem.

Finally, also indicate, self-excited vibrations are not "self-sustaining vibrations", it is not "self-sufficiency" on energy, and must rely on the external energy supply, to complement the system damping energy dissipation, and enlarge the amplitude.

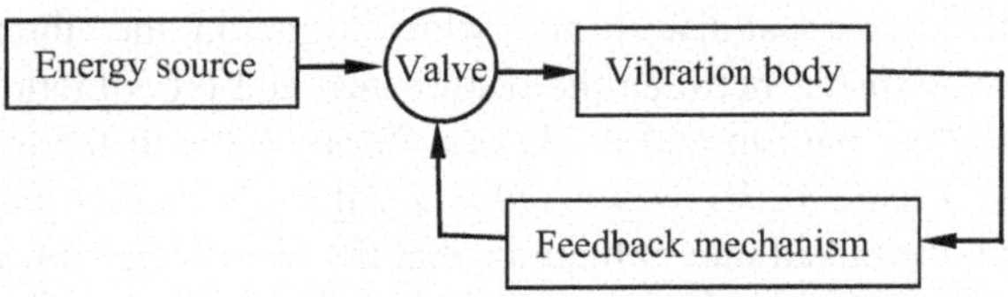

Fig. 12.2 The general organization of self-excited vibration system

Thus, the self-excited vibration system must have some mechanism to absorb energy from the outside to maintain or expand its vibration.

Free vibration and forced vibration also requires energy, the former is converting the initially stored mechanical energy in the system into alternative vibration, and while the latter is converting alternating excitation energy into alternative vibration. The self-excitation vibration is different from the above two, it is to convert a kind of dc energy into the alternating vibration. In this sense, the self-excited vibration system is much like a "converter" that converts direct current into alternating current.

Above all, a self-excited vibration system generally should be consists of the four parts: vibrating body, energy source, the "valve" adjusting the energy supply, and the feedback mechanism which according to the vibration motion of the vibration body to control energy supply, as shown in Fig. 12.2.

In metal cutting process, because of different feedback quantity and different feedback methods, we can have different types of chatter and different ways of losing stability, all types of self-excited vibration in cutting process will be described as follows.

12.2 The Cutting Chatter Caused by Cutting Speed Feedback

Considering a single degree of freedom vibration system [79], the excitation force is controlled by its own vibration velocity $\dot{x}$, and the force is the function of vibration velocity $F = F(\dot{x})$. This system is called the velocity feedback system, and its block diagram is shown in Fig. 12.3. The system consists of two parts: the vibration body and the force control mechanism, which is the feedback mechanism.

12.2.1 The Mechanism of Speed Feedback in Cutting Process

Figure 12.4a is a schematic for cylindrical turning, machine tool structure is simplified as a mass m, a spring k and a damper c of a single-degree-of-freedom system, V is the cutting velocity of the tool relative to the work-piece, F_c is the cutting force

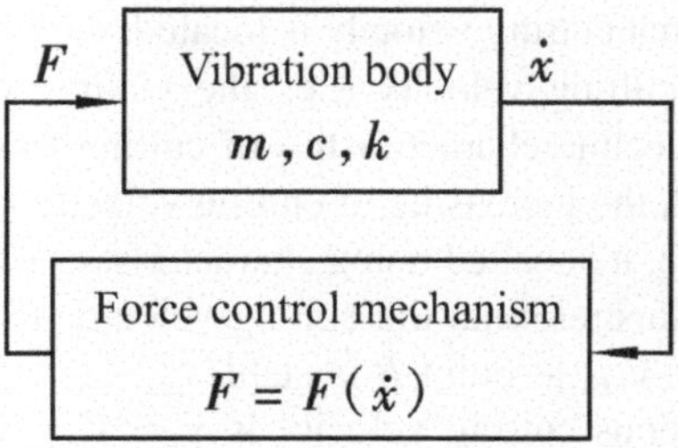

Fig. 12.3 Speed feedback system

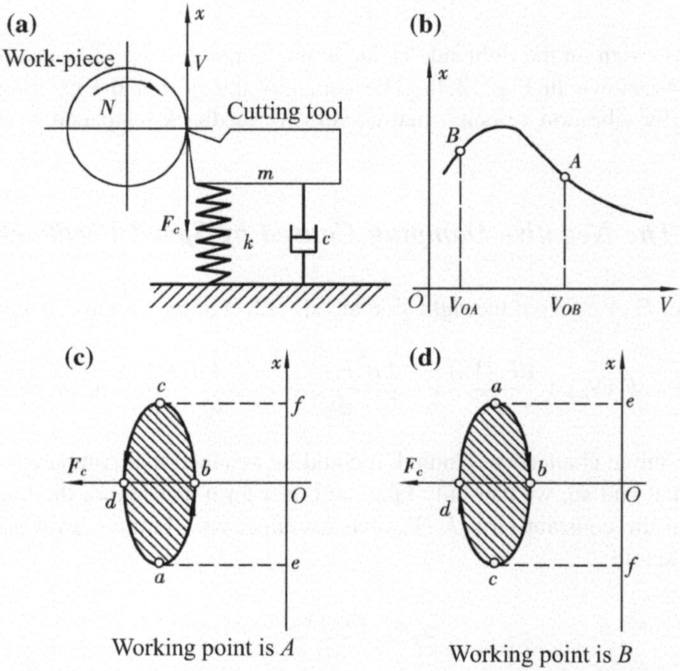

Fig. 12.4 Vibration caused by speed feedback mechanism during cylindrical turning. **a** Sketch of turning system. **b** $V - F_c$ curve. **c** Working point is at the decline characteristic region of cutting force (unstable). **d** Working point is at the rising characteristic region of cutting force (stable)

imposed on cutting tool. In general, $F_c = f(V)$, i.e., F_c is the function of F_c, the curve of F_c versus V is shown in Fig. 12.4b.

In the ideal case, the V is constant and the F_c is constant, and the compression of the spring (that is, the elastic deformation of the machine tool structure) remains unchanged. Thus, the cutting process is carried out smoothly and the cutting system is at its equilibrium point. But the stability of this equilibrium depends on the location of the system working point. Specifically, it depends on which part of the curve $V \sim F_c$ the nominal cutting velocity is located: the down part of cutting force with the increasing of cutting velocity (i.e., the vicinity of point A as shown in Fig. 12.4b, it is called decline characteristics of cutting force for short), or the rise part of cutting force with the increasing of cutting velocity (i.e., the vicinity of point B as shown in Fig. 12.4b, it is called rising characteristics of cutting force for short).

The following will illustrate that the cutting velocity V_{OA} at point A is unstable and the cutting velocity V_{OB} at point B is stable.

The actual instantaneous cutting velocity is $V = V_o + \dot{x}$, where, V_o is nominal cutting velocity and $\dot{x}$ is the vibration velocity superposed on V_o. The equation of motion for the system shown in Fig. 12.4a is

$$m\ddot{x} + c\dot{x} + kx = -F_c(V_o + \dot{x}) \tag{12.1}$$

The negative sign on the right side is due to the increase of F_c will cause x and $\dot{x}$ to decrease, as shown in Fig. 12.4a. The equation shows that the exciting force is controlled by vibration velocity, that is, velocity feedback happened.

12.2.2 The Negative Damping Caused by Speed Feedback

Developing $F_c(V_o + \dot{x})$ at the right side of Eq. (12.1) in the vicinity of V_o results in

$$F_c(V) = F_c(V_o) + \frac{dF_c(V_o)}{dV}\dot{x} + \frac{1}{2}\frac{d^2F_c(V_o)}{dV^2}\dot{x}^2 + \frac{1}{6}\frac{d^3F_c(V_o)}{dV^3}\dot{x}^3 + \cdots \tag{12.2}$$

If only the initial chatter was studied, it could be assumed that compared with V_o, $\dot{x}$ is very small and so, we can only take the linear term and ignore the higher order terms. And the constant term $F_c(V_o)$ has no effect on vibration, so it also can be ignored. Denote

$$c' = \frac{dF_c(V_o)}{dV}, \tag{12.3}$$

Equation (12.2) may be simplified as

$$F_c(V) \approx c'\dot{x}. \tag{12.4}$$

Substituting it into Eq. (12.1) and transposing term result in

$$m\ddot{x} + (c + c')\dot{x} + kx = 0. \tag{12.5}$$

This is the free vibration equation of a single-degree-of-freedom system, but unlike ordinary vibration system, the system damping is composed of two parts: a part is the damping of vibration system itself (machine tool structure), this is general damping, is a hindrance to vibration, that is, the damping coefficient c is positively, and the damping is called "positive damping"; The other part is the equivalent damping due to velocity feedback, and its damping coefficient c' can be positive or negative, depending on the location of V_o on $V \sim F_c$ curve: if V_o is located in the portion of rising characteristics of cutting force, as point B in Fig. 12.4b, then from Eq. (12.3), we know that $c' = dF_c(V_o)/dV > 0$, and the total damping $c + c' > 0$, i.e., it is positive damping; if V_o is located in the portion of decline characteristics of cutting force, as point A in Fig. 12.4b, then from Eq. (12.3), we know that $c' = dF_c(V_o)/dV < 0$, then, equivalent damping c' is negative. When the negative damping is large enough to overcome the positive damping of the system structure itself, the total damping can be negative, namely $c + c' < 0$. As stated earlier, negative damping is actually a kind of "help", which does not hinder the vibration of the system, but also increases the vibration of the system. Let

$$\omega_o^2 = k/m, \xi = (c + c')/(2\omega_o m), \tag{12.6}$$

Equation (12.5) becomes

$$\ddot{x} + 2\xi\omega_o\dot{x} + \omega_o^2 x = 0, \tag{12.7}$$

where, ξ is the total damping ratio, which may be positive, negative or zero. The general solution of above equation is

$$x(t) = Ae^{-\xi\omega_o t}\cos(\omega_d t - \psi), \tag{12.8}$$

where, $\omega_d = \omega_o\sqrt{1 - \xi^2}$. The vibration course corresponding to the three cases: $\xi > 0$, $\xi = 0$ and $\xi < 0$ are shown in Fig. 12.5a–c. The three cases are called stable, critical and unstable, respectively.

A system with negative damping is not stable, because it is in a state of "explosive", as long as there is any disturbance causing even a very small initial vibration (displacement or velocity), it will be expanding due to negative damping aid, and developed into a strong self-excited vibration. This kind of self-excited vibration is called a chatter caused by the decline characteristic of the cutting force.

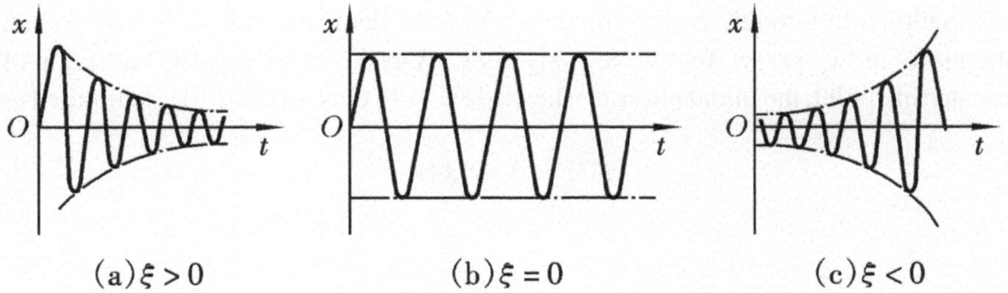

Fig. 12.5 Vibration process with different damping ratio

12.2.3 *The Energy Mechanism of Self-excited Vibration*

The mechanism of energy input and energy consumption of self-excited vibration system are now analyzed.

12.2.3.1 The Energy Mechanism of Cutting Process

When the machine tool works in the decline area of the cutting force, the cutting force will do the net positive work to the vibration system, i.e., cutting force inputs energy to the vibration system. And when the machine tool works in the rising area of the cutting force, the cutting force does a net negative work on the vibration system, that is, the energy of the vibration system is dissipated by the cutting process. To illustrate the issue, try to compare Fig. 12.4c, d, the two loops in the figures show the relationship between vibration displacement and the cutting force under the two conditions, respectively. First, see figure *c*, in the half cycle of x increase, the relationship between F_c and x is shown in curve *abc*, and in the half period of x decrease, the relationship between F_c and x is shown in the curve *cda*. The difference between the two case is caused by the difference in velocity: in the first half of the cycle, $\dot{x} > 0$, $V = V_o + \dot{x}$ is relatively larger, and cutting force F_c is relatively smaller; and in the second half of the cycle, $\dot{x} < 0$, $V = V_o + \dot{x}$ is relatively smaller, and cutting force F_c is relatively larger. Therefore, the relationship between $F_c \sim x$ is divided into two branches, and the whole cycle becomes a loop. Since the direction of the cutting force always points to the negative direction of x (see Fig. 12.4a), the cutting force is doing negative work in the first half period, and the amount of negative work is given by the area of the graph *abcfoea*; In the second half period, the cutting force does positive work, and the positive work is given by the area of the graph *cdaeofc*. Obviously, the positive work in quantity is greater than negative work, and the area of the closed curve *abcda* gives the net positive work done by the cutting force in a vibrating cycle, namely the energy inputting into the vibration system.

From the same analysis on Fig. 12.4d, the closed curve *abcda* shows the net negative work done by the cutting force in a period, namely the energy dissipation off in vibration system.

Now, let's quantitatively estimate input of the vibration energy or the energy dissipation off through cutting process. Assume damping rate ξ is very small, therefore, in Eq. (12.8) $Ae^{-\xi\omega_o t} \approx 1$, $\omega_d \approx \omega_o$. And through the proper selection of the starting point, the initial phase can be made $\psi = 0$, thus, Eq. (12.8) is simplified as

$$x(t) = A \cos \omega_o t. \tag{12.9}$$

The work done by the cutting force in a vibrating cycle is

$$W_p = -\oint F_c(V)\mathrm{d}x. \tag{12.10}$$

Substituting Eqs. (12.4) and (12.9) into (12.10) results in

$$W_p = -c'A^2\omega_o \int_0^{2\pi} \sin^2 \omega_o t\mathrm{d}(\omega_o t)$$

After integration, we have

$$W_p = -c'\pi\omega_o A^2. \tag{12.11}$$

Obviously, when the system is working in the area of the cutting force decline, due to Eq. (12.3), $c' < 0$, and so $W_p > 0$, thus, the cutting process is doing positive work and inputting energy to the system; On the other hand, when working in the area of cutting force rising, $c' > 0$ and $W_p < 0$, the cutting process is doing negative work, and the vibration energy dissipates through the cutting process.

12.2.3.2 The Energy Dissipated by the Internal Damping of the Machine Tool Structure

The vibration energy dissipated by the internal damping of the machine tool structure in a vibration cycle is

$$W_d = \oint c\dot{x}\mathrm{d}x = \oint c\dot{x}^2\mathrm{d}t.$$

Substituting Eq. (12.9) into above equation and integrating leads to

$$W_d = -c\pi\omega_o A^2. \tag{12.12}$$

Because there's $c > 0$, so $W_d < 0$. This indicates that the damping force of the machine structure can only do negative work to the vibration system and dissipate the vibration energy. When $W_p + W_d > 0$, the cutting properties of the cutting force can destabilize the cutting process and cause chatter.

12.2.4 Energy Balance and Vibration Amplitude Stability

The amplitude stabilization and stabilized amplitude of the system are discussed from the energy balance. From Eq. (12.12) we know the relationship between the

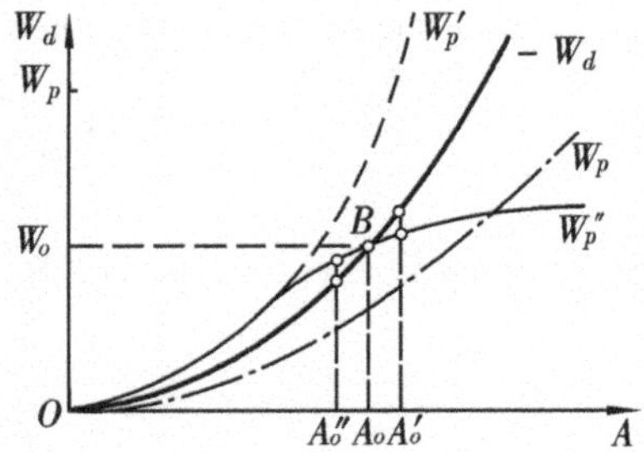

Fig. 12.6 Energy balance and amplitude stability

energy dissipated by system's damping in a vibrating period $-W_d$ and the amplitude A, as shown by thick solid lines in Fig. 12.6; The positive work W_p done by cutting force as expressed in Eq. (12.11) is also shown by a dot dash line in Fig. 12.6. In the case of, there is $-W_d > W_p$ under any amplitude A, and so it is impossible to have chatter. The above curve W_p is drawn by a certain value of c', suppose that $|c'|$ takes a larger value and gets another curve W_p' as shown by interrupted line in Fig. 12.6. At this case, because always $W_p' > -W_d$ for any amplitude A, so there will be a self-excited vibration, and its amplitude will expand continuously.

The three curves $-W_d$, W_p and W_p' are all parabolic curves, they are tangent with each other at the origin and there is no any other intersecting point, namely the energy accumulation and dissipation always can't be balance, either as curves $-W_d$ and W_p, which do not produce any self-excited vibration; or as curves $-W_d$ and W_p', which do produce self-excited vibration and the amplitude will go up infinitely. In a word, it is impossible to achieve a stable self-excited vibration for the above three curves.

In fact, the self-excited vibration system generally has amplitude stability. In order to explain this problem, the high-order term of the cutting force (i.e., the non-linear terms) in Eq. (12.2) is required to be considered. To do this, substitute Eqs. (12.2) into (12.10), and denote

$$\frac{d^2F_c(V_o)}{dV^2} = c'', \quad \frac{d^3F_c(V_o)}{dV^3} = c''', \tag{12.13}$$

the terms higher than 3-order and the constant term are ignore, then we have the positive work done by the cutting force in one period of vibration,

$$W_p = -\oint \left(c'\dot{x} + \frac{1}{2}c''\dot{x}^2 + \frac{1}{6}c'''\dot{x}^3 \right) \mathrm{d}x.$$

Substituting Eq. (12.9) into the above equation and integrating result in

$$W_p = -c'\pi\omega_o A^2 - c'''\frac{3\pi}{4}\omega_o A^4. \tag{12.14}$$

When it comes to energy balance, there should be $W_d + W_p = 0$, substituting Eqs. (12.12) and (12.14) into the this condition results in

$$-c\pi\omega_o A^2 - c'\pi\omega_o A^2 - c'''\frac{3\pi}{4}\omega_o A^4 = 0.$$

From this equation the stabilized vibration amplitude can be worked out as

$$A_o = \frac{2}{\omega_o}\sqrt{-\frac{c+c'}{3c'''}}. \tag{12.15}$$

As stated before, the prerequisite of is $c + c' < 0$, and for the above equation to have real solution, we should have $c''' > 0$. Now, let's see Eq. (12.14), in the right side, the first term must positive (because $c' < 0$), and its value increases as A^2 increasing. The second term is negative (because $c''' > 0$), this is a disincentive factor which suppresses the increase in W_p. When A is relatively small, the effect of this factor does not remarkable, while A increasing, however, the suppression effect of the factor will sharply increase. In Fig. 12.6, curve W_p'' (thin solid line) shows the relation between the amplitude A and the work done by the cutting force in a vibrating cycle, which takes the nonlinear factors into consideration. Compared with the curve W_p' (dashed line), the inhibition effect of nonlinear terms can be clearly seen. The intersection point B of W_p'' and $-W_d$ gives the stabilized amplitude A_o.

To illustrate the amplitude A_o is stable, assuming that due to accidental factors interference, amplitude from A_o suddenly increases to A_o', it can be seen from the figure that $-W_d > W_p''$, namely the consumption of energy is greater than the energy accumulation, the total energy of the system decreased, thus amplitude decreased, from A_o' drop back to A_o point; On the other hand, if the A_o falls to A_o'' due to accidental interference, the total energy of the system will be increased by similar analysis, and the amplitude A_o'' will be upturned to A_o.

12.2.5 The Threshold of Chatter

The chatter caused by the decline characteristics of cutting force is mainly happened under precision cutting condition. In this case, the friction between the tool flank and the work-piece cutting surface is the main component of the cutting force. It is generally believed that this friction force is directly proportional to contact surface area between the work-piece and the cutting tool, and the contact surface area is directly proportional to the average width of flank wear land VB, and the main cutting force can be written as

$$F_c(V) = VB \cdot p_c(V),$$

where, $p_c(V)$ is the cutting force corresponding to the unit wear width, substituted the above equation in Eq. (12.3), we have

$$c' = VB \cdot \mathrm{d}p_c(V)/\mathrm{d}V = VB \cdot c'_o,$$

where, $c'_o = \mathrm{d}p_c(V)/dV$. Consider the critical situation between stability and instability, that is

$$c + c' = c + VB \cdot c'_o = 0.$$

Therefore, the surface abrasion of the critical flank wear land can be worked out,

$$VB_{cr} = -c/c'_o. \tag{12.16}$$

It must noticed that in the decline area of the cutting force we have $c'_o < 0$, and so $VB_{cr} > 0$.

By the above analysis it can be seen that to perform precision machining in the velocity area of cutting force decline characteristics, after the amount of flank wear $VB > VB_{cr}$, cutting system will lose stability and chatter of the type of vibration velocity feedback in the cutting force decline area will happen.

12.3 The Cutting Chatter Caused by Delayed Displacement Feedback

12.3.1 Displacement Feedback, Negative Stiffness and Static Instability

The negative damping and dynamic instability caused by velocity feedback are described in the last section. Now, the feedback of vibration displacement and its effect are investigated. The block diagram of the system is shown in Fig. 12.7, and the equation of motion is

$$m\ddot{x} + c\dot{x} + kx = F(x). \tag{12.17}$$

It can be seen that the characteristic of this system is that the force itself imposed on the vibrating system is controlled by the vibration displacement. In the above equation, $F(x)$ is generally a nonlinear function, and when x is less, it can be expressed as a power series near $x = 0$, only taking the linear term, and ignoring higher order and the constant terms, we have

$$F(x) \approx -k'x, \quad -k' = \frac{\mathrm{d}F(0)}{\mathrm{d}x}, \tag{12.18}$$

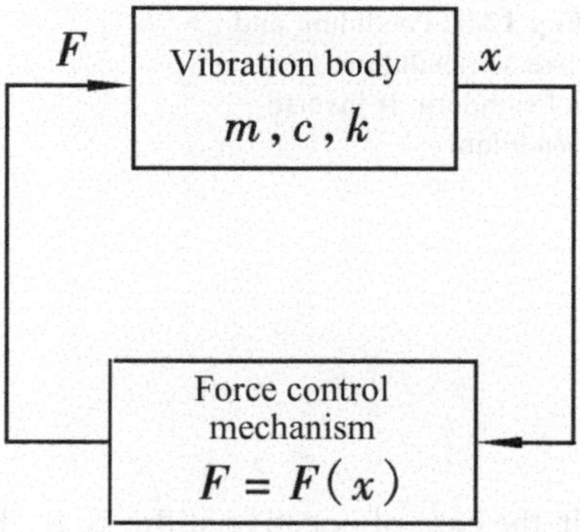

Fig. 12.7 Displacement feedback

Substituting into Eq. (12.17) and transposing terms, results in

$$m\ddot{x} + c\dot{x} + (k + k')x = 0. \tag{12.19}$$

It looks like a free vibration equation of motion for a single-degree-of-freedom system, but the difference is that the stiffness coefficient consists of two parts: the first part is the stiffness of the vibration body (k), generally is positive; the second part (k') is the "equivalent stiffness" generated by the displacement feedback, which depends on the property of the function $F(x)$. If near the $x = 0$, the function $F(x)$ decreases with the increase of x, namely $-k' < 0$, the total system stiffness (k + k$'$) > 0, then Eq. (12.19) has essentially no difference with free vibration equation of motion for a single-degree-of-freedom system, just stiffness increased, due to the displacement feedback, and system natural frequency ω_o also increased accordingly,

$$\omega_n^2 = \frac{k + k'}{m} = \omega_o^2 + \frac{k'}{m}. \tag{12.20}$$

where, ω_n is the natural frequency of the system after taking the equivalent stiffness caused by displacement feedback into consideration.

If the function $F(x)$ increases with the increase of x, namely $-k' > 0$, and if $-k' > k$, the total stiffness of the system (k + k$'$) < 0, that is, the total stiffness of the system becomes negative, which becomes a destabilizing factor.

Figure 12.8a, b respectively give two examples of systems with positive and negative stiffness, namely pendulum and inversed pendulum. Obviously, the latter is unstable. But this instability is very different from the instability caused by negative damping. Now back to the Eq. (12.20), the natural frequency of the system is

$$\omega_n = \sqrt{\frac{k + k'}{m}}, \tag{12.21}$$

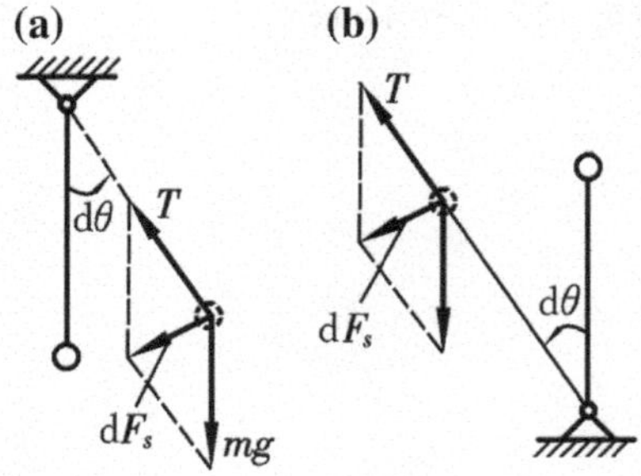

Fig. 12.8 Pendulum and inverse pendulum. **a** Pendulum. **b** Inverse pendulum

In the case of negative stiffness, ω_n becoming imaginary, namely the system natural frequency does not exist, this indicates that the system could not make the reciprocating vibration, as shown in Fig. 12.8a but as a way of displacement increases monotonously deviating from its equilibrium position. In fact, in the Eq. (12.21), if $k + k' < 0$, i.e., negative stiffness, introduce the following notation

$$p = \sqrt{\frac{k' + k}{m}} = i\omega_n, \tag{12.22}$$

$$\zeta = \frac{c}{2\sqrt{(k' + k)m}} = -i\xi, \tag{12.23}$$

P, ζ, as defined, are positive real numbers, and the Eq. (12.19) can be written using the above notation as

$$\ddot{x} + 2\zeta p\dot{x} - p^2 x = 0. \tag{12.24}$$

Let

$$x(t) = Ae^{st}, \tag{12.25}$$

Substituting into Eq. (12.24) results in characteristic equation

$$s^2 + 2\zeta ps - p^2 = 0. \tag{12.26}$$

The above formula can solved,

$$\begin{matrix} s_1 \\ s_2 \end{matrix} = \left(-\zeta \pm \sqrt{1 + \zeta^2}\right)p. \tag{12.27}$$

The general solution of Eq. (12.24) is

$$x(t) = A_1 e^{s_1 t} + A_2 e^{s_2 t}. \tag{12.28}$$

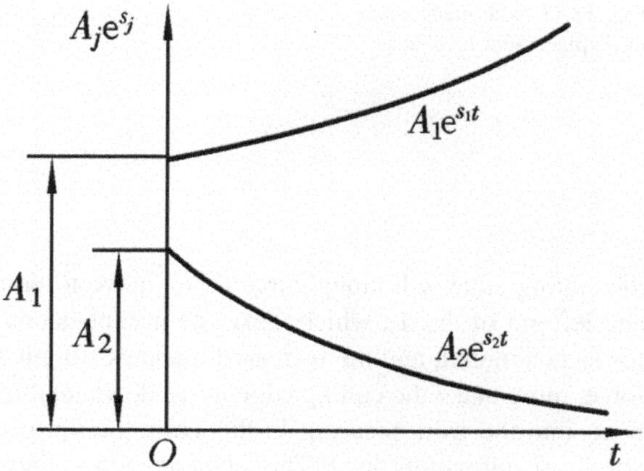

Fig. 12.9 The variations of the two terms at right side of Eq. (12.28)

Notice that $s_1 > 0$, and $s_2 < 0$, the first term will monotonically increase, while the second term will monotonically decrease approaching to zero, as shown in Fig. 12.9. The general trend when the time is large, $x(t)$ will increase monotonically (but if A_2 is larger or s_1 is smaller, $x(t)$ may drop at first, and then rise).

The instability caused by negative stiffness is called static instability to distinguish from the instability caused by negative damping as described in the previous section, which is called dynamic instability.

As an example of static instability, the negative stiffness and static instability caused by tool deformation in the process of metal cutting are analyzed. Figure 12.10 represents a simplified cutting process. The work-piece moves with the velocity V downward and a layer metal of thickness s_o is cut off by the cutting edge. The downward cutting force is F_o, and under its action, the cutter shank will have elastic deformation, while the cutting edge from its original position without cutting load, i.e., the point o moves to o'. At this point, the elastic recovery force F_s (not shown) is balanced with the cutting force F_o.

But this balance is not stable, as a result of any accidental disturbance in the cutting process, such as the cutting edge encounter a hard point in work-piece materials, cutting force F_o will get a increment dP, under the influence of dP, then

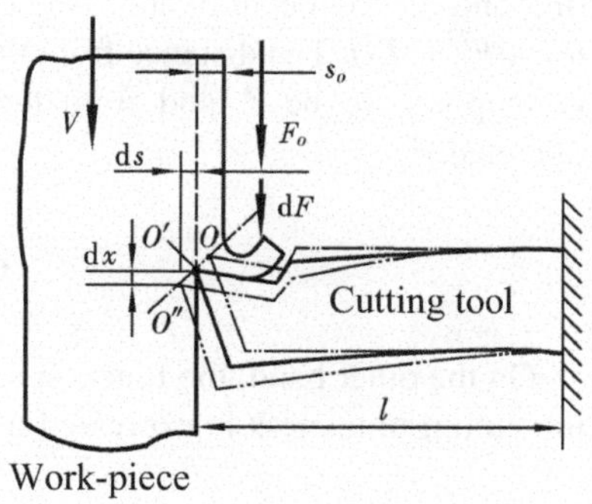

Fig. 12.10 The breaking edge phenomenon

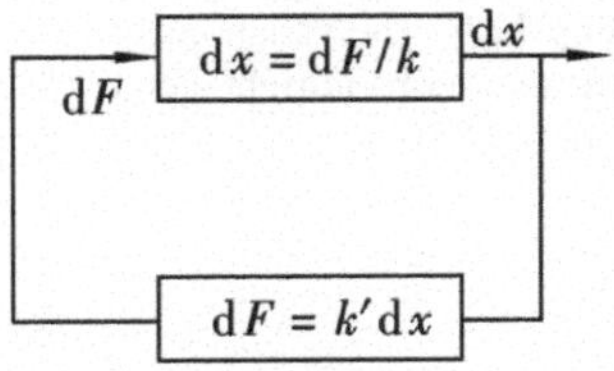

Fig. 12.11 The mechanism of displacement feedback

the cutting edge will along curve $oo'o''$ move to points o'', it moves down by dx, and left out of the ds, which make the instantaneous thickness of cut increases by the same amount, and the increased thickness of cut again cause the rise of cutting force, more make the cutting edge move down, and the cutting edge is more deeply prick into the work-piece…, In this case, the tip of the cutter can be broken by a rapidly rising cutting load. This phenomenon is known as the "breaking edge".

Block diagram of the system can be expressed in Fig. 12.11, in which the above box represents the cutter or machine tool structure, its input are the incremental of cutting force acting on the cutter dF, and output dx is the longitudinal deformation of the cutting edge; the box below represents the control mechanism of cutting force (that is, the cutting process), its input is dx, namely the movement of the cutting edge in the vertical direction, and the output is the change of cutting force dF, which is the feedback on the cutting tool, obviously, this is a system with the displacement feedback.

The tool is considered as a cantilever beam, and it is considered that the clamping of the cutter is completely rigid, and the stiffness of the cutter is known by the material mechanics,

$$k = \frac{\mathrm{d}F}{\mathrm{d}x} = \frac{3EI}{l^3}, \tag{12.29}$$

In the formula, l is the suspension length of the tool, and EI is its flexural rigidity, which is of course the positive stiffness.

The equivalent stiffness of the displacement feedback is calculated as follows: firstly, the relationship between the vertical sinkage of the cutting edge in Fig. 12.12 and its horizontal outstretched ds (that is, the increment of the cutting thickness). As a cantilever beam, the end part of the pole has a concentrated load dF, and the deflection and Angle of the end are respectively $f = \mathrm{d}Fl^3/(3EI)$, $\theta = \mathrm{d}Fl^2/(2EI)$. The distance from the neutral plane of the shank to the cutting edge is supposed to be Z, and from the geometric relationship in Fig. 12.12 can be obtained

$$\mathrm{d}s = \frac{3Z}{2l}\mathrm{d}x, \tag{12.30}$$

On the other hand, the function relation $F(s_o + \mathrm{d}s)$ between the cutting force and the cutting thickness is expressed as a power series near s_o,

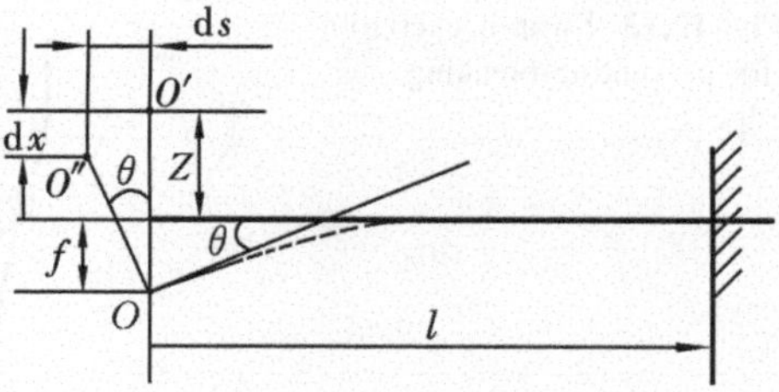

Fig. 12.12 An analysis on shank deformation and cutting edge displacement

$$F(s_o + \mathrm{d}s) = F(s_o) + \frac{\mathrm{d}F(s_o)}{\mathrm{d}s}\mathrm{d}s + \cdots,$$

And the incremental of the cutting force is

$$\mathrm{d}F = F(s_o + \mathrm{d}s) - F(s_o) = k_s \mathrm{d}s + \cdots, \tag{12.31}$$

where,

$$k_s = \frac{\mathrm{d}F(s_o)}{\mathrm{d}s}. \tag{12.32}$$

Because $F(s_o + ds)$ is the increasing function of ds, so $k_s > 0$. Substituting Eqs. (12.30) into (12.31) leads to

$$\mathrm{d}F = k_s \frac{3Z}{2l}\mathrm{d}x,$$

The equivalent stiffness coefficient is obtained

$$k' = \frac{\mathrm{d}F}{\mathrm{d}x} = k_s \frac{3Z}{2l}. \tag{12.33}$$

From the above equation and the Eq. (12.29) we get the total stiffness

$$k - k' = \frac{l^3}{3EI} - k_s \frac{3Z}{2l}, \tag{12.34}$$

Among them, $-k_s 3Z/2l$ is the equivalent negative stiffness caused by the displacement feedback. The condition of the breaking edge phenomenon is

$$\frac{l^3}{3EI} - k_s \frac{3Z}{2l} < 0. \tag{12.35}$$

An effective measures, to prevent breaking edge is change the shape of the tool shank, when the deformation of the cutting edge down, it will retreat from the work-piece at the same time, rather than plunge into the work-piece, this way, the

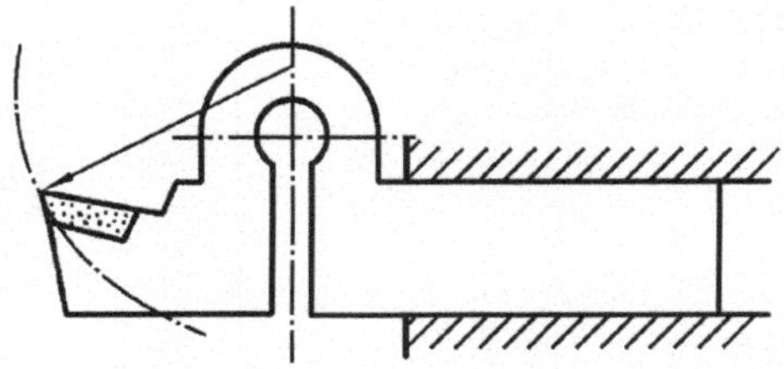

Fig. 12.13 Swan-necked tool (for preventing breaking edge)

second term in above equation will become is a positive stiffness, and of course the system won't lose its stability.

The elastic cutter bar (swan-necked tool) shown in Fig. 12.13 can meet the above requirements. The cutting edge of the cutter under the action of cutting force is generally moving along an arc, and it moves backward while moving down.

From this section you can see, that simple displacement feedback can only increase the original stiffness, or form negative stiffness and cause static instability, but it can't cause dynamic instability, that is, could not cause self-excited vibration. As we will see, the delayed feedback of displacement can cause self-excited vibration.

12.3.2 Delayed Displacement Feedback

If the instantaneous excitation forces $F(t)$ acting on the system is not controlled by the vibration displacement $x(t)$ at that time, but controlled by the vibration displacement $x(t - T)$ a time T before, then get the displacement delayed feedback system, or called "time delay system", the block diagram as shown in Fig. 12.14. Its motion equation is

$$m\ddot{x}(t) + c\dot{x}(t) + kx(t) = F[x(t - T)], \tag{12.36}$$

The function $F[x(t - T)]$ is linearized,

$$m\ddot{x}(t) + c\dot{x}(t) + kx(t) = px(t - T), \tag{12.37}$$

where,

$$p = \frac{\mathrm{d}F(0)}{\mathrm{d}x}. \tag{12.38}$$

Time-delay systems can be stable or unstable. In this paper, a kind of intermediate state between stability and instability is analyzed, that is, the possibility that the system produces such a constant amplitude vibration. For this reason, set

$$x(t) = a_o \cos \omega t, \tag{12.39}$$

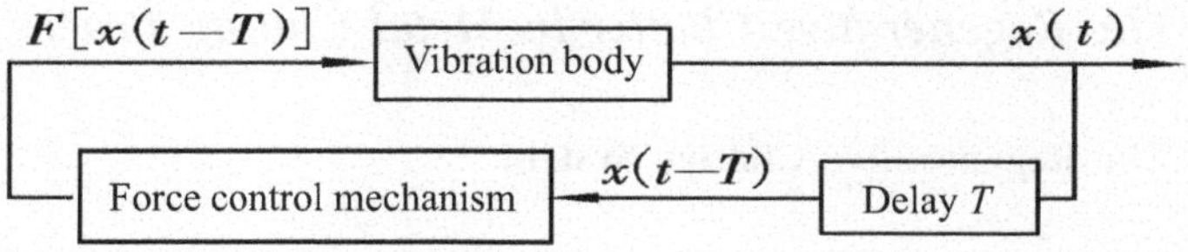

Fig. 12.14 The displacement delayed feedback mechanism

Hence,

$$\dot{x}(t) = -\omega a_o \sin \omega t, \tag{12.40}$$

And

$$\begin{aligned} x(t-T) &= a_o \cos \omega(t-T) = a_o \cos(\omega t - \phi) = a_o \cos \phi \cos \omega t + a_o \sin \phi \sin \omega t \\ &= \cos \phi (a_o \cos \omega t) - \frac{1}{\omega} \sin \phi (-\omega a_o \sin \omega t) \end{aligned} \tag{12.41}$$

where, $\phi = \omega T$ is the phase lag caused by time delay. Introduction of marks

$$-p \cos \phi = k', \tag{12.42}$$

$$p \sin \phi / \omega = c', \tag{12.43}$$

From Eqs. (12.39)–(12.43) we have

$$px(t-T) = -k'x(t) - c'\dot{x}(t). \tag{12.44}$$

Substitute this type in (12.37) and transfer it,

$$m\ddot{x}(t) + (c + c')\dot{x}(t) + (k + k')x(t) = 0. \tag{12.45}$$

Time delayed displacement feedback, therefore, is equivalent to the displacement and velocity feedback at the same time, it changed both the system damping and stiffness. Equations (12.42) and (12.43) give, respectively, the coefficients of equivalent stiffness and equivalent damping due to the delayed displacement feedback. Depending on the length of time delay, negative stiffness or negative damping can occur, resulting in static or dynamic instability.

In this chapter, the "regenerative chatter" in the metal cutting process is taken as an example to illustrate the mechanism and the law of the dynamic instability caused by displacement delayed feedback.

12.3.3 *The Regenerative Chatter in Metal Cutting Process*

12.3.3.1 The Regenerative Chatter System

Regenerative chatter occurred in metal cutting process is a kind of typical dynamic instability phenomenon, caused by time delayed feedback of vibration displacement, is also one of the main mechanisms of self-excited vibration of metal cutting machine tools.

Figure 12.15 is a schematic of radial plunge cylindrical turning, among them, the machine tool structure is simplified into a single-degree-of-freedom system, cutting motion consists of the work-piece rotation N and the feed of the tool along the work-piece radial s_o, and it is equal to the average thickness of cut in quantity. $x(t)$ is the vibration displacement of the tool relative to the work-piece in horizontal direction, namely the deformation of the machine tool structure, and $F(t)$ is the cutting force imposed on the tool. Where, the machine tool structure is the vibration body in Fig. 12.14, and $F(t)$ acting on it produces vibration displacement $x(t)$. On the other hand, $x(t)$ causes the instantaneous cutting thickness $s(t)$ to fluctuate around its mean value s_o that the change will, in turn, cause the cutting force $F(t)$ change. Therefore, the cutting process is the force control mechanism in Fig. 12.14, which controls the excitation force according to the vibration displacement to realize the displacement feedback. It must be also seen that $s(t)$ is not only related to the vibration displacement of the cutting edge at that time $x(t)$ (see the inner surface of the shaded part shown in Fig. 12.15), but also to the vibration displacement $y(t)$ before a revolution of the work-piece (see the outer surface of the shaded part shown in Fig. 12.15). Therefore, there is delayed feedback of vibration displacement.

Under the condition of smooth cutting, the surface of the work-piece is evenly cut with a layer of metal s_o that the trajectory of cutting edge with respect to the work-piece is a spiral of Archimedes, as shown by the thin solid line in Fig. 12.15; At this point, the cutting force F_o is a constant, which acts on the structure of the machine and causes the constant deformation x_o. The constant x_o, in turn, ensures that the thickness of cut s_o is constant. Theoretically, if there is no outside interference, the smooth cutting process seems to go on and on.

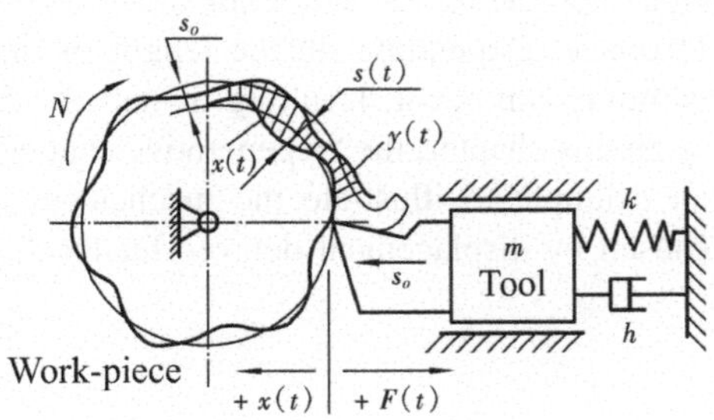

Fig. 12.15 A model of regeneration chatter

But, as explained before, in the actual case, there is always various disturbances, therefore the above stationary cutting process is doomed to be disturbed. If the cutting process can be returned to the equilibrium state after being disturbed, the process is still stable; If a disturbance makes the cutting process more and more deviated from the above equilibrium state, the cutting process is unstable.

Now suppose the cutting process suddenly encounters an interference, for example, the cutting edge encounters a hard point of work-piece materials, cutting force immediately gains a dynamic incremental $\Delta F(t)$, and the $\Delta F(t)$ acting on the machine tool structure, caused vibration $x(t)$, which in turn makes the instantaneous thickness of cut changes $\Delta s(t)$, which cause the secondary change of cutting force $\Delta F'(t)$. Under certain conditions, the amount of the cutting force will increase after a rotation of the work-piece, namely, $\Delta F'(t) > \Delta F(t)$. In the same way, once the work-piece rotates one revolution again, there will be $\Delta F''(t) > \Delta F'(t)$. And so on and so forth, $\Delta F(t)$ and $x(t)$ are rising and rising, forming a strong self-excited vibration. This kind of self-excited vibration in the cutting process is called "regenerative chatter".

Two things need to be said here: first, the "initial perturbation" that inspired this sequence of responses and developed into a self-excited vibration was "forgotten" after only one action. Since then, the response is due to the interaction between the "machine tool structure" and the "cutting process" in the system, which is not related to the disturbance of the outside world; second, the above is just for the sake of convenience, to tell the causal relationship between the various factors in order, but in fact, the various factors are in the mutual influence, and mutual restriction and are simultaneously and continuously changing, and it's not supposed to be understood that one factor changes, then, the next one changes.

The following mathematical models are used to analyze the mechanism and the rule of regenerative chatter.

12.3.3.2 The Equation of Motion

According to Fig. (12.15), the equation of motion can be written as

$$m\ddot{x}(t) + \frac{h}{\omega}\dot{x}(t) + kx(t) = -\Delta F(t), \tag{12.46}$$

where, h is the hysteretic damping coefficient. Here hysteretic damping is used only in order to make the formulae are neat. As a fact, when the damping is small, and the system vibration is a harmonic one, the difference between the models of hysteretic damping and viscous damping is not large. The negative sign on the right hand side of the equation is due to the increase of F_c will cause x to decrease (the positive direction of cutting force $F(t)$ is opposite to the positive direction of tool vibration displacement $x(t)$, as shown in Fig. 12.15).

If the variation of thickness of cut $\Delta s(t)$ is smaller, the dynamic increment $\Delta F(t)$ of the cutting force can be expressed as

$$\Delta F(t) = a_w k_s \Delta s(t), \tag{12.47}$$

where, a_w is width of cut (mm), i.e., the thickness of the work-piece (measured in the direction perpendicular to the figure surface). k_s is the coefficient of thickness of cut (N/mm^2) for cutting force, that is, the cutting force for unit of "$a_w \Delta s(t)$".

The relationship between dynamic change of the thickness of cut $\Delta s(t)$ and the relative vibrational $x(t)$ between the tool and work-piece is solved as follows. Amplify the shadow portion of Fig. 12.15, spread it out, and expressed in the Fig. 12.16. The spiral of Archimedes in Fig. 12.15 (i.e., the cutting edge trajectory while smooth cutting) was shown in Fig. 12.16 as two parallel straight lines, they are s_o apart.

The wavy lines over the two straight lines (i.e., thin lines and heavy line, respectively), are the cutting path of the edge in this revolution and last revolution of the work-piece $x(t)$ and $y(t)$, which constitute the lower surface and the upper surface of the cutting layer, the difference between the two surfaces is the instantaneous thickness of cut under the condition of vibration cutting

$$s(t) = x(t) - y(t). \tag{12.48}$$

Still consider the situation of $x(t)$ being harmonics with constant amplitude, which is the critical state between stability and instability. Suppose

$$x(t) = a_o \cos \omega t, \tag{12.49}$$

Then, from Fig. 12.16 we have

$$y(t) = x(t - T) - s_o, \tag{12.50}$$

where, T is the time (s) for one revolution of the work-piece,

$$T = \frac{60}{N}, \tag{12.51}$$

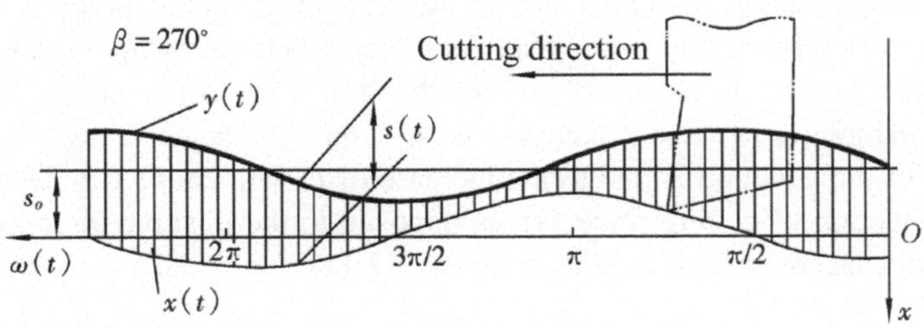

Fig. 12.16 The cutting paths of the tool during successive revolutions of the work-piece

Suppose β is the phase difference of two wave cut in successive revolutions,

$$\beta = T\omega = 60\omega/N. \tag{12.52}$$

Substituting Eqs. (12.49), (12.50) and (12.52) into (12.48) results in,

$$\begin{aligned} s(t) &= a_o \cos \omega t - a_o \cos(\omega t - \beta) + s_o \\ &= s_o + a_o \left[(1 - \cos \beta) \cos \omega t - \frac{1}{\omega} \sin \beta \omega \sin \omega t \right], \end{aligned} \tag{12.53}$$

Denoting

$$1 - \cos \beta = A, \sin \beta = B. \tag{12.54}$$

And considering Eq. (12.49) and

$$\dot{x}(t) = -\omega a_o \sin \omega t, \tag{12.55}$$

leads to

$$s(t) = s_o + Ax(t) + \frac{B}{\omega}\dot{x}(t),$$

And the dynamic variation over its average of thickness of cut is

$$\Delta s(t) = s(t) - s_o = Ax(t) + \frac{B}{\omega}\dot{x}(t), \tag{12.56}$$

Substituting into Eq. (12.47) leads to the expression of dynamic cutting force

$$\Delta F(t) = a_w k_s \left[Ax(t) + \frac{B}{\omega}\dot{x}(t) \right], \tag{12.57}$$

This formula indicates that the excitation force is controlled by vibration displacement and vibration velocity at the same time, and once again, the formula shows that the delayed feedback of displacement is equivalent to the feedback of both displacement and velocity.

Substituting Eqs. (12.57) into (12.46) and transposing result in

$$m\ddot{x}(t) + \frac{1}{\omega}(h + a_w k_s B)\dot{x}(t) + (k + a_w k_s A)x(t) = 0. \tag{12.58}$$

This is a free vibration equation of motion of the single-degree-of-freedom system, the stiffness coefficient and damping coefficient are both consist of two parts: a part is coming from the stiffness and damping of machine tool structure themselves, and another is the equivalent stiffness and equivalent damping of

Fig. 12.17 The regeneration chatter mechanism in cutting process

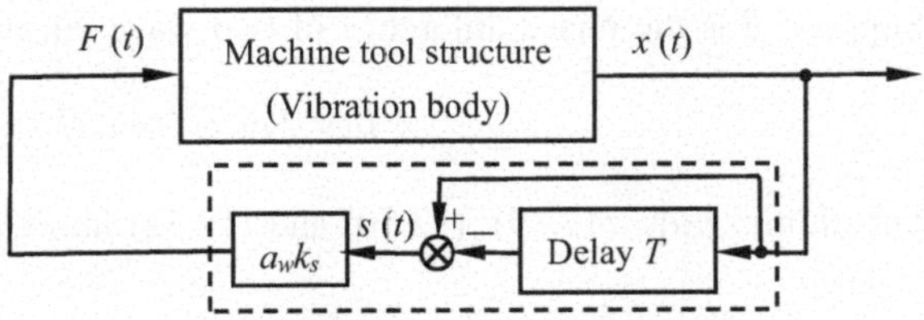

cutting process caused by delay displacement feedback, namely due to the "regenerative effect".

From Eq. (12.54) we know that $A \geq 0$, and normally $a_wk_sA \ll k$, that is, the equivalent stiffness of cutting process is positive and much smaller than the stiffness of the machine tool structure itself. Therefore, the equivalent stiffness can only increase the total stiffness of the system slightly, and have no real impact on the characteristics of the system. On the other hand, the equivalent damping a_wk_sB could make the whole cutting system lose its dynamic stability. First of all, by Eq. (12.54), B could be positive or negative, depending on Angle β, when $\beta = 180°–360°$, $B = \sin\,\beta < 0$, thus the equivalent damping a_wk_sB is negative. Secondly, if the cutting width a_w is large enough, it can make $h + a_wk_sB < 0$, i.e., the total damping of the system becomes negative, resulting in self-excited vibration.

According to the above analysis, the more detailed block diagram of the regenerative chatter system is shown in Fig. 12.17.

12.3.3.3 The Stability Equation and Stability Chart

In the above model, the work-piece revolutions N and the cutting width a_w are two main technological parameters. The influence of technological parameters on the stability of cutting system is analyzed based on motion equation.

As we mentioned earlier, we are studying the critical state between stability and instability, which is that the total damping coefficient in the Eq. (12.58) should be zero,

$$h + a_wk_sB = 0. \tag{12.59}$$

The critical width of cut can be calculated

$$a_{wcr} = -\frac{h}{k_sB} = -\frac{h}{k_s}\frac{1}{\sin\frac{60\omega}{N}}. \tag{12.60}$$

When $a_w > a_{wcr}$, the total damping of the system becomes negative, and the self-excited vibration will occur. So, a_{wcr} is called the "stability threshold."

In the condition of satisfying Eq. (12.59), the Eq. (12.58) becomes

$$m\ddot{x}(t) + (k + a_w k_s A)x(t) = 0, \tag{12.61}$$

This is the motion equation of an undammed free vibration system, its natural frequency is

$$\omega = \sqrt{\frac{k + a_w k_s A}{m}}, \tag{12.62}$$

Or,

$$\omega^2 = \omega_o^2 + \frac{a_w k_s A}{m}. \tag{12.63}$$

Here ω is the self-excited vibration frequency, and ω_o is the natural frequency of machine tool structure itself. Obviously, $\omega > \omega_o$, that is, the frequency of self-excited vibration is always slightly higher than the natural frequency of a certain unstable mode in the machine tool structure. As mentioned above, this is because the equivalent stiffness of the cutting process increases the stiffness of the entire cutting system slightly.

The Eqs. (12.59) and (12.63) compose the stability equations set of the system

$$\begin{cases} h + a_w k_s B = 0 \\ \omega^2 = (k + a_w k_s A)/m \end{cases} \tag{12.64}$$

where,

$$A = 1 - \cos\beta, \tag{12.65}$$

$$B = \sin\beta. \tag{12.66}$$

When the dynamic characteristics of machine tool structure m, h, k and the characteristics of cutting process k_s are given, by Eq. (12.64), critical width of cut a_{wcr} for chatter to occur and the chatter frequency ω under a certain revolutions N can be solved. Obviously, a_{wcr} and ω are both related with the work-piece revolutions N, figures showing these relationships are known as stability chart of the system. Figure 12.18, gives a typical stability chart. In Fig. 12.18a, the earlobe shadow areas are the unstable regions, and the area below the horizontal interrupted line is the unconditional stable region under all the work-piece revolutions. When $\sin(60\,\omega/N) = -1$ in Eq. (12.60), the minimum critical width of cut is

$$a_{wcr\min} = \frac{h}{k_s}, \tag{12.67}$$

This is the width of the unconditional stable region in Fig. 12.18a. When $a_w < a_{wcr\min}$, The cutting system is stable at any work-piece revolutions N.

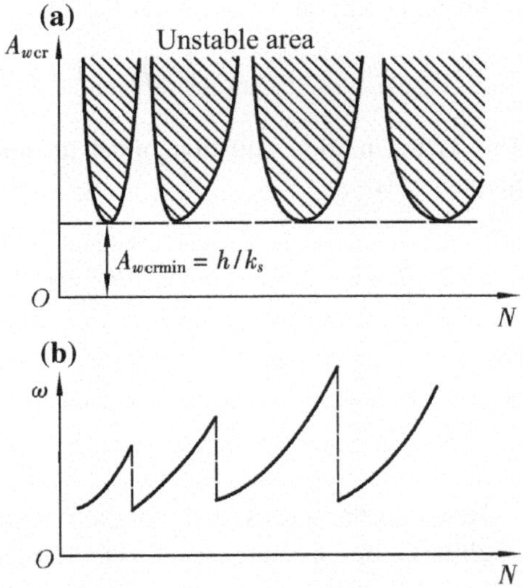

Fig. 12.18 Stability chart. **a** Critical width of cut. **b** Chatter frequency at threshold

Figure 12.18 shows that on the stability threshold ($a_w = a_{wcr}$), the relation curve between chatter frequency and work-piece revolutions is jagged.

12.3.3.4 The Cause of Negative Damping and Its Effects

When β = 180°–360°, the equivalent damping of the cutting process is negative, here the causes and effects of negative damping are analyzed in detail.

Figure 12.19a, b present the patterns of the layer to be removed by cutting under the condition of vibration, with β = 90° and β = 270°, respectively. Figure 12.19a is the case of β = 90°, the distance between the two wave lines, upper and the lower, is instantaneous thickness of cut $s(t) = x(t) - y(t)$. The $s(t)$ consists of two parts, $s(t) = s_o + \Delta s(t)$, i.e., the average s_o and the fluctuation around the average $\Delta s(t)$. Only $\Delta s(t)$ can generate vibration, and s_o produce a constant force which only can cause a variation in equilibrium position. To eliminate s_o, we can move $y(t)$ downward by a distance s_o, and get wave line $y'(t)$ as expressed by interrupted line in the figure. The distance between curves $y'(t)$ and $x(t)$ is $\Delta s(t)$. Here we need to explain that the thickness of cut $s(t)$ can only be positive, but $\Delta s(t)$ can be either positive or negative, when $y'(t)$ is above $x(t)$, $\Delta s(t) > 0$, whereas, when $y'(t)$ is below $x(t)$, $\Delta s(t) < 0$. Likewise, the cutting force $F(t)$ cannot be negative (its positive direction is shown in Fig. 12.15). However, from Eq. (12.47), the dynamic component of the cutting force $\Delta F(t)$ can also be either positive or negative because of the variable symbol of $\Delta s(t)$. According to the positive direction of the cutting force $F(t)$ specified in Fig. 12.15, the positive cutting force is pushing the cutter away from the work-piece, while the negative cutting force tends to pull the cutter to the work-piece.

Fig. 12.19 The generation mechanism of positive damping and negative damping

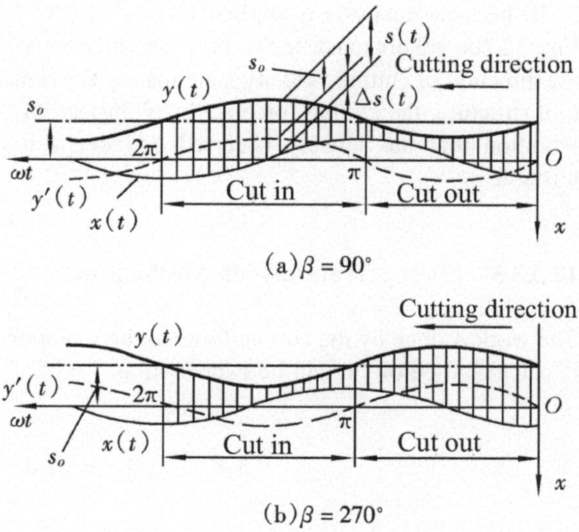

From the thin solid curve in Fig. 12.19a we can see that during one cycle, the motion of the tool can be divided into two phases: cut out ($0 \sim \pi$) and cut in ($\pi \sim 2\pi$). In cut out phase, cutting tool's tendency is leaving the work-piece, and in the cut in phase, the cutting tool's tendency is toward the work-piece. In the case of $\beta = 90°$, for most of the cut out phase, $\Delta s(t) < 0$, and thus, $\Delta F(t) < 0$, the dynamic cutting force tries to pull the cutter to the work-piece. And for most of the cut in phase, $\Delta s(t) > 0$, and thus, $\Delta F(t) > 0$, the cutting force tries to push the cutter away from the work piece.

The relationship between the direction of dynamic cutting force and the direction of the motion of the cutter is shown in Fig. 12.20a, it can be seen from the figure that when $\beta = 90°$, the direction of dynamic cutting force ΔF and the direction of cutting velocity $\dot{x}$ are mainly opposite, in this case, the cutting force acts as a drag, so the system is stable, which means that the self-excited vibration will not occur.

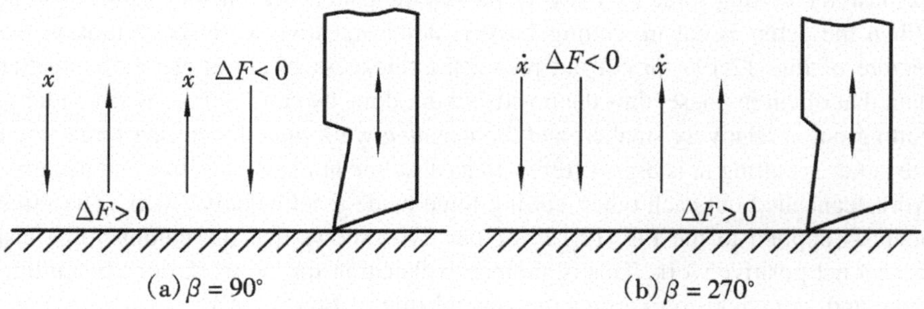

Fig. 12.20 The physical meaning of positive damping and negative damping

If the same analysis is applied to Fig. 12.19b, $\beta = 270°$, the results shown in Fig. 12.20b are obtained. In this case, the direction of dynamic cutting force ΔF and the direction of cutting velocity $\dot{x}$ are mainly the same. For the vibration of machine tool structure, the cutting force is "help" instead of "resistance". It encourages the vibration that has already occurred and keeps it expanding, so the system is unstable.

12.3.3.5 Energy Transmission Mechanism

The work A done by the cutting force to the vibration system in one vibrating cycle is calculated below, it can be expressed as

$$A = -\int_0^T \Delta F(t)\dot{x}(t)\mathrm{d}t, \quad (12.68)$$

The negative sign in the equation is due to the difference between $F(t)$ and $x(t)$, as shown in Fig. 12.15. Substituting Eqs. (12.57), (12.54) and (12.55) into above equation and integrating result in

$$A = -\pi a_0^2 a_w k_s \sin\beta. \quad (12.69)$$

From this expression it can be seen, that when $\beta = 0°–180°$, $A < 0$, the cutting force did negative work to the vibration system, that is, vibration system will feed back mechanical energy to the cutting process, as the heat energy consumed away, so the cutting process is stable; When $\beta = 180°–360°$, $A > 0$, cutting force did positive work, if the positive work is greater than the energy dissipations of the machine tool structure damping, vibration energy will accumulate, vibration will be intensified, the cutting process is unstable. This conclusion is consistent with the previous conclusion that is derived from the equivalent damping.

In fact, the above results can be intuitively seen from Fig. 12.19: first of all, remember that the cutting force is doing positive work when the cutter is cut out, because the cutting force $F(t)$ always pushes the cutter away from the work-piece. When the cutter is cut in, cutting force is doing negative work. Now look at the picture of Fig. 12.19a, in cut out phase, the thickness of cut is generally smaller than that of cut in phase, thus the positive work done by cutting force when cutter is cutting out is relatively smaller, and the negative work done by cutting force when the cutter is cutting in is bigger (refers to its absolute number), positive and negative works cancelled out each other, cutting force made a net negative work. The same analysis applied to the Fig. 12.19b of $\beta = 270°$ shows that the cutting force has done a net positive work. This is another explanation for the work done by cutting force and its conclusion verifies the general rule of Eq. (12.69).

12.4 The Cutting Chatter Caused by Modal Coupling

The last two sections discussed the phenomenon that the cutting system as a single-degree-of-freedom generates self-excited vibration due to the negative damping of velocity feedback or displacement delayed feedback. This section will introduce the phenomenon of self-excited vibration caused by the displacement feedback between the various degrees of freedom (i.e., between different modes) to prompt the vibration system to absorb energy from the outside world.

12.4.1 Modal Coupling and the Stability of Modal Coupling System

The free vibration equation of motion for two-degrees-of-freedom vibration system is

$$\boldsymbol{m}\ddot{\boldsymbol{x}} + \boldsymbol{k}\boldsymbol{x} = \boldsymbol{0}, \tag{12.70}$$

where, $\boldsymbol{x} = \{x_1, x_2\}^{\mathrm{T}}$ is displacement vector, and $\boldsymbol{m}$, $\boldsymbol{k}$ are the second order square matrixes, they are called the mass matrix and the stiffness matrix, respectively. When $\boldsymbol{m}$, $\boldsymbol{k}$ are both diagonal matrixes, Eq. (12.70) may written as

$$m_1\ddot{x}_1 + k_1x_1 = 0, \tag{12.71}$$

$$m_2\ddot{x}_2 + k_2x_2 = 0, \tag{12.72}$$

That is the equations of motion for two independent, single-degree-of-freedom systems. We call that under the coordinates x_1, x_2 these two-degrees-of-freedom have neither mass coupling nor stiffness coupling, the x_1, x_2 are the natural coordinates of the system, and Eqs. (12.71) and (12.72) are the equations of motion of the two models, respectively. The above two equations, however, may be coupled due to displacement feedback in the load. Suppose the above two-degrees-of-freedom are loaded by exciting forces $-F_1$ and $-F_2$, respectively, and F_1 and F_2 themselves are in term controlled by vibration displacements of the two modes, i.e., $F_1 = F_1(x_1, x_2)$, $F_2 = F_2(x_1, x_2)$, after linearization of the two functions, we have $F_1 = \lambda_{11}x_1 + \lambda_{12}x_2, F_2 = \lambda_{21}x_1 + \lambda_{22}x_2$, so, the two equations of motion become

$$m_1\ddot{x}_1 + k_1x_1 = -\lambda_{11}x_1 - \lambda_{12}x_2, \tag{12.73}$$

$$m_2\ddot{x}_2 + k_2x_2 = -\lambda_{21}x_1 - \lambda_{22}x_2. \tag{12.74}$$

After transposing and denoting $k_1 + \lambda_{11} = K_{11}$, $\lambda_{12} = K_{12}$, $\lambda_{21} = K_{21}$ and $k_2 + \lambda_{22} = K_{22}$ we can rewrite the above two equations as

$$m_1\ddot{x}_1 + K_{11}x_1 + K_{12}x_2 = 0, \tag{12.75}$$

$$m_2\ddot{x}_2 + K_{21}x_1 + K_{22}x_2 = 0. \tag{12.76}$$

These are the free vibration equations of motion of the two-degree-of-freedom, they are, however, no longer independent, but coupled through the new stiffness matrix $[K_{ij}]$ $(i, j = 1, 2)$. The general literature refers to this coupling as "modal coupling", in fact, rather, it should be called "load-coupling between modes".

The damping of the system itself has been omitted in the above discussion, and only displacement feedback is assumed, so if it is a single-degree-of-freedom system, the self-excited vibration will not occur as mentioned before. However, this is a two-degree-of-freedom system, and the cross feedback of vibration displacement between two-degrees-of-freedom can lead to the instability of the system, i.e., causing chatter.

To verdict the stability of the system expressed by Eqs. (12.75) and (12.76), suppose the formal solutions is

$$x_1(t) = A_1e^{pt}, \quad x_2(t) = A_2e^{pt}. \tag{12.77}$$

Substituting them into Eqs. (12.75) and (12.76) results in

$$\left(m_1p^2 + K_{11}\right)A_1 + K_{12}A_2 = 0, \tag{12.78}$$

$$K_{21}A_1 + \left(m_2p^2 + K_{22}\right)A_2 = 0. \tag{12.79}$$

For the equation set to have un-trivial solution, we must have

$$\begin{vmatrix} m_1p^2 + K_{11} & K_{12} \\ K_{21} & m_2p^2 + K_{22} \end{vmatrix} = 0,$$

The above determinant can be developed, as

$$m_1m_2p^4 + (K_{11}m_2 + K_{22}m_1)p^2 + K_{11}K_{22} - K_{12}K_{21} = 0, \tag{12.80}$$

This is the characteristic equation. Suppose that in the system expressed by Eqs. (12.75) and (12.76), we have $K_{11} > 0$, $K_{22} > 0$ (if not, the system will be statically unstable). Let

$$K_{11}/m_1 = n_1^2 > 0; \quad K_{22}/m_2 = n_2^2 > 0 \tag{12.81}$$

We can rewrite Eq. (12.80) as

$$p^4 + (n_1^2 + n_2^2)p^2 + (K_{11}K_{22} - K_{12}K_{21})/(m_1m_2) = 0, \tag{12.82}$$

It can be solved as

$$(p^2)_{1,2} = \frac{1}{2}\left[-(n_1^2 + n_2^2) \pm \sqrt{(n_1^2 + n_2^2)^2 - 4\frac{K_{11}K_{22} - K_{12}K_{21}}{m_1m_2}}\right], \tag{12.83}$$

or rewrite it as,

$$(p^2)_{1,2} = \frac{1}{2}\left[-(n_1^2 + n_2^2) \pm \sqrt{(n_1^2 - n_2^2)^2 + 4\frac{K_{12}K_{21}}{m_1m_2}}\right]. \tag{12.84}$$

Extracting $(p^2)_1$ and $(p^2)_2$ solved from above equation results in p_1, p_2 and p_3, p_4. The stability of the system depends on the values of these four numbers, which are related to the values of $K_{ij}(i, j = 1, 2)$. The influence of relationship of K_{ij} on the values of $p_1 \sim p_4$ and the stability of the system will be briefly analyzed in the follows.

12.4.1.1 Equation (12.85) Is True and $K_{11}K_{22} - K_{12}K_{21} > 0$

For ordinary two-degrees-of-freedom system (not self-excited vibration), from Eq. (12.81), we know $n_1^2 > 0$, $n_2^2 > 0$, and from reciprocal theory, we know $K_{12} = K_{21}$, so the evaluation of the part in the sign of radical (Eq. 12.84) is positive, i.e.,

$$(n_1^2 - n_2^2)^2 + 4\frac{K_{12}K_{21}}{m_1m_2} > 0, \tag{12.85}$$

So the root is a real number, and then, because $K_{11}K_{22} - K_{12}K_{21} > 0$, then from Eq. (12.83) we know the value of radical must smaller than $\left|n_1^2 + \mathrm{n}_2^2\right|$, therefore, $(p^2)_1$ and $(P^2)_2$ both are negative real numbers, which are denoted as $-\omega_1^2, -\omega_2^2$, where ω_1, ω_2 are both positive real numbers. Extracting of the root, we have

$$p_{1,2} = \pm\sqrt{(p^2)_1} = \pm i\omega_1,$$
$$p_{3,4} = \pm\sqrt{(p^2)_2} = \pm i\omega_2.$$

This corresponds to the constant vibration with the constant amplitude, and the system is stable.

12.4.1.2 Equation (12.85) Is True and $K_{11}K_{22} - K_{12}K_{21} < 0$

Suppose Eq. (12.85) is true, then the radical in Eqs. (12.84) and (12.83) is real number, and if $K_{11}K_{22} - K_{12}K_{21} < 0$, then from Eq. (12.83) we know the value of radical must bigger than $\left|n_1^2 + n_2^2\right|$, therefore, $(P^2)_1 > 0$, extracting of the root, we have

$$p_{1,2} = \pm\sqrt{(p^2)_1} = \pm\xi$$

where, ξ is a positive real number, and the term of $e^{\xi t}$ is including in the solution, which is an unstable solution of the non-periodic divergence.

From the above two points, we know that under the condition of Eq. (12.85) (that is the reciprocal theorem is satisfied, as everybody knows that ordinary linear multi-degree-of-freedom system all meet this theorem), the system either exists a stable periodic motion (actually, because of the system damping, the periodic motion will inevitably be attenuation), or there will be a static instability, but won't produce dynamic instability, i.e., the self-excited vibration will not occur.

12.4.1.3 Equation (12.86) Is True

Suppose

$$(n_1^2 - n_2^2)^2 + 4\frac{K_{12}K_{21}}{m_1 m_2} < 0, \tag{12.86}$$

Then from Eq. (12.84), we know the two solutions of the equation are a pair of conjugate complexes,

$$(p^2)_{l,2} = -h \pm il,$$

Extracting of the root, we have

$$p_{1,2,3,4} = \sqrt{(p^2)_{1,2}} = \pm(\xi \pm i\omega),$$

where, h, l, ξ *and* ω are all positive real numbers. And from Eq. (12.77) we know the solution of the system is (take only the x_1 as an example):

$$x_1(t) = Ae^{\xi t}e^{i\omega t} + Be^{\xi t}e^{-i\omega t} + Ce^{-\xi t}e^{i\omega t} + De^{-\xi t}e^{-i\omega t}.$$

The first two terms are the self-excited vibration with constantly increasing amplitude

The two-degrees-of-freedom (or multi-degree-of-freedom) system has a self-excited vibration under the condition of (12.86), which is called modal coupled self-excited vibration. It's obviously different from the self-excited vibration due to negative damping that we talked about earlier, because we're not dealing with damping here.

Let's look back Eq. (12.81), n_1 and n_2 are the natural frequency of a system of two-degrees-of-freedom under the condition of no mutual coupling ($k_{12} = k_{21} = 0$), and the (12.86) showed that the closer the n_1 and n_2 are, the easier the self-excited vibration of modal coupling to cause.

12.4.2 The Self-excited Vibration of Modal Coupling in Metal Cutting Process

The self-excited vibration of modal coupling in metal cutting process was analyzed in [80] with an example of a boring bar which has a rectangular cross section. The cutter and boring tool can vibrate in the two perpendicular directions (two vibration modes). The stiffness coefficients of the boring bar in the two perpendicular directions are k_1 and k_2, and $k_1 < k_2$.

The two vibration modes are coupled through cutting force, i.e., this is "load-coupling between modes". The modal coupling analyzing method used are almost the same as described above.

The analyzing results give us some practical and unexpected implications:

First, only when the direction of smaller stiffness on the section of the boring bar is orientated within the angle included between the directions of cutting force and the normal to the work-piece surface, the chatter due to modal coupling could happen, and beyond that range no chatter happens. So the orientation of the direction of smallest stiffness of the boring bar should avoid this range;

Second, at the bisector of the angle included between the directions of cutting force and the normal to the work-piece surface, the critical width of cut (beyond which, chatter will happen), reaches its minimum and so the orientation of the smallest stiffness direction of the boring bar should as far as possible avoid this bisector direction;

Third, the critical width of cut is proportional to the difference of the stiffness of $k_2 - k_1$, so the circular section of the boring bar ($k_1 = k_2$) is the easiest to generate the self-excited vibration of modal coupling. For boring bar in the opposite direction to cut flat, makes the stiffness coefficients of the boring bar in the two directions are different, this seems to be weaken the stiffness of the boring bar system, but in fact, as long as reasonable decorate the stiffness axis direction, this improve the system ability of resisting the self-excited vibration of the modal coupling.

Chapter 13
Non-linear Theory of Regenerative Chatter in Cutting Processes (I)

Firstly, a nonlinear chatter model proposed by other researchers was introduced, which reflects the nonlinear stiffness of machine tool structure. The achievement of this model was discussed and the difficulties and problems were pointed out. Then, a new nonlinear chattering model established by the author was introduced. The model is based on a nonlinear factor of tool jumping out of work-piece material, when vibration is severe enough. The characteristics and conclusions of this model, such as amplitude stability and the influence of the cutting parameters on the stabilized amplitude, were discussed in detail.

Since the advent of the theory of regenerative chatter in the metal cutting process in the 1940s [80–86], the study of it has been largely confined to the category of linear theory. As mentioned earlier, the linear chatter theory successfully explain and predict the critical conditions for chatter to occurrence, namely the stability threshold, but after crossing the threshold, for the overall process of chatter further development and finally to be stabilized, linear theory can't explain and predict. This is due to the assumption of "small displacement" the linear model is based on, and neglecting the influence of non-linear factors. Because of this, the linear model creates an illusion it seems that when the chatter occurs, the amplitude will rise infinitely and the unstable region is "a monolithic block" that cannot be explored. Therefore, the linear theory naturally treats the unstable region as the restricted zone and is deterred by the "stability threshold". Also because of this, the linear theory failed to reveal the whole process of cutting chatter growth and finally stabilized at a certain lever, it did not reveal how the various non-linear factors control the stabilized chatter amplitude, and did not explore the ways and measures for suppression of the chatter amplitude.

These limitations and shortcomings of linear theory have been known by some researchers. As early as the 1970s, some scholars in the world tried to develop the nonlinear theory of the machine tool chatter [87–91], and obtained some results. The most typical of them is the chatter model based on non-linear stiffness of the structure of machine tool [90, 91] established by S. A. Tobias and N. A. Hanna. This chapter will firstly introduce this nonlinear model, pointing out its

H. Shi, *Metal Cutting Theory*, Springer Series in Advanced Manufacturing,
https://doi.org/10.1007/978-3-319-73561-0_13

achievements and problems, and then systematically discusses a non-linear theory of regenerative chatter developed by the author.

In 1980s, 1990s and 2000s, the author is committed to the study of nonlinear theory of cutting chatter, after a large amount of experiments and theoretical analyses, found the basic nonlinear factors which restricting the chatter process, and based on these factors, a nonlinear model describing the cutting chatter is established. The nonlinear theory with minimal, and ubiquitous assumptions, has given a series of important phenomena about machine tool chatter a satisfactory explanation and correct prediction, these phenomena have been unable to explain before. And also the theory reveals some important rules of practical significance and little-known, such as the extremely sensitive dependence of chatter amplitude on cutting parameter, and the way for outside disturbance to effect the stability of the cutting process and so on.

It is amazing that behind an unstable area of intense chatter, this nonlinear theory has found a "peaceful" area of cutting parameters, which is a stable cutting area with high productivity and high stability. This suggests that it is not always the case that, as linear theory asserts, preventing chatter must be at the expense of productivity.

A large number of theoretical results are confirmed by the experiment, which indicates that the established nonlinear model can reveal the physical mechanism and mathematical laws of machine tool chatter more deeply, and have a richer connotation.

The research achievements in the area of nonlinear theory of cutting chatter have provided a theoretical basis and technical guidance for chatter on line monitoring and automatic control, which is an unavoidable key technology in computer-aided manufacturing or artificial intelligence manufacturing.

13.1 A Chatter Model Based on Non-linear Stiffness of Machine Tool Structure

13.1.1 The Main Points of the Model and Its Achievements

A non-linear cutting chatter model was presented in [90, 91], which is based on the non-linear stiffness of the machine tool structure. Now the model is introduce as follows.

The term reflecting elastic restoring force $kx(t)$ in Eq. (12.46) is replaced by a cubic polynomial (For the sake of simplicity, here the term x^2 in the polynomial has been omitted, because the calculation shows that this one has very little effect),

$$F_s = k[x(t) + \beta x^3(t)], \tag{13.1}$$

Then Eq. (12.46) becomes (the cutting method is supposed to be milling),

$$m\ddot{x}(t) + \frac{h}{\omega}x(t) + k[x(t) + \beta x^3(t)] = -a_p k_s z_c \Delta s(t). \tag{13.2}$$

This is a Duffing equation [79], where, z_c is the number of the teeth of milling cutter that are concurrently working. The width of cut a_w is replaced by depth of cut a_p and this has no substantial change because for the same cutter the two are proportional. The possibility of the existence of the critical condition of constant amplitude self-excited vibration is still discussed, for this purpose, let

$$x(t) = a_o \cos \omega t, \tag{13.3}$$

Substituting the above equation into Eq. (13.2), and using harmonic balance method [79] lead to

$$h + a_p k_s z_c B = 0; \tag{13.4}$$

$$a_o = 1.1547\sqrt{D/\beta}, \tag{13.5}$$

where,

$$\mathrm{D} = \frac{hA}{kB} + (\frac{\omega^2}{\omega_o^2} - 1) \tag{13.6}$$

$$\omega_o^2 = k/m, \tag{13.7}$$

where, A and B are still given by Eqs. (12.65) and (12.66). Equations (13.4) and (13.5) are the stability equations of the system. When the parameters of the dynamic characteristics of the machine tool ω_o, k, β, h and the parameter of the cutting process k_s are given, from Eqs. (13.4) and (13.5) the amplitude and the frequency of self-excited vibration can be solved as the functions of cutting parameters,

$$a_o = \varphi(N, a_p), \tag{13.8}$$

$$\omega = \psi(N, a_p). \tag{13.9}$$

Equation (13.8) defines a curved surface of zero damping in space (a_o, N, a_p), as shown in Fig. 13.1 (only a piece of the curved surface is shown in the figure). Any a point (N, a_p) on the plane N-a_p is corresponding to a point A on the surface, the height of the point A is the stabilized amplitude a_o under the technological parameters (N, a_p). Equation (13.9), then marks every point on the surface with a value of ω, the frequency of stabilized chatter vibration.

Let a_o = 0, then, Eqs. (13.4) and (13.5) get back to Eq. (12.64), that is, when vibration amplitude is very small, the stability equations of a non-linear model will degrade into that of a linear model. And the section of surface $a_o = \varphi(N, a_p)$ at

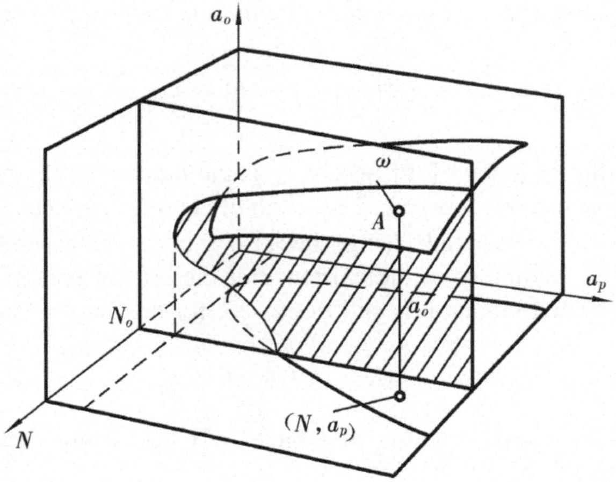

Fig. 13.1 *WS* stability chart

plane $a_o = 0$ (i.e., the plane N-a_p) are the stability chart of a linear model as shown in Fig. 12.18 (in Fig. 13.1 only one earlobe, one of the unstable areas, is given). The figure is called *WS* stability chart (where, *W* stands for width of cut or depth of cut, and *S* stand for cutting speed or spindle revolutions) [90, 91].

In Fig. 13.2a, the *WS* stability chart, one earlobe is shown, through the point ① on the earlobe making a section of N = constant (200 r/min), we get a *AW* stability chart ① on the plane a_o–a_p, as shown in Fig. 13.2b. The curve ② is another *AW* stability char corresponding to the point ② (N = 187.5 r/min) on the earlobe in Fig. 12.2a (where, *A* stands for amplitude). In graphing, the data in British system of literature [90, 91] have been converted into international system: $h = 13694.61$ N/mm, $f_o = 173.25$ Hz, $k = 0.3273 \times 10^6$ N/mm, $z = 24$, $z_c = 4$ and $\beta = 410$ mm^{-2}.

Now let's focus to the curve ① and the area surrounded by the curve (shaded area) in Fig. 13.2. The ordinate expresses the stabilized amplitude and the abscissa expresses the dimensionless depth of cut. The figure shows the relationship of the

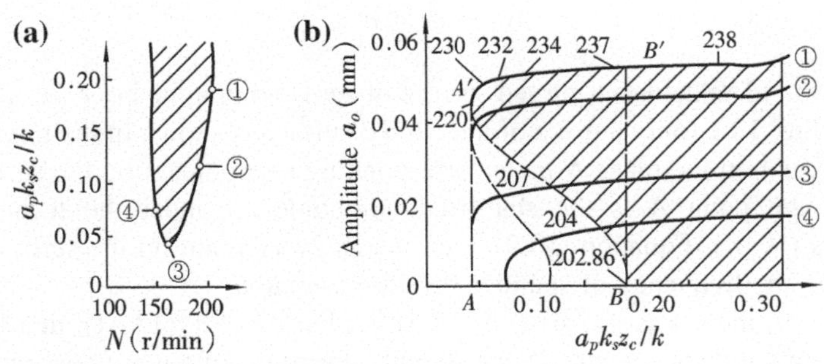

Fig. 13.2 *WS* stability chart and *AW* stability chart

depth of cut a_p and the amplitude a_o at steady state, It reveals some of the essential differences between nonlinear models and the aforementioned linear models. First of all, to be sure, the curve in the graph (solid line and interrupted line) represents the critical situation, i.e., total damping is zero, the shaded area represents the negative damping area, and namely the area of self-excited vibration, the rest is positive damping area. When depth of cut a_p is increased from zero, the system is stable because the working point (a_p, a_o) is in the area of positive damping, and so no self-excited vibration occurred. When depth of cut increases to the point B on the abscissa, it comes in the area of negative damping, and vibration immediately excited, the amplitude jumps up to point B', and is stabilized at the point. If a_p continue to increase, then the vibration amplitude will increase slowly along the solid line in the figure. If, on the other hand, the depth of cut a_p is gradually decreasing from a larger value, then the amplitude a_o will slowly decrease along solid line, to point A', and gets into the area of positive damping, self-excited vibration vanished, and the amplitude drops from point A' to point A on the abscissa.

The depths of cut corresponding to points of A and B are denoted as a_{pcr1} and a_{pcr2}, respectively. When $a_p < a_{pcr1}$, the cutting system is stable for any disturbance, and this is equivalent to the case that the origin of the phase plane (x, $\dot{x}$) is globally stable; When $a_p > a_{pcr2}$, the cutting system is under soft exciting status, any a minor disturbance can make the system unstable, this case is equivalent to that the origin of the phase plane (x, $\dot{x}$) is unstable, and there is a stable limit cycle surrounding the origin, as shown in Fig. 12.1a; When $a_{pcr1} < a_p < a_{pcr2}$, the system is stable if the disturbance is small (not crosses the interrupted line in the figure); A large disturbance (it crosses the interrupted line and gets into the area of negative damping), will, however, makes the system lost its stability, and the amplitude will jump to the solid line representing steady state chatter. This case corresponding to the situation as shown in Fig. 12.1b, the origin is stable, it surrounded by an unstable limit cycle and a stable limit cycle. The cutting system is under hard excitation status.

It can be seen that this nonlinear model does show very different features than the aforementioned linear model. In summary, there are three different characteristics:

(1) **The self-stabilization of vibration amplitude**

When the amplitude is fully developed from self-excited vibration, it will stabilize itself at a certain level, as shown in Fig. 13.2b; in linear theory, the amplitude of self-excited vibration is considered to increase indefinitely.

(2) **The separation of vibration jump threshold and vibration drop threshold**

These two thresholds correspond to $A(a_{\mathrm{pcr1}})$, $B(a_{\mathrm{pcr2}})$ on the abscissa axis in Fig. 13.2b, and generally $a_{pcr1} < a_{pcr2}$. In linear theory, these two thresholds are considered to overlap.

(3) The finite amplitude instability

This is the condition of the above stated "hard excitation", which does not exist in the linear theory.

The physical origin of the above characteristics is that the equivalent damping of the system varies with the amplitude. In the above linear model, however, the equivalent damping coefficient of the cutting process a_pk_sB (see Eq. 12.58) has nothing to do with the amplitude a_o. In the nonlinear model, the equivalent damping is still $a_pk_sz_cB$, see Eq. (13.4), here, there is one more variable is z_c, the number of the teeth of milling cutter that are simultaneous working, the variable z_c is omitted in the linear model simply because it is for turning, where $z_c = 1$, here, the model is based on the milling, so z_c need to be considered, however, it does not brings substantial change. However, because in the expression of equivalent damping there is a parameter B which is related to chatter frequency ω, and ω is related to amplitude a_o (see Eqs. 13.5 and 13.6), thus, equivalent damping $a_pk_sz_cB$ will changes with amplitude a_o, as shown in Fig. 13.3. In the figure, three curves of $-a_pk_sz_cB \sim a_o$ are given, which are corresponding to three depths of cut a_{p1}, a_{p2} and a_{p3}, respectively, and $a_{p1} < a_{p2} < a_{p3}$. The horizontal line h represents the damping coefficient of the machine tool structure, which has nothing to do with the amplitude a_o and the depth of cut a_p. First, see curve $-a_{p1}k_sz_cB \sim a_o$, it has no intersect point with the horizontal line h, and this means that no matter what the value a_o is, it has $-a_{p1}k_sz_cB < h$, i.e., $h + a_{p1}k_sz_cB > 0$, the total damping is positive, so there is no self-excited vibration. Then, see curve $-a_{p3}k_sz_cB \sim a_o$, which intersects with horizontal line h at point C, at this point we have $-a_{p3}k_sz_cB = h$, i.e., $h + a_{p3}k_sz_cB = 0$, this is the case that total damping is zero. The amplitude a_{o3} corresponding to point C is a steady state amplitude. The stability of this amplitude can be analyzed as follows: If the amplitude a_{o3} is slightly increased by accidental interference, it can be seen in the figure that there is $-a_{p3}k_sz_cB < h$, and the positive damping $h + a_{p3}k_sz_cB > 0$, which is positive, the amplitude will decrease, and then return to a_{o3}; If the accidental interference causes the amplitude to drop slightly, then there is $-a_{p3}k_sz_cB > h$, total damping $h + a_{p3}k_sz_cB < 0$, which is negative damping, and the amplitude will return to a_{o3}, so the amplitude of a_{o3} is stable. Finally, then analysis curve $-a_{p2}k_sz_cB \sim a_o$, it has two intersection points A, B with horizontal line h, and the same analysis will reveal that the amplitude a_{o2}

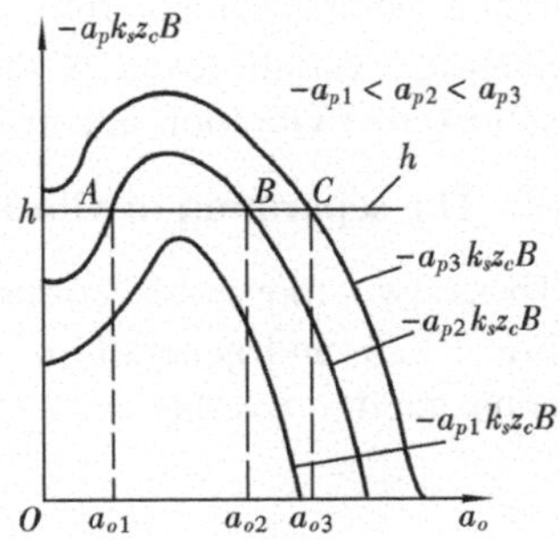

Fig. 13.3 The equivalent damping coefficients and the stability

corresponding to point B is stable, and the amplitude a_{o1} corresponding to point A is unstable, once disturbed, it either drops to zero or jumps to a_{o2}.

The characteristics of the above nonlinear model have been proved to some extent by experimental facts.

13.1.2 The Difficulties and the Problems of the Non-linear Model

Further experimental research and theoretical analysis show that these models have serious difficulties and problems.

13.1.2.1 The Nonlinear Factor on Which the Model Established Has not Been Verified Experimentally

The nonlinear factors on which the model is established, i.e., the nonlinear stiffness of the machine tool structure has not been verified experimentally. Electromagnetic excitation technology and its corresponding analytical calculation methods, which are maturely applied in linear model study, are helpless in non-linear model research. The reasons are as follows:

First, the amplitude of electromagnetic excitation force is too small to reveal the nonlinearity of the machine tool structure under chatter condition. The milling machine and the electromagnetic vibrator the author used in his study are exactly the same as in [90, 91], the capacity of the power amplifier used as high as 1200 VA, but the biggest exciting force amplitude is only 209 N, compared with the dynamic cutting force under machine tool chatter condition, the force is too small. According to the characteristics of amplitude and machine structure given in literature [90, 91], the amplitude of dynamic cutting force under the condition of chatter vibration is at least 900 N. In the literature [90, 91], the frequency response of the machine tool structure obtained by the excitation test is shown in Fig. 13.4a (the author obtained very similar results under the same experimental conditions). In the figure, the bend of the constant frequency curve shows the weak non-linearity of the machine tool structure under the excitation condition. Figure 13.4b is a theoretical frequency response curve calculated based on the nonlinear theoretical model and the relevant parameters of the machine tool structure adopted in the literature [90, 91]. The model can be expressed as:

$$m\ddot{x}_o + \frac{h}{\omega}\dot{x}_o + k(x_o + \beta_1 x_o^2 + \beta x_o^3) = F\cos^2 \pi f t, \tag{13.10}$$

Where, F is the amplitude of the excitation force, the value is calculated from 100 to 1100 N, f is the excitation frequency, and its value is from 166 to 240 Hz. β_1 = 18.87 mm^{-1}. The values of the remaining parameters are as described above.

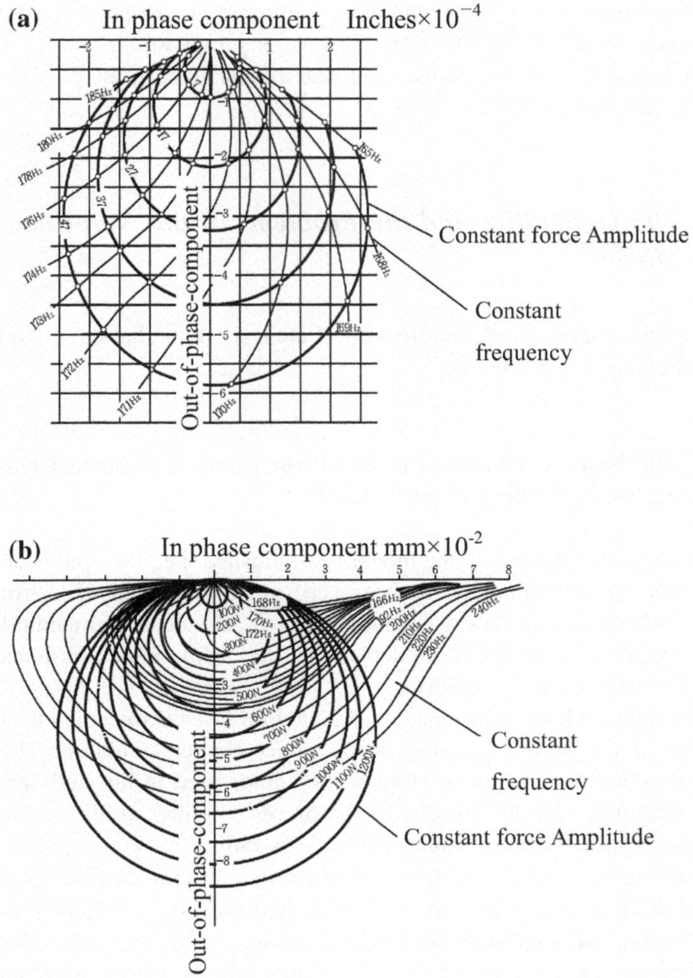

Fig. 13.4 Non-linear structure response of experimental machine. **a** Experimental harmonic response loci of chatter mode in small range. **b** Calculated harmonic response loci for force amplitude range up to 1200 N

In calculation, only the first harmonic is taken, and the higher harmonics are omitted.

In Fig. 13.4b the circles indicate the constant force amplitude, among them, the two circles corresponding to the 200 N (maximum excitation force) and 900 N (dynamic cutting force during chattering) are expressed by interrupted circle and dot dash circle, respectively. In the range of excitation test (within the interrupted circle, which is the equivalent of Fig. 13.4a, the bending of the constant frequency lines is very small, and the bending direction indicates that the structure has a soft spring property. The obvious bending of the constant frequency lines are outside the scope of vibration test, under the exciting amplitude equal to the dynamic cutting force (dot dash circle), the constant frequency lines have bended very seriously, and the bending direction shows the machine tool structure with strong hard spring characteristics.

Obviously, the method from the nonlinear characteristics of machine tool structure obtained by the vibration tests with exiting force amplitude is only 200 N to extrapolate the nonlinear dynamic characteristics of machine tool structure under the condition of real chatter, where dynamic cutting force amplitude is 900 N, is not advisable, the result is not credible.

Secondly, the electromagnetic excitation method cannot simulate the relative size relation between the average cutting force and the dynamic cutting force under real chatter condition. We know that in order for an electromagnetic exciter to work normally, and the waveform distortion is small, the DC part of the excitation current should be much larger than the amplitude of AC part. This characteristic makes the device very suitable for simulating the stability threshold in linear theory, which is to superimpose small fluctuations on considerable static cutting forces. However, under the condition of the large amplitude vibration, tool vibration can in some time away from the work-piece, the dynamic cutting force amplitude can be equal to or even greater than the static cutting force, so that the cutting force is zero for part of a vibration period. The real chatter condition is impossible to simulate by electromagnetic excitation device.

The third, in the linear theory, there are already quite perfect methods to calculate the "operative receptance" (the meaning of this concept will be expressed in next chapter) of the system based on the experimental results of direct and cross receptances. This method is based on "superposition principle". However, this principle is not suitable for nonlinear systems, so it is impossible to calculate the non-linearity of the operative receptance in the non-linearity from the direct and the cross receptances. The literature [90, 91] are actually used a direct receptances to replace the operative receptance. Of course it is not accurate.

In fact, the results of our study showed that the nonlinear stiffness of machine tool structure is not necessarily the most basic, universal non-linear factor that has to be considered.

13.1.2.2 The Results from the Model Are Inconsistent with the Results of the Chatter Test

First, if the nonlinear stiffness of machine tool structure is indeed so strong as required by the model in literature [90, 91], then the amplitude change will result in great changes of vibration frequency, as shown in Fig. 13.2b, the curve ① shown: When the vibration occurs at first, the frequency is about 202.86 Hz, when amplitude is stabilized in 53 $\times$ 10^{-3} mm, the frequency increased to 237 Hz. When $a_p k_s z_c / k$ increases from 0.0418 to 0.2675, the steady-state chatter frequency increased from 220 to 238 Hz. However, the actual chatter test shows that the chatter frequency changes only a few Hertz in a considerable range of amplitudes.

Second, according to the model in literature [90, 91], only working under a speed corresponding to the right side of a unstable lobe in *WS* stability, can appear finite amplitude instability phenomenon, the corresponding *AW* stability charts are shown in Fig. 13.2b as curves ① and ②. Working at the left side of an unstable lobe (as points ③, ④ in Fig. 13.2a) there is no possibility of a finite amplitude instability at all. Their *AW* stability charts are shown in Fig. 13.2b as curve ③ and ④.

As shown in Fig. 13.3, in this model, the physics roots of self-stabilization of vibration amplitude, the separation of vibration jump threshold and vibration drop threshold and the finite amplitude instability are as follows: due to the system stiffness increases with the increasing of amplitude, and chatter frequency also increases along with the amplitude; From Eqs. (12.65) and (12.66), the change of frequency ω changed the equivalent damping $a_p k_s z_c \sin(\omega/Nz)$ and equivalent stiffness $a_p k_s z_c[1 - \cos(\omega/Nz)]$ of the cutting process, where, z is number of milling cutter teeth. The equivalent damping coefficient $B = \sin\beta = \sin(\omega/Nz)$, corresponding to the unit value of "$a_p k_s z_c$", is a function of amplitude a_o, as shown in Fig. 13.5. In the figure there are two curves labeled ① and ②, they are corresponding to the curves ① and ② in Fig. 13.2b and the points ① and ② in Fig. 13.2a, respectively. The curves ① and ② in Fig. 13.5 are first up and then down, and it is this tendency that causes these phenomena above described.

But according to this model, when the machine working on the left side of an unstable earlobe, the equivalent damping coefficient of cutting process is monotonic decline, as shown by curves ③ and ④ in Fig. 14.5, and don't have the kind of

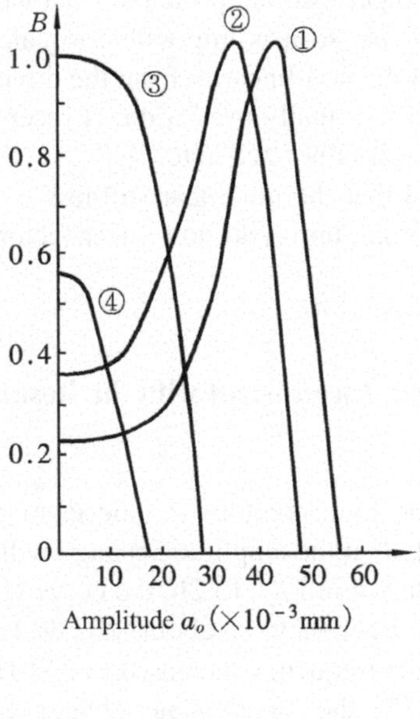

Fig. 13.5 Equivalent damping as the function of amplitude. ① $N = 200$ r/min; ② $N = 187.5$ r/min; ③ $N = 160$ r/min; ④ $N = 150$ r/min

features, that is, first up and then down, as a result, it can't cause the finite amplitude instability and threshold separation phenomenon.

Unfortunately, this is just the opposite with the experimental facts: with the cutting speed corresponding to the left side of an unstable earlobes during machining processing, the range of depth of cut (or width of cut) which produces the finite amplitude instability is much bigger than work at the right speed.

13.2 A Chatter Model Based on the Non-linearity of Cutting Edge Leaving the Work-Piece Material Due to Vibration

In this section, a factor being bound to appear when the vibration amplitude is big enough, is taken into consideration, that is, the movement pass of the cutting edge goes beyond the work-piece, and based on this, a non-linear differential equations describing the cutting process is established. The solution of the equation explained the reason of amplitude stabilization, and the sizes of stabilized amplitude under various conditions.

13.2.1 Differential Equations Describing Cutting Chatter

13.2.1.1 The Case of Small Amplitude (Linear Theory)

Figure 13.6 shows the dynamic cutting process under different vibration amplitudes. Among them, Fig. 13.6a shows the circumstances of a comparatively small amplitude a_o ($a_o = s_o/2$), the thin solid line in the figure shows the cutting pass $x_o(t)$ of a cutting tooth, and the thick solid line shows the top surface $y_o(t)$ of the layer of material to be removed, here, it is cutting pass $x_1(t)$ of the last cutting tooth. Note that the cutting pass of the tooth $x_o(t)$ is all in the work-piece material because of the smaller amplitude. The linear theory of machine tool dynamics is based on this situation.

The instantaneous thickness of cut $s(t)$ can be expressed as

$$s(t) = x_o(t) - y_o(t), \tag{13.11}$$

or,

$$s(t) = x_o(t) - x_1(t). \tag{13.12}$$

Because the $x_o(t)$ and the $x_1(t)$ are relatively small, the whole process can be linearized, and we only consider steady chatter on the stability threshold, so there is

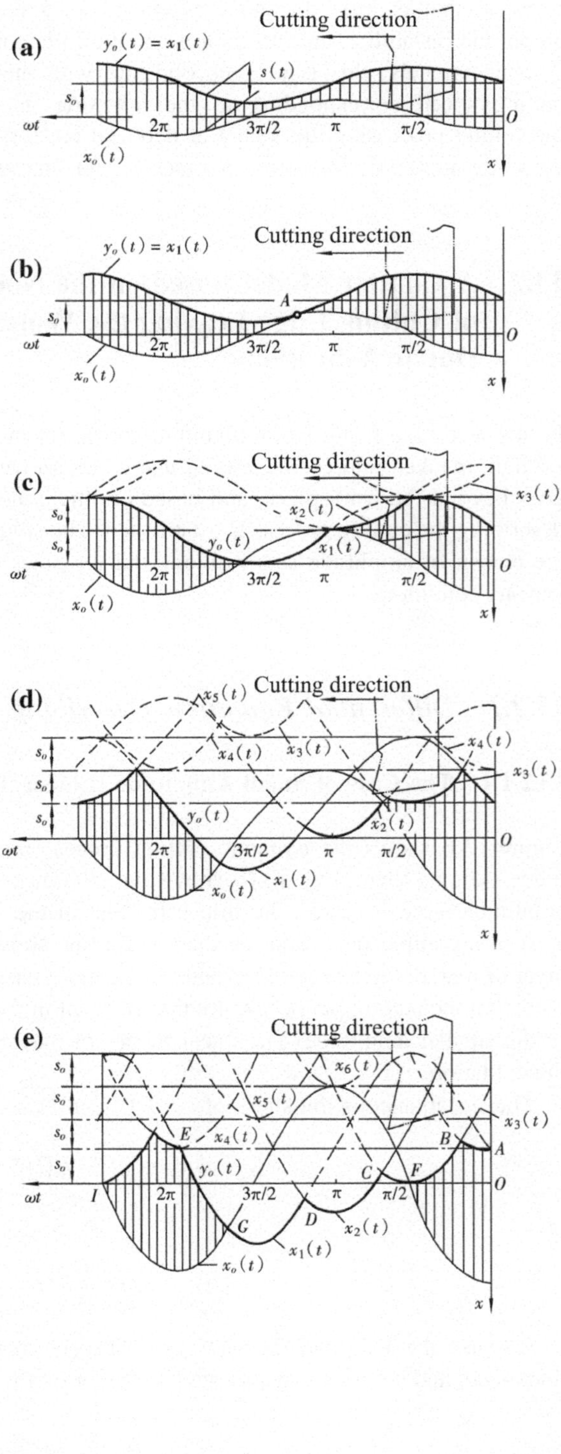

Fig. 13.6 The patterns of dynamic thickness of cut under different vibration amplitude ($\beta = 270°$). **a** $a_o = s_o/2$, **b** $a_o = s_o/\sqrt{2}$, **c** $a_o = s_o$, **d** $a_o = 2s_o$, **e** $a_o = 3s_o$

good reasons for the assumption that the $x_o(t)$ and the $x_1(t)$ are the same harmonic with equal amplitude a_o,

$$x_o(t) = a_o \cos \omega t, \tag{13.13}$$

$$x_1(t) = a_o \cos(\omega t - \beta) - s_o, \tag{13.14}$$

Where, β is the phase difference between the two wave passes of successive cutting teeth, which can be expressed as,

$$\beta = \frac{T\omega}{z} = \frac{60\omega}{zN} = \frac{2\pi\omega}{z\Omega}. \tag{13.15}$$

Substituting Eqs. (13.13) and (13.14) into (13.12) and rearranging result in the dynamic incremental thickness of cut superposed on the static thickness of cut s_o

$$\Delta s(t) = s(t) - s_o = Ax_o(t) + \frac{B}{\omega}\dot{x}_o(t), \tag{13.16}$$

where,

$$A = 1 - \cos \beta, \tag{13.17}$$

$$B = \sin \beta, \tag{13.18}$$

And $\dot{x}_o(t)$ is the instantaneous vibration velocity of the cutting edge with respect to work-piece,

$$\dot{x}_o(t) = \frac{dx_o(t)}{dt} = -a_o\omega \sin \omega t. \tag{13.19}$$

The physical meaning of coefficients A and B is that they are equivalent stiffness coefficient and the equivalent damping coefficient, respectively. Equations (13.17) and (13.18) show that they only depend on β, and from Eq. (13.15), they depend on ω, but have nothing to do with amplitude a_o. This means that when a_o is relatively small, the chatter process is linear.

Because $\Delta s(t)$ is small, it can be regarded that $\Delta F(t)$ to be in direct proportion to $\Delta s(t)$:

$$\Delta F(t) = a_p k_1 z_c \Delta s(t), \tag{13.20}$$

This is the exciting force acting on the machine tool structure.

Machine tool structure is multi-degree-of-freedom system, in order to simplify the analysis, we are here to study the unstable mode only, and thus the machine tool structure is treated as a single-degree-of-freedom linear system, its differential equation of motion can be written as:

$$m\ddot{x}_o(t) + \frac{h}{\omega}\dot{x}_o(t) + kx_o(t) = -\Delta F(t), \tag{13.21}$$

The negative sign before $\Delta F(t)$ shows that the increase of $\Delta F(t)$ will push the cutter away from the work-piece, and thus to decrease the $x_o(t)$ (see positive direction of axis x in Fig. 13.6).

Substituting Eqs. (13.16) into (13.20), and then into (13.21) results in a linear ordinary differential equation. The solution of this equation gives the critical value a_{pcr} of the depth of cut for the machine to start chattering, i.e., the "Stability threshold".

Geometrically, the cutting process represented by Fig. 13.6a has two characteristics: first, after the cutting through of every cutting edge of tooth a wave surface is left on the work-piece, which is periodic, such as

$$y_o(t + T/z) = y_o(t); \tag{13.22}$$

Second, the relationship between the wave surfaces left on the work-piece by two successively teeth, such as $y_o(t)$ and $y_{-1}(t) = x_o(t)$, than $y_{-1}(t)$ with respect to $y_o(t)$ advance a phase angle $\beta = 270°$ (that is, to move towards the negative direction of the axis ωt, and translate on the positive direction of axis x by a distance s_o, namely

$$y_{-1}(t) = y_o(t + T/z) + s_o. \tag{13.23}$$

The above is the description of the chatter process by linear theory. When the amplitude a_o is less than the threshold given by the following formula, the above description is correct:

$$a_{cr} = s_o/[2\sin(\beta/2)]. \tag{13.24}$$

Figure 13.6b represents the limit of $a_o = a_{cr}$, when the two passes of $x_o(t)$ and $x_1(t)$ are tangent to each other at point A, and the instantaneous thickness of cut at the point is zero.

13.2.1.2 The Case of Large Amplitude (Non-linear Theory)

When the amplitude is greater than the critical value identified by Eq. (13.24), as shown in Fig. 13.6c–e, in a vibration cycle, part of the pass of the cutting tool will go beyond the work-piece material, so that the cutting process is interrupted. This fact has a substantial effect on the cutting process. Let's take a closer analysis on Fig. 13.6e, which represents the situation: $\beta = 270°$, $a_o = 3s_o$. The curves in the figure represent the cutting passes of the milling cutter's teeth, in which the

$x_o(t)$ (fine line) is the pass of the n-th tooth, and $x_1(t)$ is the pass of the $(n-1)$-th tooth, and $x_2(t)$ is the pass of the $(n-2)$-th tooth…, etc. It should be noticed, that compared with $x_i(t)$, $x_{i-1}(t)$ has a time delay $\frac{T}{z} = \frac{\beta}{\omega}$, and its central line has a shift of s_o towards the negative direction of axis x, i.e.,

$$x_1(t) = x_o(t - \frac{T}{z}) - s_o, \tag{13.25}$$

$$x_2(t) = x_o(t - \frac{2T}{z}) - 2s_o, \tag{13.26}$$

$$x_m(t) = x_o(t - \frac{mT}{z}) - ms_o. \tag{13.27}$$

Before the n-th cutting tooth cuts through, the upper surface of the work-piece (solid line) is a periodic wave line formed by the curves of *AB*, *BC*, *CD* and *DE*, they are parts of the curves $x_4(t)$, $x_3(t)$, $x_2(t)$ and $x_1(t)$, respectively. When n-th cutting tooth is cutting, the part *FG* of its pass $x_o(t)$ is outside of the work-piece material. The shadow area in the figure is the material cut off by the n-th cutting tooth, which is composed by $x_o(t)$, $x_1(t)$, $x_3(t)$ and $x_4(t)$, but has nothing to do with $x_2(t)$. It can be seen that the variation of instantaneous cutting thickness is very complicated when the cutting edge left the work-piece material during part of a vibration cycle. In contrast to the small amplitude, where $s(t)$ is only related to $x_o(t)$ and $x_1(t)$ (see Eq. 13.12), here, the previous motion passes of several cutting teeth $x_o(t)$, $x_1(t)$, $x_2(t)$, …, $x_m(t)$ are likely to have their impact on the cutting thickness of the different time, while the way of influence is specified by the time delay T/z (i.e., the angle β) and the ratio of a_o/s_o.

Before the n-th cutting tooth cuts through, the upper surface of the work-piece (solid line in Fig. 13.6e) may be expressed by the following equation,

$$y_o(t) = \max[x_1(t), x_2(t), \ldots, x_m(t)]. \tag{13.28}$$

The above formula should be understood this way: for any moment of t, take the largest one as $y_o(t)$, among the values of $x_1(t)$, $x_2(t)$… $x_m(t)$, because the portion, the values are not the maximum, has been cut off by the subsequent cutting teeth and will not remain on the surface of the work-piece.

The instantaneous thickness of cut of the *nth* tooth can be expressed as:

$$s(t) = \begin{cases} x_o(t) - y_o(t), & \text{when } x_o(t) > y_o(t) \\ 0, & \text{when } x_o(t) \le y_o(t) \end{cases}, \tag{13.29}$$

The dynamic variation of thickness of cut is:

$$\Delta s(t) = s(t) - s_0. \tag{13.30}$$

After the n-th tooth cutting though, the waveform of the work-piece surface is formed by the sine curves of *FC*, *CD*, *DG* and *GI*. The shape of the cutting graphics is far more complicated than the linear conditions though, looking closely at Fig. 13.6e is not hard to find, that the waveform on the surface of the work-piece after the n-th tooth cutting though, i.e., $y_{-1}(t) = FCDGI$ ant that before the n-th tooth cutting though, i.e., $y_o(t) = ABCDE$ still have the relations under linear conditions expressed by Eqs. (13.22) and (13.23).

At this point, Eq. (13.16) failures, and Eqs. (13.20), (13.21) and (13.25)–(13.30) composite a very complex nonlinear differential equations for the finite amplitude chatter of the machine tool. Its non-linearity is expressed in the fact that the exciting force $\Delta F(t)$ is non-linearly relayed on the vibration displacement $x_o(t)$ itself. Here we do not intend to seek exact solutions of the equation, but based on observation and analysis on the phenomenon of machine tool chatter, to adopt the following basic assumptions: only the fundamental wave of exciting force is important for the anti-chatter stability of the system, and the effect of high order harmonic can be ignored. Its reasons are as follows:

First, it is well known that the chatter frequency is always close to the natural frequency of a certain mode of the machine structure, while the high-order harmonic in the excitation force is much higher than the natural frequency of the mode. As long as the higher harmonic in high frequency area did not encounter the other mode, or other modes are relatively far away from the unstable mode, then the response of the machine tool structure on such higher order harmonic exciting force is small, so that they can be ignored.

Second, the steady vibration signal recorded in the chatter test is very close to the sine wave. Figure 13.7a is a vibration signal recorded when a test milling machine takes a severe chatter that the cutter leaves the work-piece in some time, and Fig. 13.7b is the spectrum of the signal. It can be seen that the waveform of vibration is very close to the sine wave, and the base wave component of 200 Hz

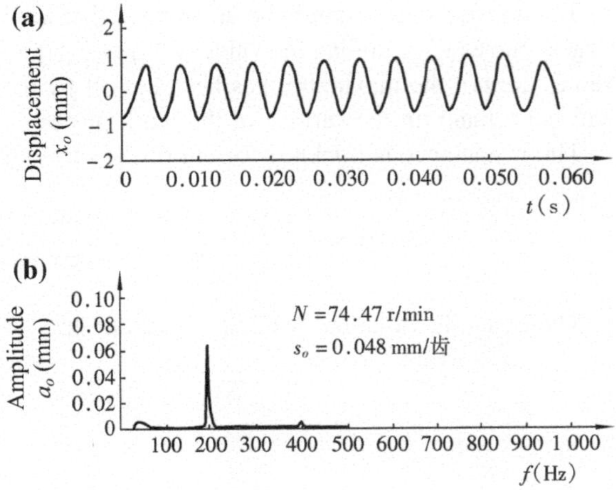

Fig. 13.7 The signal of machine tool chatter and its spectrum. **a** Time-domain signal. **b** Spectrum

occupies the absolute advantage, while the second harmonic of 400 Hz is small and can be ignored.

Based on the above assumptions, we can only study the base wave of vibration, and the steady-state vibration passes of the cutting teeth in Fig. 13.6e are expressed as harmonic:

$$x_o(t) = a_o \cos \omega t, \tag{13.31}$$

$$x_1(t) = a_o \cos(\omega t - \beta) - s_o, \tag{13.32}$$

$$x_2(t) = a_o \cos(\omega t - 2\beta) - 2s_o, \tag{13.33}$$

$$\ldots,$$

$$x_m(t) = a_o \cos(\omega t - m\beta) - ms_o, \tag{13.34}$$

where, m takes the integral part of $\left(\frac{2a_o}{s_o} + 1\right)$.

Figure 13.8 is a waveform that is actually measured from the cutting surface of the work-piece with a profiler, which is very similar to the harmonic waveform (solid line) assumed in Fig. 13.6d, e.

On the other hand, we only take the base wave for the excitation force, and we get:

$$\Delta F(t) \approx W k_s z_c [A' x_o(t) - \frac{B'}{\omega} \dot{x}_o(t)], \tag{13.35}$$

where,

$$A' = \frac{1}{a_o \pi} \int_0^{2\pi} \Delta s(t) \cos \omega t d\omega t, \tag{13.36}$$

$$B' = \frac{1}{a_o \pi} \int_0^{2\pi} \Delta s(t) \sin \omega t d\omega t, \tag{13.37}$$

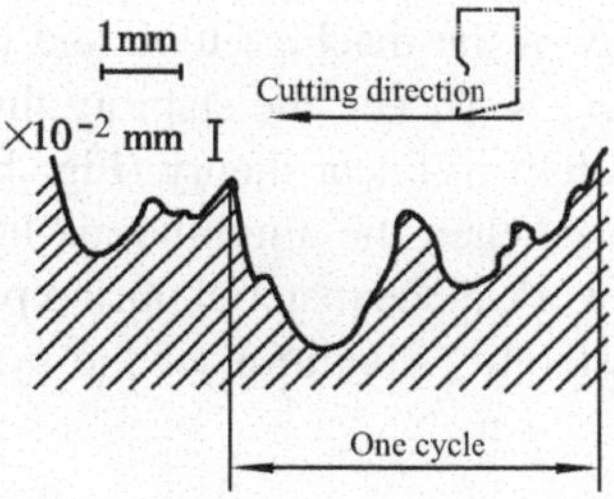

Fig. 13.8 Wave cut on machined surface. $N = 74.47$ r/min, $s_o = 0.060$ mm, $a_o = 0.080$ mm

Are the coefficients of the terms of $\cos \omega t$ 和 $\sin \omega t$ in Fourier series expansion of $\Delta s(t)/a_o$.

13.2.2 The Stability Conditions and the Stability Charts

Substituting Eq. (13.35) into (13.21) leads to

$$m\ddot{x}_o(t) + \frac{1}{\omega}[h - a_p k_s z_c B']\dot{x}_o(t) + [k + a_p k_s z_c A']x_o(t) = 0. \tag{13.38}$$

This is a free vibration differential equation of a nonlinear system, and its hysteresis damping coefficient ($h - a_p k_s z_c B'$) and the stiffness coefficient ($k + a_p k_s z_c A'$) vary with the amplitude a_o, and the reason is that both A' and B' are functions of a_o.

By solving the above equation with harmonic balance method, the stable solution can be obtained to satisfy the following equations:

$$h - a_p k_s z_c B' = 0, \tag{13.39}$$

$$\omega^2 - \omega_o^2 - a_p z_c k_s \omega_o^2 A'/k = 0. \tag{13.40}$$

It should be noticed that, where, A' and B' both are related to a_o and β, and the latters are in turn the function of ω and N (see Eq. 13.15). This set of equations describe the stability conditions of the system,

When the amplitude is small and cutting edge does not leave the work-piece material, the above equations are simplified to:

$$h + a_p k_s z_c B = 0, \tag{13.41}$$

$$\omega^2 - \omega_o^2 - a_p z_c k_s \omega_o^2 A/k = 0. \tag{13.42}$$

This is the stable solution of linear differential equations composed of Eqs. (13.16), (13.20) and (13.21) that is the well-known stability equation in linear theory (penetration rate factor does not included in). The dynamic characteristic parameters of the machine tool structure ω_o, k, h and characteristics of the cutting process k_s, z_c are known, from the above two equations the critical depth of cut a_{pcr} and the chatter frequency ω can be solved as the function of the spindle revolutions N of the machine tool, and thus determined the border of stable area on the plane $a_p \sim N$, i.e., the stability threshold, as shown in Fig. 13.9a, which is the stability chart in linear theory (Fig. 13.9b will be expained in Sect. 13.2.4.3).

When the amplitude is larger and so that the cutting edge of the teeth left the work-piece material during part of a cycle, the Eqs. (13.39) and (13.40) describe the stability condition of the system in a 3-dimensional space of a_p-N-a_o. When the

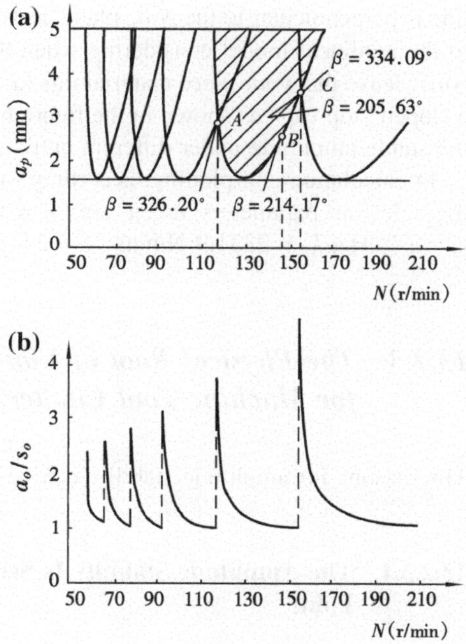

Fig. 13.9 Critical depth of cut and the amplitude at critical depth of cut. **a** Critical depth of cut a_{pcr}. **b** Amplitude at critical depth of cut a_o

parameters ω_o, k, h and k_s, z_c are known, from the two equations $a_o = \varphi(a_p, N)$ and $\omega = \psi(a_p, N)$ can be worked out, and thus a threshold surface and the stable frequency at every point on the surface are determined. Thus, it is possible to determine which areas are stable on the $a_p \sim$ N plane, and which regions are unstable and the chatter amplitude and frequency in the unstable region. In this way, we get the 3-D stability diagram as shown in Fig. 13.10 (for clarity, there is only one unstable lobe is presented in the picture).

The intersection of the graph with the plane $a_o = 0$, as shown in Fig. 13.9a, is the two-dimensional stability chart in the linear theory, which represents the relationship between critical depth of cut and cutting speed, namely, the *WS* stability chart. In this theory, the amplitude is uncertain because the stability equation of linear theory Eqs. (13.41) and (13.42) do not contain factor a_o.

In the 3-D space of a_p-N-a_o, shown in Fig. 13.10, the stability chart should be some infinitely high cylindrical surface A, as shown in the figure, and its generatrix

Fig. 13.10 The 3-D stability chart (*A-S-W*)

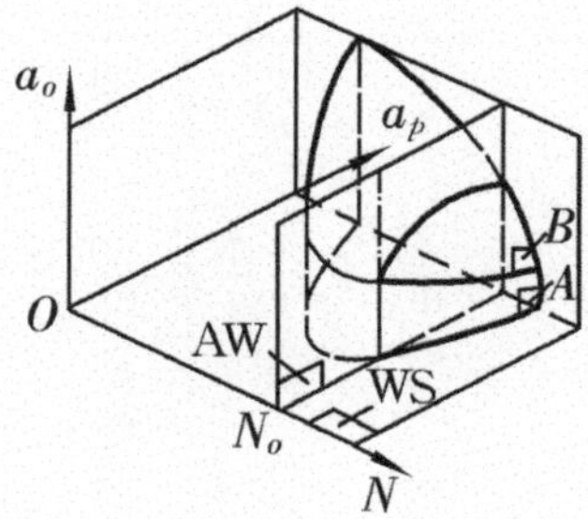

line is perpendicular to the N-a_p plane, as shown in the figure by dot dash line. Due to the nonlinear model considering when the amplitude is large enough, tool pass partly leave the work-piece material this factor, the top of each cylinder is cut off by a sloped "top cap" as shown in the figure by surface B. This top surface represents the stable amplitude under different cutting conditions (a_p, N).

In calculation and plotting, according to the excitation test and the milling test, the relevant parameters taken are: $h = 1488\ 1$ N/mm, $k = 1.232 \times 10^5$ N/mm, $f_o = 178$ Hz, $k_s = 2331.9$ N/mm^2, $z = 24$, $z_c = 4$.

13.2.3 The Physical Root of Vibration Amplitude Stability for Machine Tool Chatter

The reasons for amplitude stability can be explained in two ways.

13.2.3.1 The Amplitude Stability Is Seen from the Energy Balance Point

Figure 13.11 indicates the relationship between the vibrating displacement $x_o(t)$ (abscissa) and the instantaneous cutting thickness $s(t)$ (ordinate) of different amplitude. For each amplitude a_o, we get a loop that is represented by a solid line,

Fig. 13.11 The loop of $x_o(t) \sim s(t)$ showing the relation of vibration displacement and the thickness of cut

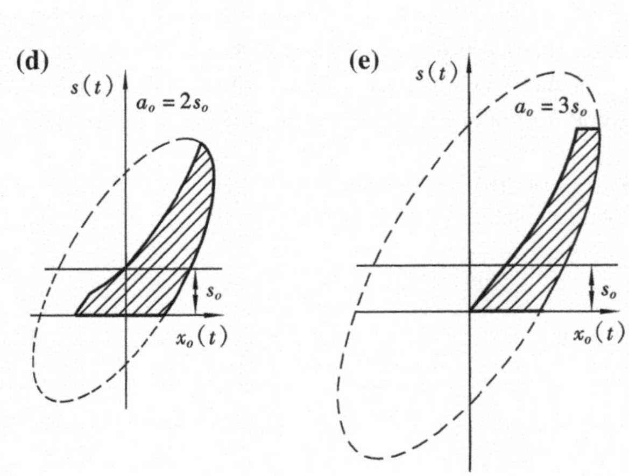

and the area (the shaded region) is proportional to the positive work A_+ that is done by the cutting force to the vibration system in a vibrating cycle.

Note that before the cutting tool edge leaving the work-piece material (Fig. 13.11a, b), the loops are ellipses, and when the tool edge pass partly exceeds the work piece material, the loop is just a part of the ellipse (Fig. 13.11c–e), and the ellipses in figure drawn by interrupted lines indicate the loop of $x_o(t) \sim s(t)$ if the edge never leave the work-piece material.

By the figure it can be seen, that when amplitude increases, the fact that "the cutting edge being out of work-piece material" significantly reduced the positive work done by cutting force to the vibration system, and this factor have the obvious effect of the inhibition of chatter amplitude continues to grow.

In a vibration period, the positive work done by the cutting force to the vibration system can be expressed as:

$$A_+ = -\int_0^{\frac{2\pi}{\omega}} \Delta F(t)\dot{x}(t)dt. \tag{13.43}$$

If the cutter does not leave the work-piece, then substituting Eqs. (13.16)–(13.20) into above equation leads to the expression of A_+, which will be proportional to the square of a_o,

$$A_+ = \pi a_p k_s z_c a_o^2 \sin\beta. \tag{13.44}$$

The relationship between A_+ and a_o is shown in Fig. 13.12 in the interrupted line *OPB*. In fact, at the point *P* on the curve, the amplitude has reached its critical value, so the cutting edge starts to leave the work-piece in part of the time, Eq. (13.16) lost efficacy, and $\Delta s(t)$ should be calculated according to Eqs. (13.28)–(13.34). As a result, the relationship between A_+ and a_o is actually shown in the solid line *OPC* (the calculation takes a_p = 2 mm).

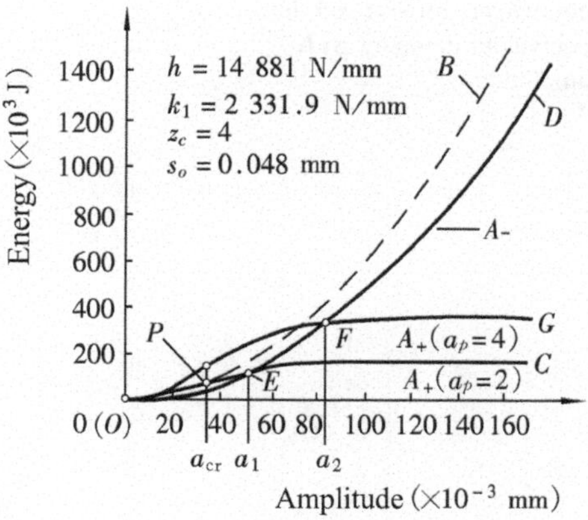

Fig. 13.12 The amplitude stability and energy balance

On the other hand, because of the damping of the machine tool structure, the system dissipates energy during a vibration period is

$$A_- = \pi h a_o^2, \tag{13.45}$$

A_- is also proportional to the square of a_o. It's worth noting that as long as it's within the linear range of the machine tool structure, A_- will always increase in proportion to a_o^2, regardless of whether or not the cutting edge has left the work-piece in part time. This is A_- obviously different from A_+. The relationship between A_- and a_o is shown in Fig. 13.12 in the parabolic *OED* (solid line).

The two curves $a_o \sim A_+$ and $a_o \sim A_-$ intersect at point *E*. The corresponding amplitude $a_1 = 0.050$ mm is the steady-state amplitude of this condition.

If another cutting depth $a_p = 4$ mm is taken in plotting, another corresponding curve $a_o \sim A_+$ will be gotten, as shown in Fig. 13.12 of the *OFG* curve (solid line). It intersects with curve $a_o \sim A_-$ at point *F*, and it corresponding to another steady-state amplitude $a_2 = 0.085$ mm.

The above analysis is for regenerative chatter, if the similar analysis is used on mode coupling chatter, the same results can be obtained: the tool leaving from the work-piece will play a role of stabilizing amplitude.

13.2.3.2 The Amplitude Stability Is Seen from the Damping Property of the System

In Eq. (13.35), $a_p k_s z_c A'$ and $-a_p k_s z_c B'$ are the coefficients of the equivalent stiffness and the equivalent hysteresis damping of the cutting process, respectively. Figure 13.13 presents the relationships of A' and B' with amplitude a_o, for $\beta = 260°$, $\beta = 270°$ and $\beta = 280°$. Before the critical value a_{cr} of amplitude is reached, A' and B' are both constants, and $A' = A = 1 - \cos\beta$, $B' = -B = -\sin\beta$;

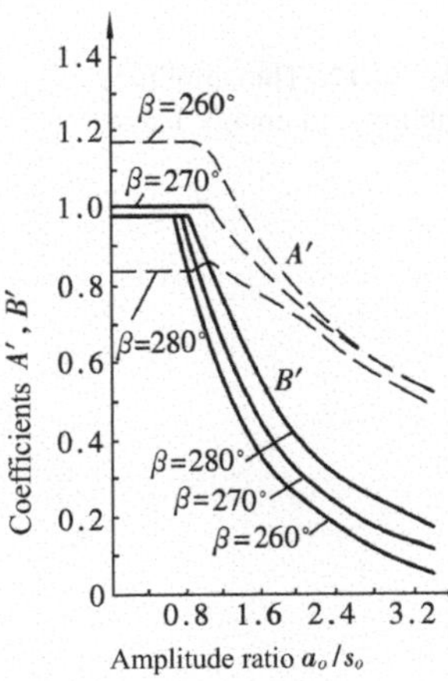

Fig. 13.13 The variation of equivalent stiffness and the equivalent damping with amplitude

once a_o exceeds a_{cr}, Both A' and B' are falling rapidly and the system becomes nonlinear. Notice that B' is proportional to the absolute value of the negative damping of the cutting process. The reason why the cutting process lost its stability, is that the negative damping produced by the cutting process surpass the positive damping of the machine tool structure and turn the total damping of the system into negative. It is not difficult to understand that the decrease of B' (i.e. the decrease of negative damping of the cutting process) will certainly play a role in stabilizing the amplitude and inhibiting the growth of the chatter.

13.2.4 The Influences of Various Factors on Stabilized Amplitude

The cutting parameters have a very direct impact on the amplitude.

13.2.4.1 Depth of Cut (a_p) or Width of Cut (a_w)

The intersection of an $N = N_o$ (constant) with the threshold surface (see Fig. 13.10) results in a stability chart of the a_p-a_o plane, as shown in the solid line in Fig. 13.14. The chart shows the relationship between the depth of cut a_p and the vibration amplitude a_o, i.e., an AW stability chart. The shaded part represents the region where the total damping coefficient is negative, namely the unstable region. The chart shows that when $a_p < a_{pcr}$, the process is stable; When $a_p = a_{pcr}$, the system loses its stability, its amplitude jumps to the critical amplitude a_{cr}, and when $a_p > a_{pcr}$, the stable amplitude a_o will slowly increase along the curve AD. The curve AD represents the stable amplitude of various a_p. This is because any disturbance causes the work point to leave the line AD and eventually returns to the curve.

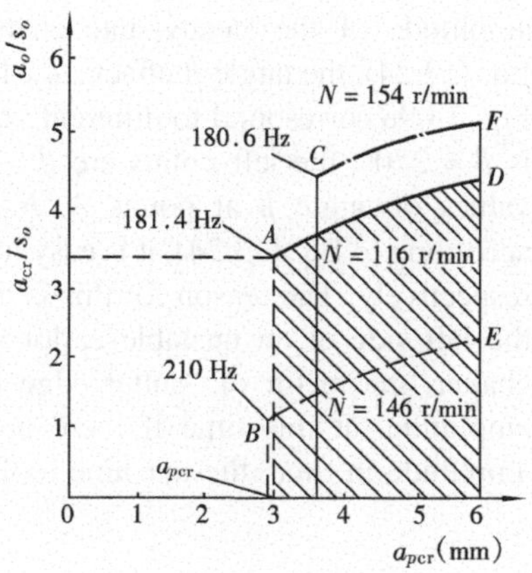

Fig. 13.14 The AW stability chart and the factors influencing stabilized amplitude

13.2.4.2 Thickness of Cut (s_o)

Notice that the ordinate in Fig. 13.14 is the ratio a_o/s_o, obviously under the same conditions, the bigger the s_o is, the larger the value of a_o is. In this model, s_o is actually the function of "scale", because the critical amplitude a_{cr} is proportional to s_o (see Eq. 13.24). As we know, in this model, only when the amplitude increasing enough so that part of the pass of the cutting edge exceeds the work-piece material, can it be stabilized. And the larger the s_o is, the more difficult for the edge to leave the work-piece material. Therefore, it is not difficult to understand that the stabilized amplitude increases with the increasing of s_o or, of the feed rate.

It needs to be explained that the amplitude increases with s_o is only true in a certain range. When s_o is greater than a critical value, the non-linearity of the cutting force will bring about fundamental changes, which will be discussed in next chapter.

13.2.4.3 Spindle Revolutions (N)

In Fig. 13.14, the solid line *AD*, interrupted line *BE* and the dot dash line *CF* are the *AW* stability charts corresponding to three different spindle speed, and the section plane is respectively through *A*, *B* and *C* in Fig. 13.9a. From comparing the three stability charts in Fig. 13.14, it can be seen that when the spindle speed *N* increases from 116 (*A*) to 146 r/min (*B*), i.e., from the left side of an earlobe shifts to the right side, the amplitude decreased to less than a third of the original. If the speed continues to increase to 154 r/min (*C*), that is to go to the left of the next earlobe, the amplitude will suddenly jump to more than four times the amplitude of point *B*. It is visible that spindle speed has a very large impact on vibration amplitude. Figure 13.9b expresses the change of the critical amplitude a_{cr} with the spindle revolutions. Linking Fig. 13.9a, b, it is visible that in the left side region of each unstable earlobe a_{cr} is larger, move to the right region, a_{cr} decreases, and gets into left side of the next earlobe, a_{cr} skyrocket up again.

The reason why the critical a_{cr} is so sensitive to spindle speed, and so the amplitude of the steady-state chatter, because it changed β, and according to Eq. (13.24), the angle β affects a_{cr}. In fact, the points of the earlobe curve shown in Fig. 13.9a correspond to different values of angle β: the lowest point of the earlobe is $\beta = 270°$, the left points are $\beta > 270°$, and the right points are $\beta < 270°$. And values of angle β at points *A*, *B* and *C* are 326.2°, 214.17° and 334.09°, and according to Eq. (13.24), it is easy to calculate the a_{cr} as $3.44s_o$, $1.046s_o$ and $4.46s_o$, respectively. The reason for this is that, because to work at a spindle revolutions at the left side of an unstable earlobe (e.g., Fig. 13.9a point A), if in the event of chatter, the factor of "cutter edge out of work-piece" to appear only in a large amplitude, at this time if there are no other factors in the system to stabilize amplitude in time, the machine tool will endure extremely violent chatter.

Other factors, such as the dynamic characteristics of the machine tool and the mechanical characteristics of the cutting process, mainly affect the chatter frequency, and then change the Angle β, and finally affect the amplitude.

The above rules are of practical significance to the choice of cutting speed in the actual machining process, which will be discussed in the next chapters.

13.2.5 Remarks

13.2.5.1 Amplitude Stability Comes from the Limitations of Feed Rate

In the cutting, when an chatter occurs, after the amplitude rises to large enough, there will certainly be an amplitude stabilizing factor to stabilize it rather than continued to rise, which is part of the cutting edge pass goes out of the work-piece materials and the tool is only part of the time in participating in cutting, the energy input into the system by the cutting process thus sharp decline, so as to inhibit the chatter amplitude and stabilize it.

As in any machine and any machining process, the selected feed (or thickness of cut) is always limited, when the vibration amplitude of the tool relative to the work-piece full increased, the part of its pass is bound to run out of the work-piece material, so this factor is widespread, essential elements for chatter amplitude limiting. Therefore, it can be said that machine vibration amplitude stability comes from the limitations of machine tool feed rate.

13.2.5.2 The Effect of s_o Plays as a "Scale"

If chatter amplitude is stabilized due to the factor of tool edge leaving the work-piece, so, because this factor began to play a role, only when the amplitude of a_o is bigger than the critical value of a_{cr} determined by Eq. (13.24), therefore, the stability of amplitude a_o is always slightly larger than the critical a_{cr}; As for how much larger, it is related to the difference between the depth of cut a_p and its critical value a_{pcr}.

The critical amplitude a_{cr} is proportional to the nominal cutting thickness s_o and related to the phase difference β between the passes of the two successive cutting teeth; the angle β is in turn determined by the vibration frequency ω and the spindle speed N of the machine tool. In this model, the effect of s_o plays as a "scale", it increased, then the a_{cr} increases proportionally, and so, the steady amplitude a_o increases. Other factors affect the chattering amplitude by altering ω or β and then affect amplitude.

13.2.5.3 Only the Fundamental Wave of the Cutting Force Is Important to the Stability of the Process

In the analysis and establishment of the model, one hypothesis is used: when dynamic cutting, only the fundamental wave of the cutting force is important to the stability of the process, and the effect of high harmonics can be ignored. This hypothesis greatly simplifies the theoretical model, making it possible to use simpler mathematical tools to describe and analyze rather complex phenomena. This assumption is acceptable as long as in the higher frequency range than the chatter mode there is no other significant mode of the machine structure, or when other modes are further away from the instability mode.

The nonlinear model proposed in this chapter explains the phenomenon of amplitude stability. The variation of the amplitude predicted is consistent with the experimental facts qualitatively.

However, the experiment shows that the real chatter is not only as simple as the linear theory, but also more complicated than the nonlinear model proposed in this chapter. For example, the phenomenon of "finite amplitude instability" of machine tool is neither explained by linear theory and nor nonlinear model proposed in this chapter. To explain this phenomenon, in addition to that the motion pass of the cutting edge partly exceeds the work-piece material reflected by the model presented in this chapter, another important nonlinear factors, i.e., the nonlinear dependence of cutting force on the thickness of cut under the large amplitude must be also put into account, and establish a more perfect theoretical model.

Chapter 14
Non-linear Theory of Regenerative Chatter in Cutting Processes (II)

In this chapter, two basic nonlinear factors in cutting process are taken into consideration at the same time: the pass of the cutting edge partly exceeds the work-piece material and the cutting force nonlinearly dependents on the thickness of cut, and a more perfect theoretical model about self-excited vibration of machine tool is established. This model can explain and predict the two important phenomena of cutting chatter: amplitude stability and the phenomenon of machine tool chatter induced by external shock. The latter is known as "finite amplitude stability". These phenomena is unexplainable by traditional linear theory. This model leads to some meaningful conclusions, one of the most important is: under certain conditions, increase of the machine load (rather than decrease load), is helpful to suppress chatter, or even completely eliminate the flutter, and the cutting process is stabilized. The theoretical prediction results are in accordance with the experimental data.

14.1 The Nonlinear Cutting Chatter Model Based on the Cutting Edge Leaving the Work-Piece Due to Severe Vibration and Non-linearity of Cutting Force

14.1.1 Actual Chatter Process of Machine Tools

Figure 14.1 is a schematic diagram of the chatter test on a vertical milling machine. The top surface of the work-piece of the soft steel has a slope of 1°, so that the milling depth of cut can be continuously increased or reduced in the milling process. The accelerometer mounted on the milling head monitors the vibration signals of the milling head in the direction perpendicular to the cutting surface of the

H. Shi, *Metal Cutting Theory*, Springer Series in Advanced Manufacturing,
https://doi.org/10.1007/978-3-319-73561-0_14

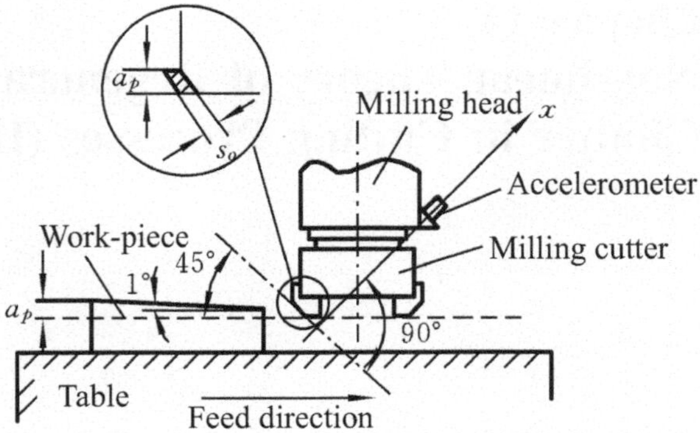

Fig. 14.1 A schematic diagram of chatter test device

work-piece. The vibration signal is recorded by a tape recorder and then processed on a Fourier analyzer.

The typical test results are shown in Fig. 14.2, and the figures *a* and *b* correspond to different mean thickness of cut per tooth (the former is larger and the latter is smaller), and the other conditions are identical. The "×" in the figure shows the corresponding steady-state amplitude when the depth of cut a_p increases, and the

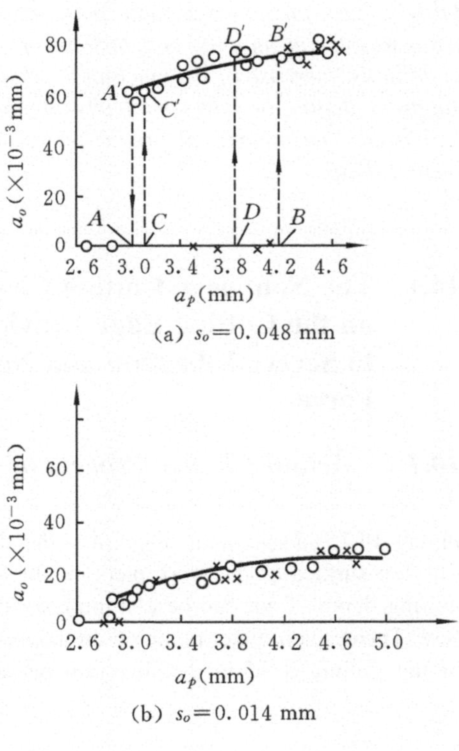

Fig. 14.2 The relationship between vibration amplitude and the depth of cut

"O" represents the steady-state amplitude when the depth of cut a_p decreases, and the curve in the figure represents the experimental relationship between a_p and a_o.

Let's take a look at Fig. 14.2a. The test starts from the a_p = 2.2 mm (beyond the picture) to continuously increase the depth of cut (see the "×" point in the figure). The milling process is stable at the beginning and the amplitude is zero. At point *B*, when a_p = 4.15 mm, the machining process suddenly loses its stability, the amplitude rapidly increases, and then stabilizes at about $a_o = 75 \times 10^{-3}$ mm, which is the *B*′ point in the figure. Continue to increase the depth of cut, the amplitude will increase along curve *A*′*B*′.

In the second experiment, the depth of cut was reduced continuously from approximately a_p = 4.6 mm (see "O" points in the figure), the amplitude is decreasing along the curve *A*′*B*′, it passing through point *B*′ (note that the amplitude does not drop), until point *A*′ is reached, process restored stability, and then the amplitude fell sharply to zero, i.e., the point *A* at abscissa axis.

The amplitude of *B*′ is called "jump amplitude", and the amplitude of *A*′ is called "drop amplitude". The depth of cut *B* at the abscissa corresponding to "jump amplitude" is "jump depth of cut" or "jump threshold", and the depth of cut *A* at the abscissa corresponding to "drop amplitude" is "drop depth of cut" or "drop threshold". In linear theory and the non-linear model I presented in last chapter, the two depths of cut, or the two thresholds, are overlapping, in fact, as shown in Fig. 14.2, for the chatter amplitude jumping and dropping, "thresholds" are different.

The two points *A*, *B* on the abscissa, representing drop threshold and jump threshold divide the cutting processes into three types: when depth of cut is below point *A*, the cutting process is stable of the conventional sense, namely, for infinitesimal perturbation or finite large disturbance the process is stable; when depth of cut is higher than *B*, the process is unstable under normal meaning, and any small disturbance can destabilize the system;

Between two points *A* and *B* will be the so-called "the finite amplitude instability", namely the cutting process for small disturbance is stable, whereas a large enough disturbance can destabilize the system.

The finite amplitude stability of the cutting process can be illustrated in Fig. 14.3. The vibration signal in the time domain (it is in a condensed tine scale) is shown in the figure, and its cutting depth is constant, and is within the range of *A*–*B* on the abscissa of Fig. 14.2a, which is in the region of finite amplitude instability.

Figure 14.3a for cutting depth a_p = 3.05 mm corresponding to point C on the abscissa of Fig. 14.2a, due to the depth of cut is less than the jumping depth of cut, so the process is stable for a variety of small disturbance in environment. Then, it was disturbed by a wooden hammer blow (1st Hammer blow) imparted to the cutter head which set up a damped transient. A second blow (2nd Hammer blow) was sufficiently large for the transient to build up into a steady-state chatter. And then the amplitude jumps from point *C* on the abscissa to the point *C*′. This is so-called jumping phenomenon.

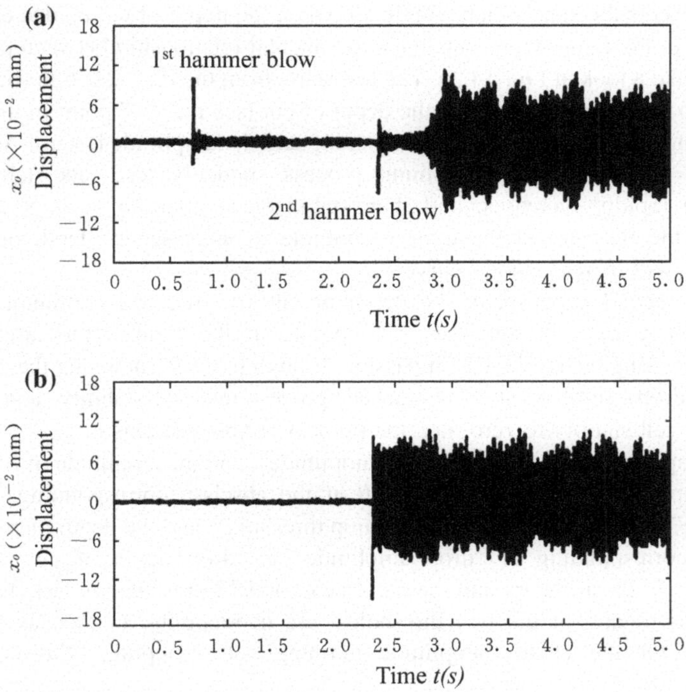

Fig. 14.3 The time course of the finite amplitude instability process. **a** $N = 74.47$ r/min, $s_o = 0.047$ mm/tooth, $a_p = 3.05$ mm. **b** $N = 74.47$ r/min, $s_o = 0.047$ mm/tooth, $a_p = 3.81$ mm

Finite amplitude instability can arise also by impulses generated by interrupted cutting. In the test presented in Fig. 14.3b the work-piece contained a 20 mm hole. The depth of cut was $a_p = 3.81$ mm, corresponding to point D in Fig. 14.2a. Initially the process was stable. However, as the cutter entered the surface discontinuity, a sufficient large disturbance was created which induced the system to chatter, at an amplitude corresponding to point D' in Fig. 14.2a. It is important to note that these vibration persisted even after the cutter had passed the 20 mm hole.

The range of the depth of cut over which finite amplitude instability can arise (range A–B in Fig. 14.2a) may be quite narrow. Figure 14.2b corresponds to a much smaller mean chip thickness of $s_o = 0.014$ mm but otherwise identical conditions. Note that the decrease of s_o greatly reduced the amplitude level at which the chatter vibration settled themselves. According to [94, 95], when finite amplitude instability arises then the chatter amplitudes will grow until the effective relief angle of the cutting tool has been eliminated. This effect did not play an essential part in the tests described; even with fully developed chatter amplitudes the effective clearance angle of the tool was still larger than 5.7°.

The finite amplitude instability of machine tool is actually a self-excited vibration induced by forced vibration. People used to study the forced vibration and self-excited vibration in machining process separately, and did not pay attention on the interaction between them. The traditional linear theory and the non-linear model introduced in previous chapter can't explain this phenomenon. But the understanding and the application of this phenomenon and its rule makes it possible to avoid the effect of various impact factors in the machining, and to improve the stability of the machining process, so it is with practical significance.

14.1.2 The Analysis of Finite Amplitude Instability

The procedure to be adopted is the same as that normally applied to the analysis of conventional linear chatter. It consists of three stages as follows: first, analysis of the dynamic cutting process, as arising under chatter conditions; second, estimation of the dynamic cutting coefficients with the aim of determining the magnitude of the dynamic cutting force; third, determination of the operative receptance locus of the machine tool/work-piece system. The mathematical linking of these three stages leads to the dynamic stability conditions of the process which are conveniently represented in graphical form as a stability chart.

14.1.2.1 Dynamic Cutting Process Under Chatter Conditions

In the previous chapter (Chap. 13), the analysis of the cutting process in the case of the cutting edge leaving the work-piece due to vibration is fully effective, and can be found in Eqs. (13.24)–(13.34).

14.1.2.2 Dynamic Cutting Force Under Chatter Conditions

A well-known fact in the cutting principle is that the change of cutting force is generally non-linear in the larger range of cutting thickness. Figure 14.4 indicates the experimental relationship between the milling force and the thickness of cut s_o per tooth.

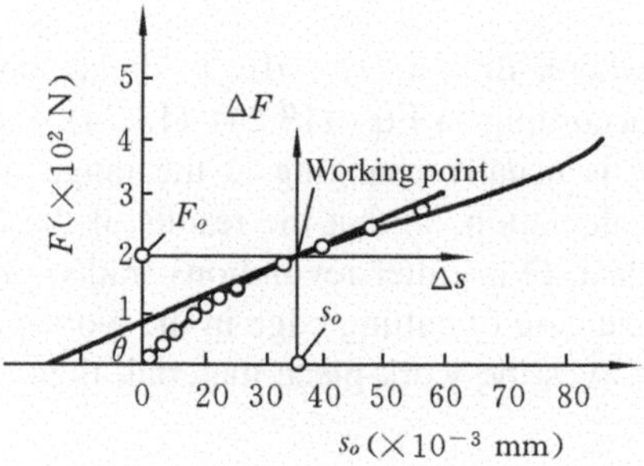

Fig. 14.4 Experimental relationship between mean cutting force and mean thickness of cut per tooth

Fitting the data into a cubic polynomial, we have

$$F = a_p z_c\left(ks + ls^2 + \mathrm{m}s^3\right), \tag{14.1}$$

The coefficients are:

$$\begin{aligned} k &= 6109.6\,N/mm^2, \\ l &= -54141.6\,N/mm^3, \\ m &= 203769.0\,N/mm^4 \end{aligned}$$

To move the origin to the working point, which is determined by the mean thickness of cut s_o and the corresponding average cutting force F_o, the dynamic increment of the cutting force is obtained

$$\Delta F = a_p k_s z_c\left[(\Delta s) + c_1(\Delta s)^2 + c_2(\Delta s)^3\right], \tag{14.2}$$

where,

$$\Delta F = F - F_o;\ \Delta s = s - s_o \tag{14.3}$$

$$F_o = a_p z_c\left(ks_o + ls_o^2 + ms_o^3\right), \tag{14.4}$$

Coefficients k_s, c_1 and c_2 were calculated according to the following formulae,

$$k_s = k + 2ls_o + 3ms_o^2, \tag{14.5}$$

$$c_1 = (l + 3ms_o)/k_s, \tag{14.6}$$

$$c_2 = m/k_s. \tag{14.7}$$

The penetration rate factor is introduced to reflect the resistance of work-piece material to cutting tool's cut in [80], and the dynamic increment of the cutting force caused by vibration $x_o(t)$ can be expressed as:

$$\Delta F = a_p k_s z_c\left[(\Delta s) + c_1(\Delta s)^2 + c_2(\Delta s)^3 + \frac{2\pi c}{z\Omega}\mathrm{d}r\right], \tag{14.8}$$

where, $\mathrm{d}r = \dot{x}_o(t) = \mathrm{d}x_o(t)/\mathrm{d}t$ is penetration rate [80, 90, 91], Δs is calculated according to Eqs. (13.24)–(13.34), c is penetration rate factor. When end milling, c is usually changing at the range of 0.0l–0.03, and $c = 0.0167$ is taken in the calculation, so that the results of the analysis are consistent with the experimental data. Ω is cutter revolutions (rad/s), and $\Omega = 2\pi N/60$. The above is certainly for the case of cutting edge in the work-piece material, and for the time when the edge leaves the work-piece material, then $c = 0$.

Just to be sure, Fig. 14.4 only shows the relationship between the cutting force F_o under static cutting test and the mean thickness of cut s_o, the coefficient of k_s is actually the slope of curve F_o–s_o. Strictly speaking, in Eq. (14.8) the coefficient from dynamic cutting experiments should be used, but because of the difference is small, here we are only going to make an approximate analysis, therefore in the equation, the dynamic cutting coefficient is approximately replaced by k_s, cutting force coefficient.

In the previous chapter, it has been demonstrated that in most cases, only the base wave of the excitation force is important, and the effect of its high harmonics can be ignored, so taking the base wave of Eq. (14.8) leads to

$$\Delta F \approx a_p k_s z_c \left[G x_0(t) - \frac{H}{\omega} \dot{x}(t) \right], \tag{14.9}$$

where,

$$G = \frac{1}{a_o \pi} \int_0^{2\pi} \left[(\Delta s) + c_1 (\Delta s)^2 + c_2 (\Delta s)^3 + \frac{2\pi c}{z\Omega} \mathrm{d}r \right] \cos \omega t \mathrm{d}\omega t; \tag{14.10}$$

$$H = \frac{1}{a_o \pi} \int_0^{2\pi} \left[(\Delta s) + c_1 (\Delta s)^2 + c_2 (\Delta s)^3 + \frac{2\pi c}{z\Omega} \mathrm{d}r \right] \sin \omega t \mathrm{d}\omega t. \tag{14.11}$$

G and H can be explained respectively as equivalent stiffness and equivalent damping coefficient for the cutting process of the unit ($a_p k_s z_c$). Figure 14.5 represents the change of G and H with amplitude a_o. Note that the initial G and H increases with a_o, and after the critical amplitude a_{cr} is reached then they decreases dramatically. This change reflects the strong non-linearity of the cutting process.

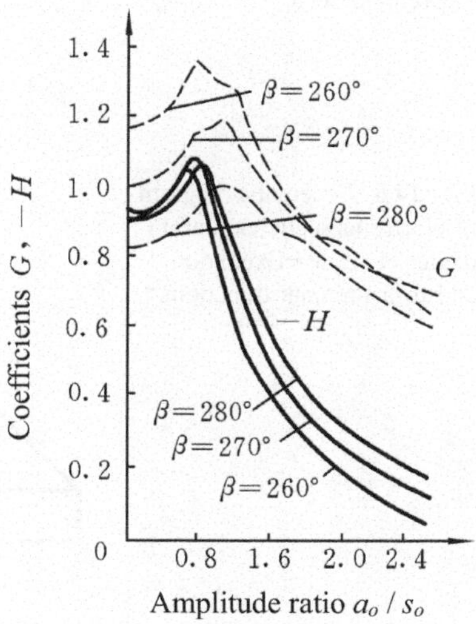

Fig. 14.5 The variation of coefficients of G, $-H$ with amplitude

14.1.2.3 Operative Receptance of the Machine Tool Structure

In this model, the machine tool structure is still assumed to be linear. The operative receptance of the milling machine is obtained by using electromagnetic excitation method.

The three-dimensional electromagnetic exciter is installed on the milling machine working-table, it is able to excite the simulated milling cutter (armature) mounted on the spindle in the directions of three axes 1, 2, 3, as shown in Fig. 14.6; and to measure the relative vibration between the simulated cutter and the table in the three directions. The excitation test is carried out when the simulated cutter is rotating so as to simulate the real cutting process. The Dynamic receptances obtained through excitation test are presented in Fig. 14.7, which include three direct receptances R_{11}, R_{22} and R_{33} and six cross dynamic receptance R_{12}, R_{21}, R_{23}, R_{32}, R_{13} and R_{31}, where, R_{ij} is the complex amplitude of vibration in direction i due to unit amplitude of exciting force on the direction of j (i, j = 1, 2, 3). All nine dynamic receptance components make up a receptance matrix:

$$\boldsymbol{R}(\omega) = \begin{bmatrix} R_{11} & R_{12} & R_{13} \\ R_{21} & R_{22} & R_{23} \\ R_{31} & R_{32} & R_{33} \end{bmatrix}, \tag{14.12}$$

Each of these elements is a complex function of frequency ω, knowing $R(\omega)$, the dynamic receptance of the response in any direction and excited in any direction can be calculated out by means of the superposition principle.

The unit vector in the direction of excitation force can be expressed as

$$\boldsymbol{F} = (F_1, F_2, F_3)^{\mathrm{T}}, \tag{14.13}$$

And the unit vector of the direction in which the vibration is measured can be expressed as

$$\boldsymbol{x} = (x_1, x_2, x_3)^{\mathrm{T}}, \tag{14.14}$$

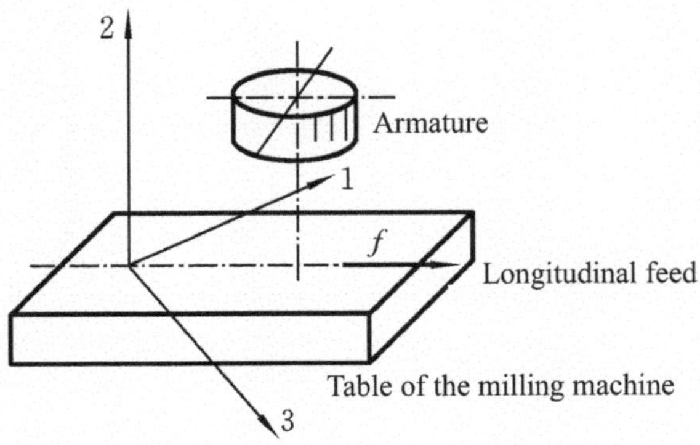

Fig. 14.6 Schematic diagram of electromagnetic excitation device. 1, 2, 3—Excitation and measurement directions

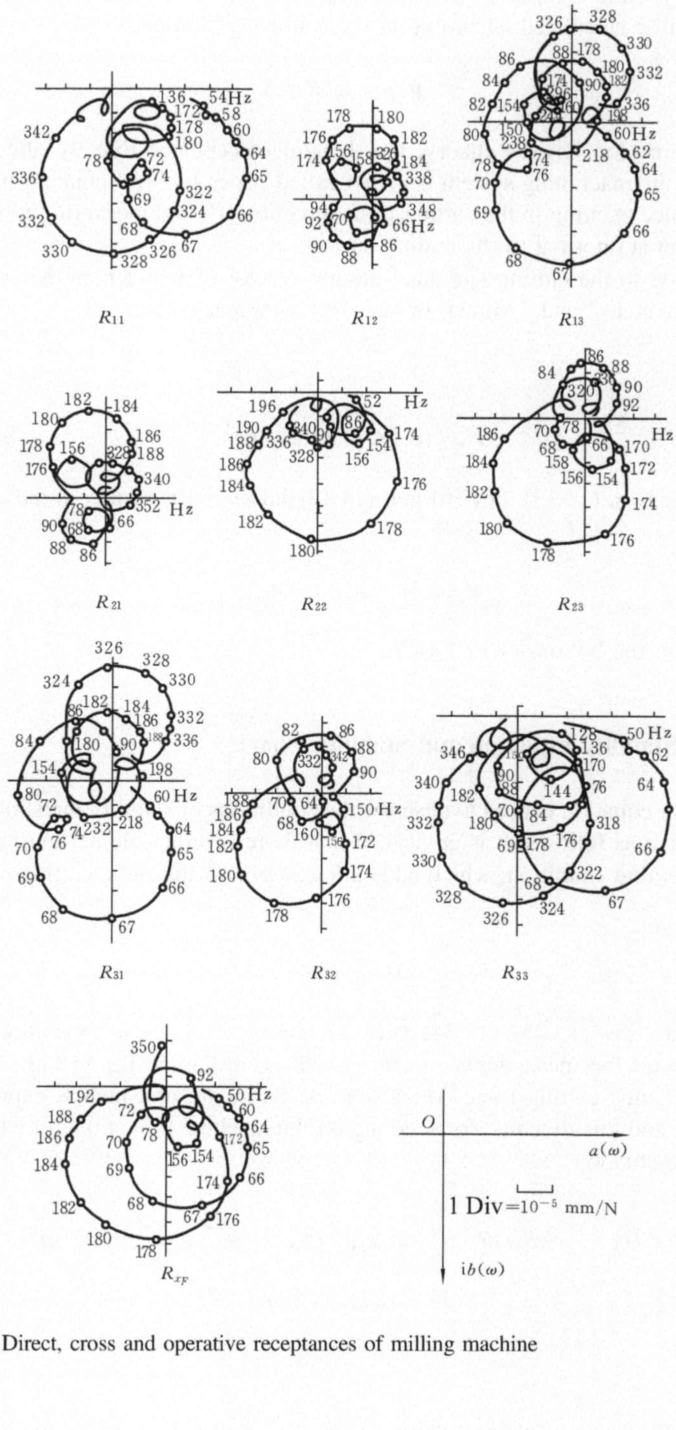

Fig. 14.7 Direct, cross and operative receptances of milling machine

Then the dynamic receptance of the response in the x direction due to excitation force F can be calculated according to the following formula,

$$\boldsymbol{R}_{\boldsymbol{xF}} = \boldsymbol{x}^{\mathrm{T}}\boldsymbol{RF}. \tag{14.15}$$

In machine tool chatter theory, the dynamic receptance directly affecting the stability of the machining system is the so-called "operative receptance", namely in the receptance exciting in the cutting force direction ($\boldsymbol{F}$) and measuring vibration in the direction ($\boldsymbol{x}$) normal to the cutting surface.

According to the cutting test data, the unit vector of $\boldsymbol{x}$ and $\boldsymbol{F}$ in the coordinate system of axes 1, 2 and 3 shown in Fig. 14.6 is respectively:

$$\boldsymbol{x} = (0.50,\ 0.71,\ 0.50)^{\mathrm{T}}; \tag{14.16}$$

$$\boldsymbol{F} = (0.28,\ 0.42,\ 0.86)^{\mathrm{T}}. \tag{14.17}$$

Substituting Eqs. (14.12), (14.16) and (14.17) into (14.15) results in the operative receptance

$$\boldsymbol{R}_{\boldsymbol{xF}}(\omega) = \boldsymbol{a}(\omega) + \mathrm{i}\boldsymbol{b}(\omega), \tag{14.18}$$

As shown in the bottom of Fig. 14.7.

14.1.2.4 Stability Equation and Stability Chart

The stability equation is used to describe the conditions of stability threshold, which can be derive as follows. It is envisaged that there is an accidental interference in the static cutting condition, which adds a fluctuation to the static cutting motion of the cutter

$$x_o(t) = a_o e^{\mathrm{i}\omega t}, \tag{14.19}$$

According to Eqs. (13.25)–(13.34), such a fluctuation will cause a variation of Δs(t) superposed on the mean depth of cut s_o, and according to Eq. (14.8), Δs(t) will cause a dynamic cutting force $\Delta F(t)$, the first harmonic of which is expressed by Eq. (14.9), and this dynamic force acting on the machine tool structure will lead to a second fluctuation,

$$\begin{aligned} x_o'(t) &= -\Delta F(t) R_{zF} = -a_p k_s z_c \left[G x_o(t) - \frac{H}{\omega} \dot{x}_o(t) \right] \cdot (a + \mathrm{i}b) \\ &= -a_p k_s z_c [G + \mathrm{i}H] a_0 e^{\mathrm{i}\omega t} (a + \mathrm{i}b) \end{aligned}$$

The negative sign in the above equation indicates that a positive increment of the cutting force will push the cutter away from the work-piece so that to make $x_o'(t)$ decreases.

Since we only study the constant amplitude and continuous chatter at the stability threshold, the $x_o(t)$ and $x_o'(t)$ should be the same harmonics, i.e.,

$$a_0 e^{i\omega t} = -a_p k_s z_c [G + iH] a_0 e^{i\omega t}(a + ib),$$

Cancel $a_0 e^{i\omega t}$ and separate the real part and the imaginary part, we have

$$Ga + Hb = -1/(a_p k_s z_c), \tag{14.20}$$

$$Gb - Ha = 0. \tag{14.21}$$

This is the stability condition. It must be noted that this condition is related to the amplitude a_o, because G and H are all functions of a_o.

The above two equations constitute a very complex system of simultaneous equations. A special computer program had been developed, after importing the relevant parameters of cutting process k, l, m, c, z_c and s_o and the dynamic characteristics of the machine tool $a + ib$, the program can automatically work out the amplitude of vibration a_o of the system and chatter frequency f as functions of the depth of cut a_p and the spindle revolution N,

$$a_o = \varphi(a_p, N); \tag{14.22}$$

$$f = \psi(a_p, N). \tag{14.23}$$

The above two equations have defined a threshold surface in the 3-D space a_p-N-a_o as shown in Fig. 14.8, and labeled every point on the surface with a chatter frequency.

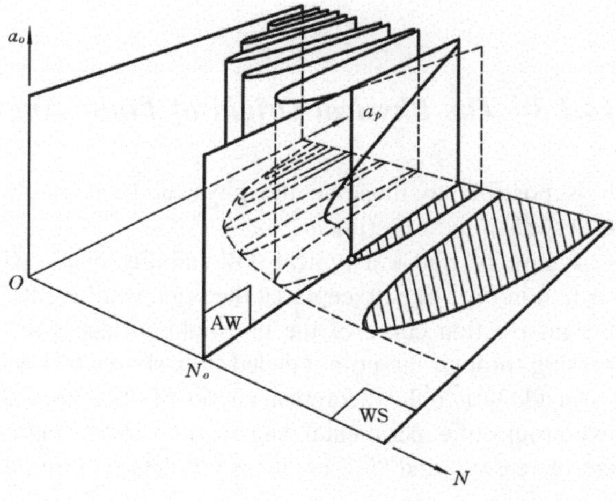

Fig. 14.8 3-D stability chart

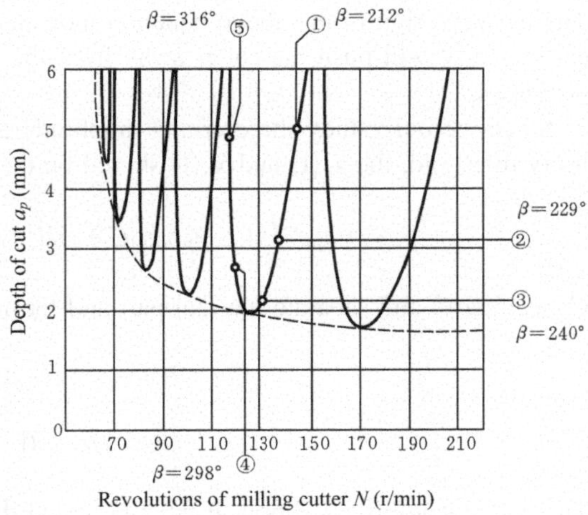

Fig. 14.9 The *WS* stability chart on section plain $a_o = 0$

Let $a_o \approx 0$, then G and H degrade into $(1 - \cos\beta)$ and $-(\sin\beta + c\beta)$, and Eqs. (14.20) and (14.21) become

$$a(1 - \cos\beta) - b\sin\beta - bc\beta = -1/a_p k_s z_c, \tag{14.24}$$

$$b(1 - \cos\beta) + a\sin\beta + ac\beta = 0. \tag{14.25}$$

This is known as the stability equation in linear theory.

In the geometric sense, the normal stability graph is obtained by cutting the threshold surface with the plane $a_o = 0$, as shown in Fig. 14.9. This stability chart is in the plane of the depth of cut a_p and revolutions N, that is, the horizontal plane in Fig. 14.8, which is called *WS* stability chart.

Taking an N = constant (N_o) plane, (the vertical plane in Fig. 14.8), the relationship between the depth of cut a_p and the amplitude a_o is obtained, i.e. the *AW* stability chart.

14.1.3 The Physical Origin of Finite Amplitude Instability

It is possible to illustrate the physical roots of the finite amplitude instability observed in the experiment.

Figure 14.10a is a typical *AW* stability chart, which is the curve of the *AW* section in Fig. 14.8, except that the scale is different. This curve actually represents the intersection curve of the threshold surface with an *AW* section cutting plane passing through the point labeled 2 as shown in Fig. 14.9.

In addition to the theoretical curves (dashed lines and solid lines), in Fig. 14.10a, two groups of experimental data are drawn: "×" indicates the data with the depth of cut increasing, and "O" indicates the data with the depth of cut decreasing.

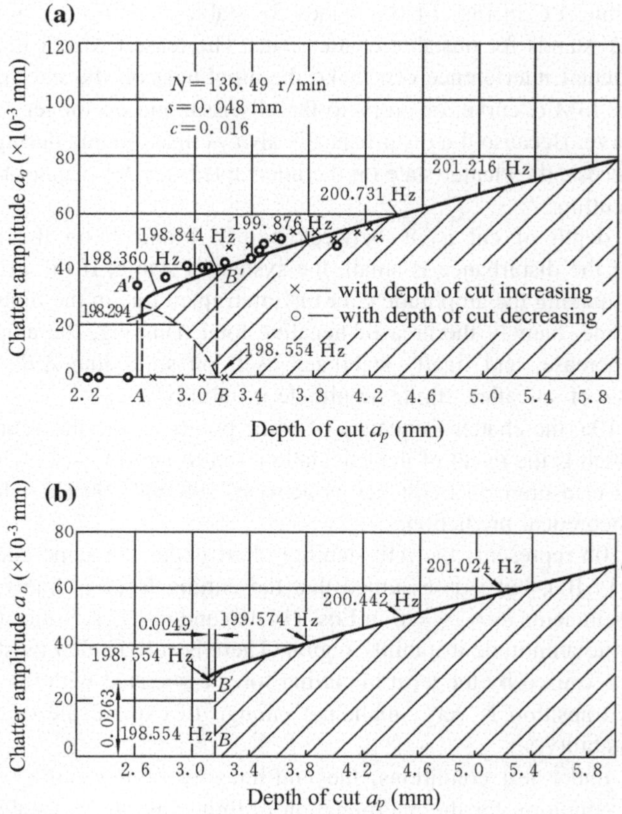

Fig. 14.10 *AW* stability chart. **a** Non-linear cutting force characteristics. **b** Linear cutting force characteristics

Now let's look at the theoretical curve, we know that the theoretical curve represents the stability threshold, the total damping is exactly zero, in the lower right side of the curve (shadow area) is negative damping area, and in the upper left area the total damping is positive.

When the depth of the cutting increases from a small value, initially the system is stable, because the working point is in the area where the total damping is positive. After the point of *B* is reached, the total damping turns negative, the system loses its stability, the amplitude increases dramatically, "jumps" to *B*′ point, and continuous increase of depth of cut, only cause the amplitude slowly rise along the *A'C* curve (the solid line). When the depth of cut from a larger value reduces gradually, at he beginning due to the total damping is negative, the amplitude will rise to *A'C* curve, and then decreases along the curve, through point *B*′, until *A*′ point, total damping into positive, amplitude dropped to point *A*.

You can also see that the experimental results expressed by ×s and Os are very close to the theoretically predicted curve.

The solid line $A'C$ in Fig. 14.10a stands for stable chatter state, and the interrupted line BA' stands for unstable chatter state. The reason why it is unstable, is that any accidental interference can make the amplitude on the interrupted curve whether jumps to $A'C$ curve, or drops to the abscissa, and no longer return to the interrupted curve. Because the disturbance is always unavoidable during the actual machining process, the chatter state on the interrupted line BA' cannot be observed in the actual cutting.

When the depth of cut a_p is between points A and B on the abscissa of Fig. 14.10a, if the disturbance is small, the system is stable; If the disturbance is large enough, making the amplitude over the interrupted line in the figure, i.e., into the shadow zone, namely the area of negative total damping, the amplitude will continue to increase, and finally stabilizes itself on solid line $A'B'$, this is the physical source of so-called "finite amplitude instability".

In Fig. 14.10a, the chatter frequency of some points on the theoretical curve is calculated, which is the result of the calculation according to Eqs. (14.20), (14.21), and the actual measurement frequency is between 198 and 200 Hz, which is very close to the theoretical prediction.

Figure 14.10b represents the AW stability chart under the same cutting conditions as Fig. 14.10a, but it is assumed that the cutting force is linear, that is, to assume the coefficients $c_1 = c_2 = 0$ in Eqs. (14.10) and (14.11). Notice that there is actually no finite amplitude instability region. There is still a small twist on the top curve, which is caused by the term in cutting force associated with the penetration rate. But this transition is very small, not enough to explain the cause of finite amplitude instability.

Therefore, under test conditions, the non-linearity of the cutting force is an indispensable condition for the interpretation of finite amplitude instability.

14.1.4 The Influence of Cutting Parameter on the Nonlinear Chatter Process

Based on the above non-linear cutting chatter model, this section analyzes the influence of cutting parameters on chatter process.

14.1.4.1 Influence of Machine Tool Spindle Revolutions on AW Stability Chart

There are five points labeled with digits 1, 2, …, 5, respectively, on an lobe of the WS stability chart shown in Fig. 14.9, they correspond to five revolutions. The WA stability charts in the cross section passing through the five points is shown in Fig. 14.11 and the corresponding number is marked.

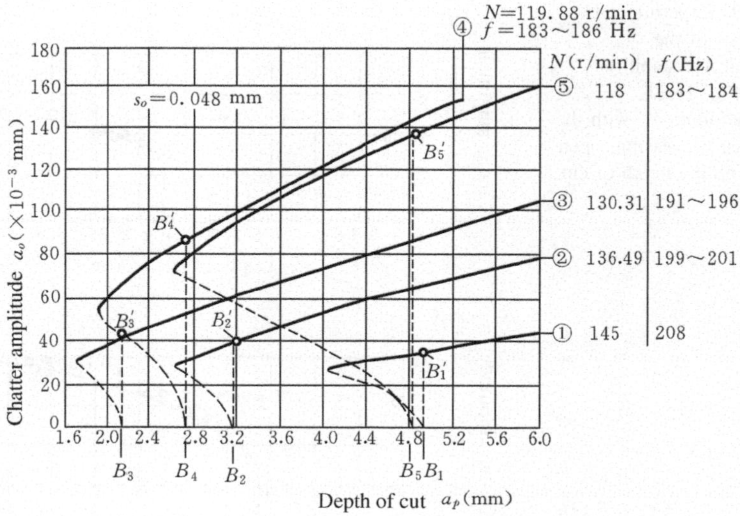

Fig. 14.11 Theoretical *AW* stability charts for different rotation speeds

From the curve 1 to the curve 5, the jump amplitude is gradually increasing. Comparing curve 1 and curve 5, the jump depth of cut of the two *WA* stability charts are almost the same, but the jump amplitudes of the latter is about four times that of the former.

This result provides us some meaningful enlightenment: under the condition of not have larger impact (continuous cutting), should choose the cutting speed corresponding to the right part of a stability lobes, so that in the event of chatter, its amplitude is small; Under the condition of the larger impacts are inevitable (discontinuous cutting), you should choose the cutting speed corresponding to left part of a stability lobes, then very large disturbance will be required to make the process unstable, though if it does become unstable it will chatter with very large amplitude.

In the former case, a small increase, in the latter a small reduction, of the rotation speed will have a marked improvement on stability, provided, of course, that there is a sufficiently large gap between the lobes and they do not overlap.

In Fig. 14.9 point 1 and 5 represents the cutting conditions is difficult to test by experiment, because the jump threshold is too high, and the experimental results of the cutting speed corresponding to point 2, 3, 4 are given in Fig. 14.12, compared them with the relevant theoretical curves in Fig. 14.11, you can see that they coincide with each other pretty good. The results of this study indicate that the unstable cutting area is indeed not "monolithic". The amplitude of different working points is quite different.

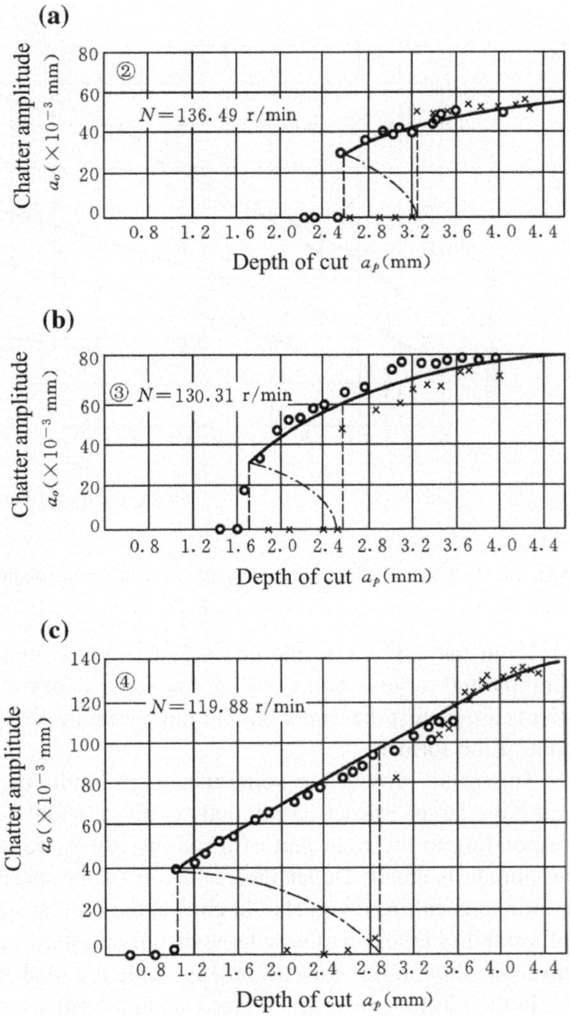

Fig. 14.12 Experimental verifications of the Theoretical *AW* stability charts in Fig. 14.11. Multiplication sign—with the depth of cut increasing, open circle—with the depth of cut decreasing

14.1.4.2 Influence of Mean Thickness of Cut on *AW* Stability Chart

Figure 14.13a shows the *AW* stability charts with a series of different mean thickness of cut. From these charts, it can be concluded that the cutting process can be stabilized by increasing the mean thickness of cut per tooth (i.e., increasing the feed rate) and making it exceed a certain threshold.

Let's fix a_p and increase s_o to see how a_o changes. For this reason, draw a vertical line passes the point of $a_p = 4.7$ mm on the abscissa. When $s_o < 50 \times 10^{-3}$ mm/teeth, the jump depth of cut in the *AW* stability charts are

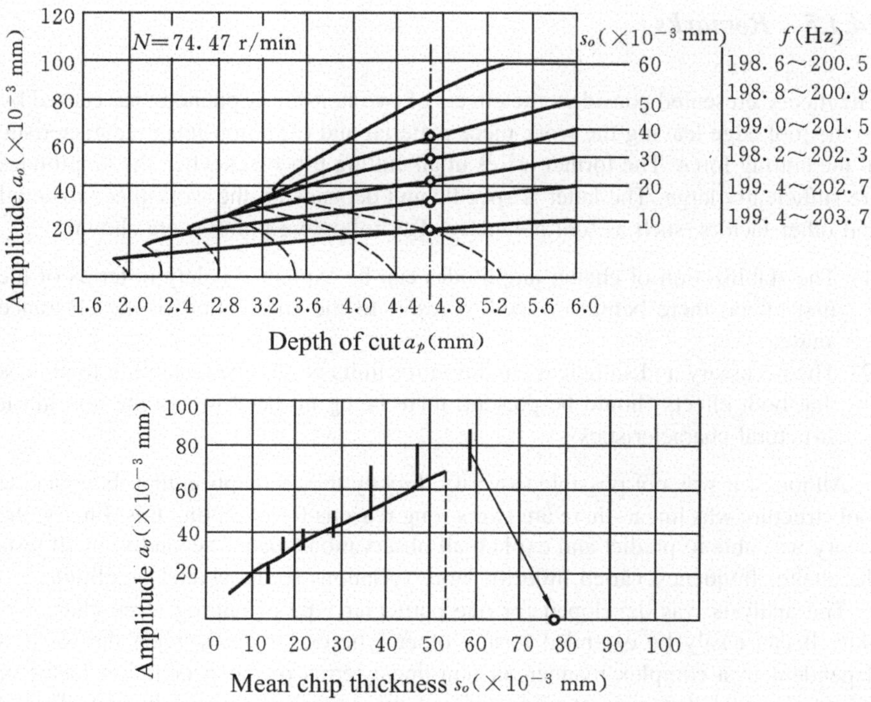

Fig. 14.13 Chatter amplitude as a function of depth of cut and mean chip thickness. **a** *AW* stability charts under different s_o. **b** The relationship of s_o and a_o

lower than 4.7 mm, so the system will be unstable and the chatter amplitude will be indicated by "Os" like the corresponding curve. However, for the curve of $s_o = 60 \times 10^{-3}$ mm/teeth, the jump depth of cut is greater than 4.7 mm, so the cutting process should be stable. If there is no greater impact, its amplitude should be zero.

Figure 14.13b represents the relationship between the amplitude a_o and s_o in the case of $a_p = 4.7$ mm. At first a_o is increasing roughly proportional to s_o, and when s_o exceeds 54×10^{-3} mm/teeth, the chatter will suddenly disappear and the amplitude will drop to zero. The amplitude variation range of actual measurement is also given on the same chart. The maximum thickness of cut with chatter is $s_o = 59.3 \times 10^{-3}$ mm/teeth, with a larger thickness of cut $s_o = 78.5 \times 10^{-3}$ mm/teeth, the amplitude drops to zero (see Fig. 14.13b). The findings suggest that there is an area of "calm" and high-productivity behind the intense chatter region.

The increase in the thickness of cut of each tooth (or the feed rate per tooth) has the effect of eliminating the chatter, which is of course due to the nonlinear dependence of the cutting force on s_o.

14.1.5 Remarks

The theory presented considers the effects of two non-linear phenomena, caused by (i) the tool edge leaving the work-piece material and (ii) a non-linear characteristic of the cutting force. The former arises in all cutting processes when the amplitudes are sufficiently large. The latter is specific and depends on the work-piece material and other factors, such as tool geometry. The following points were shown:

(1) The stabilization of chatter amplitudes can be explained solely in terms of the first effect; there being no need to resort to the elimination of the clearance angle.
(2) The necessary and sufficient condition for finite amplitude instability to arise is that both effects should be present; there being no need to assume non-linear structural characteristics.

Although it was not possible to verify directly the assumption that the machine tool structure was linear, there are two strong reasons for accepting this. Firstly, the theory was able to predict and explain all observations made. Secondly, in all tests the chatter frequency varied little for large variations of the chatter amplitude.

The analysis was developed for one particular type of cutting force characteristic. It can easily be extended to any other type or for cases when the force is dependent in a complex manner on non-linear terms of the penetration factor or mixed terms of the penetration factor and the chip thickness coefficient. This is made possible by the recognition that, because of the separation of the modes of vibration, first order approximations are adequate. In other words, when the chatter vibration is reasonably harmonic then only the fundamental components of the Fourier expansions of force variations need to be considered. By using this principle, it was possible to describe complex phenomena with the aid of relatively simple mathematics.

The theory presented has led to some important practical conclusions. The most interesting one confirms and explains what is well known in the workshop, that is, that in some cases chatter can be eliminated by a substantial increase of the mean chip thickness, by an increase of productivity.

14.2 Digital Simulation of the Cutting Chatter Process

By means of non-linear digital modeling and simulation method to study the mechanism of the cutting chatter process, and to reveal whole process for chatter to occurrence, to development and finally to be stabilized, and analyzes the influence of cutting parameter on this process, in order to achieve the purpose of the forecast and control chatter is a very meaningful work. The non-linear theory of machine tool chatter developed and the wide application of digital computer in engineering have provided theoretical basis and technical conditions for this purpose.

In this section, based on the non-linear theoretical model of machine tool chatter, described above, the dynamic characteristics of metal cutting process are studied by microcomputer numerical simulation. The simulating results show a series of new conclusions, such as the amplitude stability and the finite amplitude instability, etc., which are different from traditional linear theory. The conclusions of practical significant is that the stabilized vibration amplitude is closely related to cutting parameters, so the reasonable adjustment of cutting parameters can effectively suppress the chatter. It is particularly important to find that the suppression of chatter does not necessarily require sacrificing productivity, and sometimes increasing the cutting load can actually inhibit the chatter.

14.2.1 Equation of Motion and Theoretical Model

The two common non-linear factors in the metal cutting process were analyzed previously, these are: part of the pass of the cutting edge is out of the work-piece material, and the nonlinear dependence of the cutting force on thickness of cut. Based on the two non-linear factors, and the simplification of the machine tool structure to a single degree of freedom system, the following non-linear differential —difference equation of machine tool chatter can be derived:

$$\ddot{x}_o(t) + 2\omega\xi\dot{x}_o(t) + \omega_o^2 x_o(t) = -\frac{\omega_n^2}{k}\Delta F(t), \tag{14.26}$$

where, $x_o(t)$ shows the relative motion between the cutting edge and the work-piece in the direction normal to the cutting surface, as shown by thin solid line in Fig. 14.14 (which is the same as Fig. 13.6e, here the figure is given again is for the

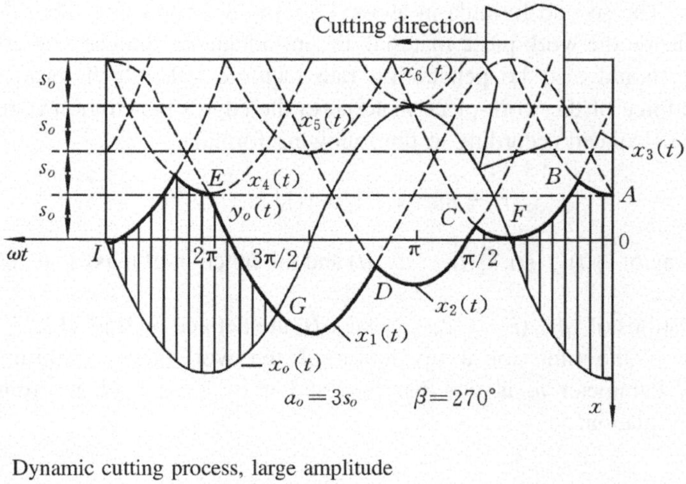

Fig. 14.14 Dynamic cutting process, large amplitude

convenience of reading with the text). ω_o is the natural frequency of the unstable mode in machine tool. ξ is the damping factor of the mode and k is stiffness coefficient of the mode. $\Delta F(t)$ is the dynamic incremental cutting force. The reason why there is a negative sign in the right side of the above equation is the same as in the Eq. (13.21).

$$\Delta F(t) = Da_p\left[s^{\mu}(t) - s_o^{\mu} + \mu s_o^{\mu-1}c\frac{60\dot{x}_o(t)}{zN}\right], \tag{14.27}$$

where, D, μ are constants, and normally $\mu = 0.7$–0.8, a_p is depth of cut (or width of cut), $s(t)$ is the instantaneous thickness of cut, s_o is mean thickness of cut, c is penetration rate factor, N is machine tool spindle revolutions, z is the number of the teeth of the cutter, here, the case is supposed to be turning, so $z = 1$. The first term of Eq. (14.27) shows the total cutting force caused by the instantaneous thickness of cut $s(t)$, it is expressed as polytrope relationship widely used in the theory of metal cutting as empirical formulae; The second term is the mean cutting force caused by the mean thickness of cut s_o, which should be taken out from the total cutting force so as to get the dynamic incremental cutting force; The third term expresses the resistance of work-piece material to the cutting edge's penetration, among which the part of $Da_p\mu s_o^{\mu-1}$ is the slope of the curve of total cutting force versus instantaneous thickness of cut $s(t)$ at working point (s_o, F_o). Compared with Eq. (14.8), it can be seen that this part is just the cutting force coefficient k_s in general literature [80].

The instantaneous thickness of cut $s(t)$ is calculated according to the following formula,

$$s(t) = \begin{cases} x_o(t) - y_o(t) & (x_o(t) > y_o(t)) \\ 0 & (x_o(t) \le y_o(t)) \end{cases}, \tag{14.28}$$

where, $y_o(t)$ is top surface of the work-piece as shown by thick solid line in Fig. 14.14. The second formula in above Eq. (14.28) shows that when the cutting edge is outside the work-piece material, the instantaneous thickness of cut is zero, and there should also be penetration rate factor c = 0, which means that the resistance force of the work-piece material to the edge's penetration is zero.

$y_o(t)$ is calculated according to the following formula,

$$y_o(t) = max[x_1(t),\ x_2(t),\ \ldots,\ x_m(t)], \tag{14.29}$$

The meaning of $x_o(t)$, $x_1(t)$, $x_2(t)$, …, $x_m(t)$ and the function of max[.] are the same as in Fig. 13.6e.

The relations of $x_i(t)$, ($i = 1, 2, \ldots$) and $x_o(t)$ are defined by Eqs. (13.25)–(13.27). Where, T is the time for a revolution of the work-piece or milling cutter, $T = 60/N$. Parameter m is time-varying, and it is determined according to the following equation:

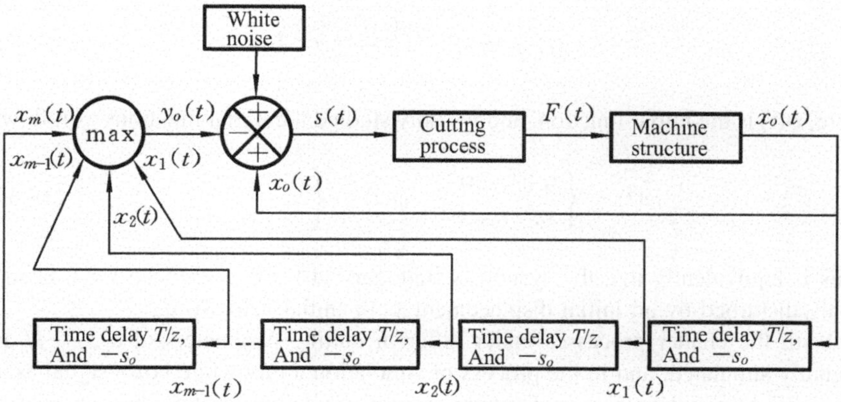

Fig. 14.15 A theoretical model for digital simulation of non-linear chatter processes

$$m(t) = \mathrm{Int}\left[\frac{|2\min x_o(\zeta)|}{s_o}\right] + 1, \ (\zeta \leq t), \tag{14.30}$$

where, function Int[·] means to take the integer part of the value in square brackets, "min" means to take the minimum value of function $x_o(\zeta)$ before $t = \zeta$. It is easy to know when $i > m$, $x_i(t)$ and $x_o(t)$ have no intersection point, that is, $x_i(t)$ is all over the $x_o(t)$. From Fig. 14.14 we know, that $|\min x_o(\zeta)| = a_o = 3s_o$, and according to Eq. (14.30), $m = 7$, i.e., $x_7(t)$ and $x_o(t)$ have no intersection point. [Only $x_6(t)$ is plotted in the figure, which just happens to be tangent to the $x_o(t)$].

The above model is expressed by the block diagram shown in Fig. 14.15. This is a non-linear and variable time delay model. This model accurately describes the physical nature of the machine tool chatter, which has a March more rich connotation than the traditional linear model. But due to its complexity, it is almost impossible to study the system with strict analytical methods.

We have adopted previously the method of taking the base wave for both the excitation force and the vibration displacement, and thus simplified the system model. Therefore, the stability equation was introduced and the amplitude and frequency of the stable chatter were obtained. Here, in order to reveal the whole process of chatter occurrence, development and finally being stabilized, the simplified methods above mentioned cannot be used, and instead, digital simulation technology were used to study the characteristics of this system.

14.2.2 Digital Simulation Method

The numerical solution of the above differential equation is obtained by the fourth-order Rung-Kutta method. The initial conditions are,

$$x_o(t) = \varphi(t), -\frac{m(t)T}{z} \leq t \leq 0, \tag{14.31}$$

Here, φ(t) is the initial function, and we only studied the initial functions as follows:

$$\varphi(t) = \begin{cases} 0 & t<0 \\ x_o & t=0 \end{cases}, \quad \dot{\varphi}(t) = \begin{cases} 0 & t<0 \\ \dot{x}_o & t=0 \end{cases}. \tag{14.32}$$

This is equivalently that the system is stationary at t < 0, and when t = 0 is suddenly disturbed by an initial displacement x_o or initial velocity $\dot{x}_0$.

With the chatter processes under different cutting parameters of a_p, s_o, N are digitally simulated, and in the process of simulation a tiny white noise signal as an incentive, to simulate the actual cutting process of all kinds of disturbances (such as described in Chap. 12, accidental disturbance from the outside world is necessary for a dynamic system which is in unstable state to start vibration, therefore, it is necessary to add white noise signals in the simulation).

In order to verify the correctness of the numerical simulation results, the chatter cutting experiments were conducted on a vertical milling machine with an end milling cutter, and according to the measured data assigned the related parameters in Eqs. (14.26) and (14.27) as follows:

$\omega_o = 1118.4\ s^{-1}$, $\xi = 0.0393$, $k = 188\ 203.6$ N/mm, $z = 24$, $D = 6254.8$ N, $\mu = 0.727$, $c = 0.0120$.

14.2.3 The Numerical Simulation Results

The results of digital simulation are well matched with the experimental facts and come up with some rules of practical significance, most of which are not predicted and explained by linear models.

14.2.3.1 The Relationship Between Depth of Cut a_p and Chatter

Gradually increase the depth of cut a_p, and the simulation results are shown in Fig. 14.16a (on a condensed time scale). When the depth of cut increases continuously to reach a critical value a_{pcr2}, the cutting process suddenly loses its stability, the amplitude rapidly increases, and then, stabilized at a fixed amplitude. If the initial depth of cut is larger and gradually decreases continuously, the time history is shown in Fig. 14.16b, only when the depth of cut decreases to reaches an other critical value a_{pcr1}, vibration can disappear, system restore stability. a_{pcr2} is called the jump threshold, and a_{pcr1} is called the drop threshold, they are not the same as the linear theory describes, but are separate, and $a_{pcr2} > a_{pcr1}$.

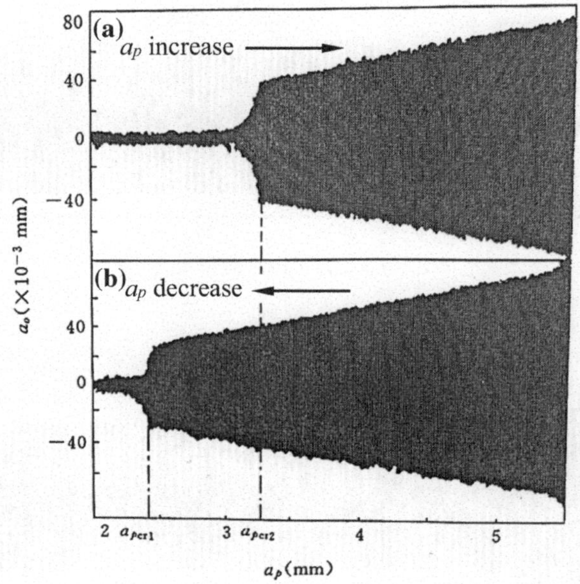

Fig. 14.16 The separation of vibration jump threshold and vibration drop threshold (simulation results)

This is in good match with the experimental results. And it is well known that the separation of jump threshold and drop threshold is one of the particular characteristics of non-linear system.

When $a_p < a_{pcr}$, the system is asymptotically stable in the sense of Lyapunov, no matter how much initial disturbance x_o is unable to destabilize the system. The time history is shown in Fig. 14.17.

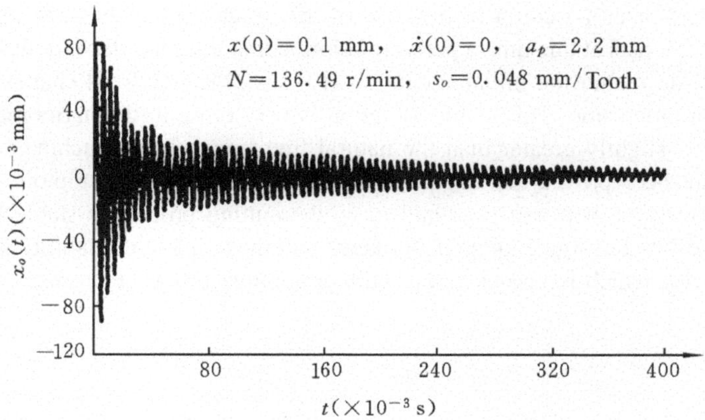

Fig. 14.17 The simulated unconditional stability of the cutting system ($a_p < a_{pcr1}$)

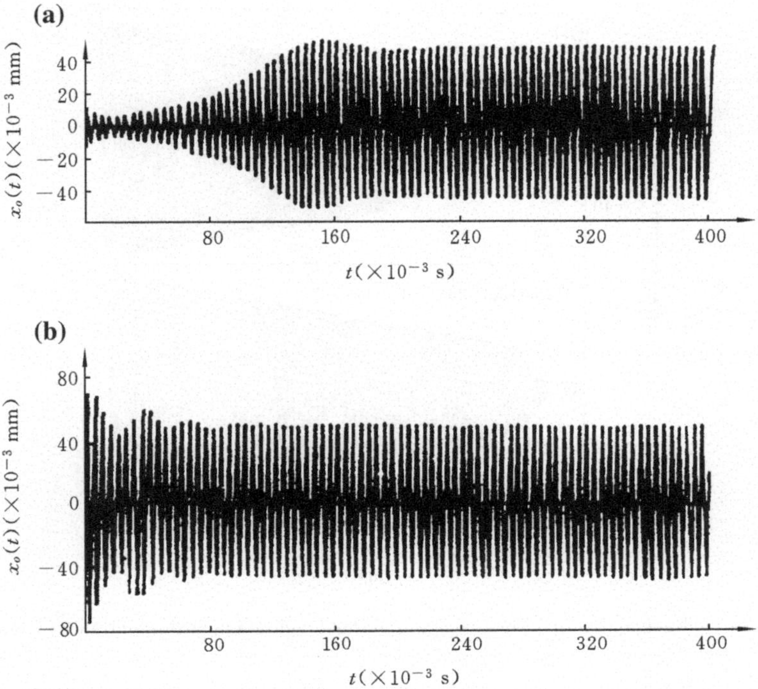

Fig. 14.18 The simulated unconditional instability of the cutting system ($a_p > a_{pcr2}$). **a** $x(0) = 0.01$ mm, $\dot{x}(0) = 0$, $a_p = 3.5$ mm, $N = 136.49$ r/min, $s_o = 0.048$ mm/teeth. **b** $x(0) = 0.08$ mm, $\dot{x}(0) = 0$, $a_p = 3.5$ mm, $N = 136.49$ r/min, $s_o = 0.048$ mm/teeth

When the $a_p > a_{pcr2}$, the system is unstable in the sense of Lyapunov, at this moment, any tiny initial disturbance, will provoked a chatter, make the amplitude increases rapidly, and then, it is stabilized on certain amplitude. The time history is shown in Fig. 14.18a. The Fig. 14.18b represents the chatter history of a larger initial disturbance. To compare *a* and *b* of this graph, it can be seen that in the system there exists at this time a periodic solution that does not depend on the initial conditions, no matter the initial excitation $x(0)$ size, the stabilized chatter size and the form are the same. This stable vibration is very close to the harmonic, and its frequency is slightly greater than the natural frequency of the machine's unstable mode. It has been proved that this periodic solution has orbital asymptotic stability.

When the a_p is between a_{pcr1} and a_{pcr2}, the cutting process is stable for small disturbances, and a large enough disturbance will make it lost its stability, as shown in Fig. 14.19, which is known as the finite amplitude instability zone.

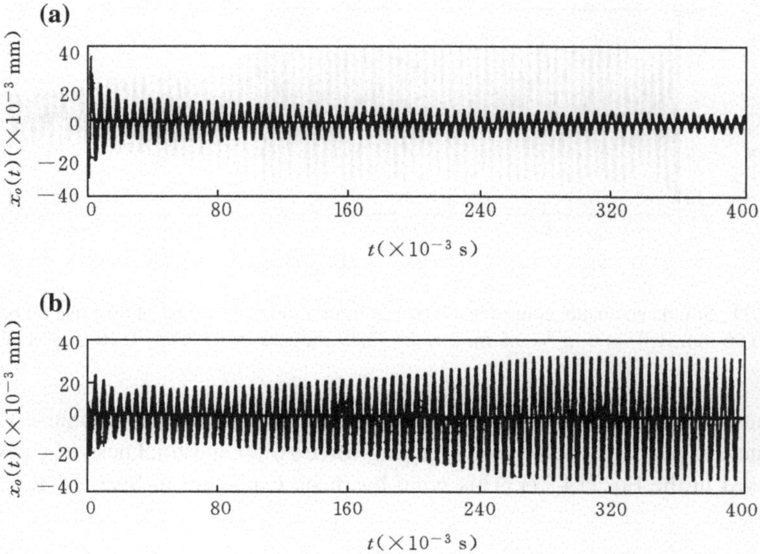

Fig. 14.19 The simulated finite amplitude instability of the cutting system ($a_{pcr1} < a_p < a_{pcr2}$). **a** $x(0) = 0.023$ mm, $\dot{x}(0) = 0$, $a_p = 2.6$ mm, $N = 136.49$ r/min, $s_o = 0.048$ mm/teeth. **b** $x(0) = 0.025$ mm, $\dot{x}(0) = 0$, $a_p = 2.6$ mm, $N = 136.49$ r/min, $s_o = 0.048$ mm/teeth

14.2.3.2 The Influence of Mean Thickness of Cut s_o on Chatter

As can be seen from the Fig. 14.20, the amplitude begins to increase gradually when s_o is increased, but when s_o exceeds a critical value of s_{cr}, the chatter suddenly disappears. The reason has been explained in the last section. This gives us a meaningful lesson: to increase the feed rate, so that it exceeds a certain threshold value, and can help to suppress the flutter.

Figure 14.21 is the simulation result of a sudden increase in the feed rate during the cutting process from $s_o = 0.048$ mm per tooth to $s_o = 0.10$ mm per tooth (at the point A).

Fig. 14.20 Simulated relationship between thickness of cut s_o per tooth and chatter amplitude a_o

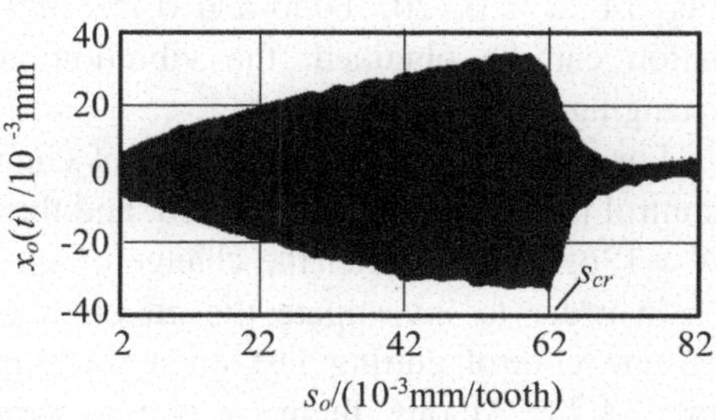

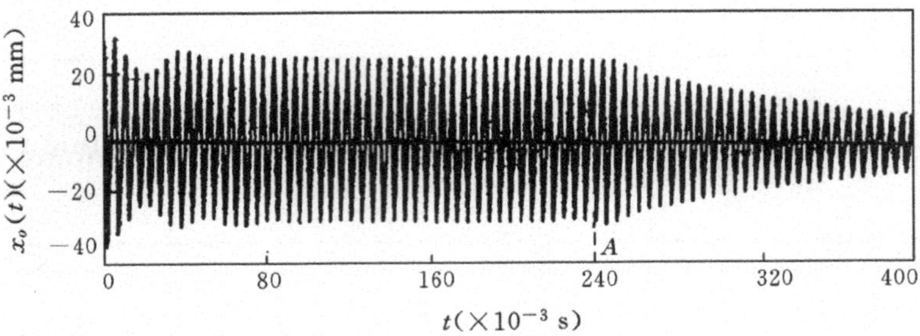

Fig. 14.21 Simulated chatter control result by means of a sudden increase of feed rate (at point *A*). $x(0) = 0.06$ mm, $\dot{x}(0) = 0$, $a_p = 3.5$ mm, $N = 136.49$ r/min, $s_o = 0.048 \rightarrow 0.10$ mm/teeth

When s_o is beyond a critical value, the chatter disappears because of the non-linear dependence between the cutting force $\Delta F(t)$ and thickness of cut $s(t)$ as expressed in the Eq. (14.27). This point has been explained in Sect. 14.1.4.2.

14.2.3.3 The Influence of Machine Tool Spindle Revolutions *N* on Chatter

The chatter is very sensitive to the spindle revolutions *N*. Small changes in *N* can cause dramatic changes in chatter amplitude, as shown in Fig. 14.22.

As can be seen from the figure, the rotation speed is different, the phase difference β ($\beta = T\omega/z = 60\omega/N$) between the two adjacent teeth is different. As shown in Fig. 14.22b, when the amplitude is increased to approximately equal to s_o, the cutting pass of the cutting edge has been partially out of the work-piece material, the excitation energy has been thus reduced, and the continuous increasing in chatter amplitude is thus suppressed.

In Fig. 14.22a, c, however, the amplitude has been risen to respectively about equal to $2s_o$ and $3s_o$ when the cutting edge passes are partly out of the work-piece material, therefore, the steady-state amplitude is much larger than the Fig. 14.21b. The simulation results show that the steady-state amplitude of the corresponding rotational speed is different, and the amplitude of the three rotating speed of Fig. 14.22 is 0.120, 0.050 and 0.150 mm respectively. Thus, a meaningful revelation can be obtained: the vibration amplitude can be controlled online by changing the rotation speed.

For this reason, the simulation of changing speed *N* in the cutting process to control the chatter was performed, and the result was shown in Fig. 14.23 (original $N = 136.49$ r/min, after the change of speed, $N = 147.1$ r/min).

In order to investigate the effect of chatter control, we performed a chatter on-line control cutting test on a medium size lathe. The results are shown in Fig. 14.23. Among them, *a* is the measured vibration signal when suddenly increased in feed rate s_o during the cutting process; *b* is a measured vibration signal

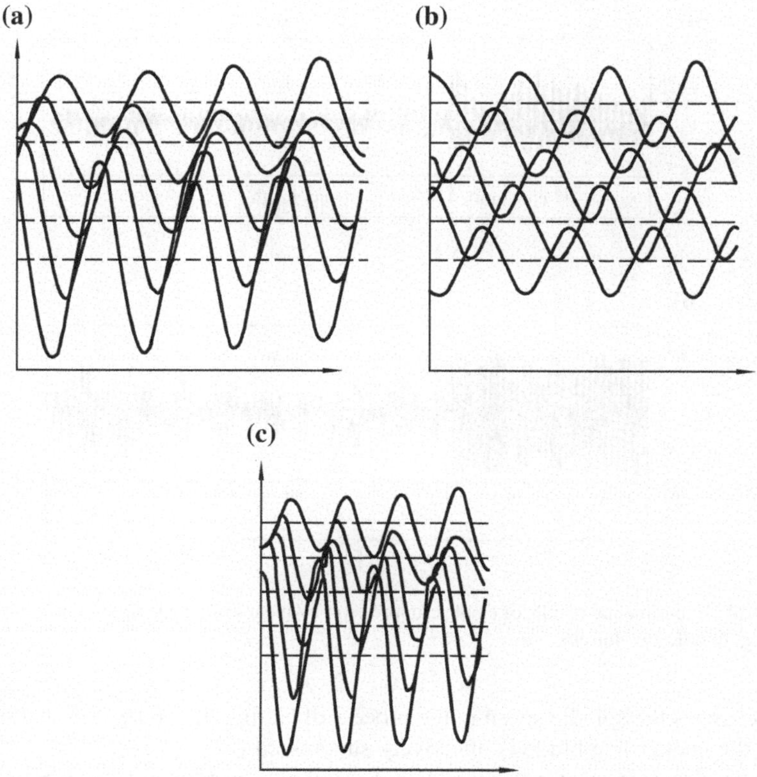

Fig. 14.22 Simulated results of the motion passes of successive cutting teeth under different revolutions of the milling cutter. **a** a_p = 3 mm, N = 119.88 r/min, s_o = 0.048 mm/teeth, β = 303°. **b** a_p = 3 mm, N = 136.49 r/min, s_o = 0.048 mm/teeth, β = 221°. **c** a_p = 3 mm, N = 158.00 r/min, s_o = 0.048 mm/teeth, β = 300°

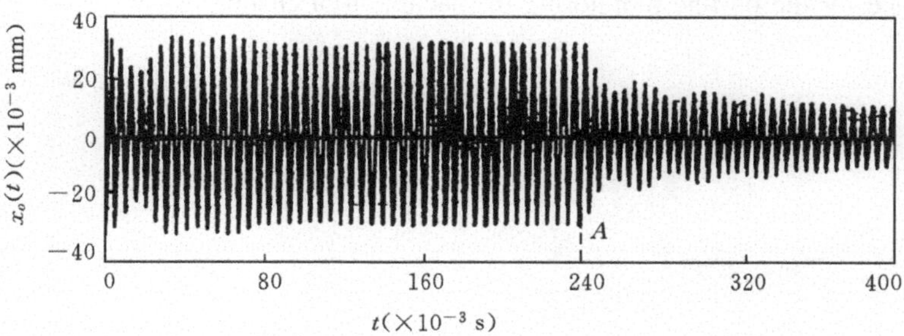

Fig. 14.23 Simulated chatter control result by means of a sudden changing spindle revolutions (at point A). $x(0)$ = 0.06 mm, $\dot{x}(0) = 0$, a_p = 3.5 mm, N = 136.49 → 147.10 r/min, s_o = 0.048 mm/teeth

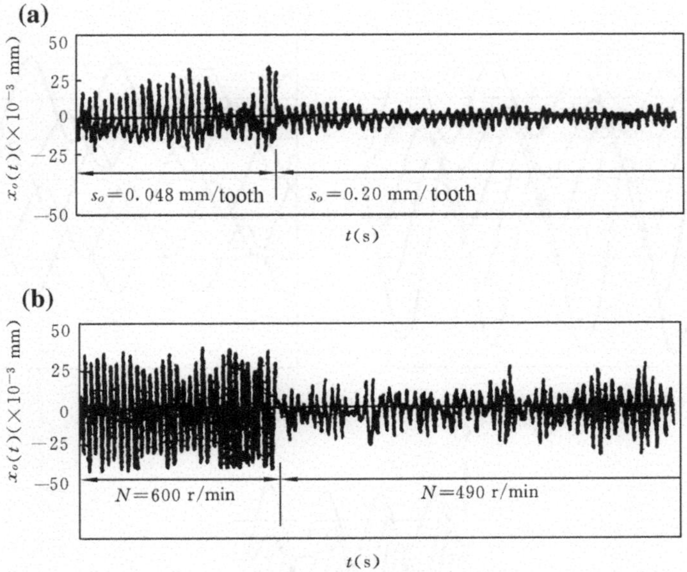

Fig. 14.24 Experimental results of chatter suppression by means of changing thickness of cut and changing spindle revolutions

when changes the spindle speed in the process of cutting. It can be seen that in both cases, the chatter amplitude is effectively suppressed (Fig. 14.24).

The study shows that computer digital simulation technology is a kind of economic and feasible method of studying the stability of machine tools in various operating conditions, and it may develop in the future become an important part of computer aided manufacturing.

The conclusions obtained in this section have been fully demonstrated by the experimental facts, and thus can be used as the theoretical basis and technical guidance for the on-line monitoring of machine tool chatter.

Chapter 15
Early Diagnosis and Online Monitoring of Machine Tool Chatter

Based on the nonlinear theory of machine tool chatter described in last two chapters, this chapter studies the early diagnosis and online control of chatter. First, through the experimental study, the appropriate monitoring signal is selected to monitor the stability of a cutting system. Then, the characteristic changes of the monitoring signal in the transition process from stable to unstable of a cutting system are analyzed, and the characteristics quantities of the system reflecting the degree of tending to instability are selected and their quick algorithm are established. The methods and strategies of the early diagnosis of chatter were discussed. The effectiveness of on line adjusting the cutting parameters to suppress the chatter is experimentally verified and the adjustment strategy established. Finally, an experimental system of on-line cutting chatter monitoring and controlling is established, and good experimental results are obtained.

The research of the previous chapter shows that online adjusting the cutting parameters is a very effective way to suppress chatter. However, this kind of adjustment must be timely, otherwise, the chatter will not be put out early, and it will still leave a chatter marks on the machined surface, affecting the surface quality of the work-piece. Therefore, the best thing to do is to early predict the chatter in the cutting process, and to adjust cutting parameters timely, and restrain chatter before it become violent, so as to ensure the quality of machining.

The top priority of early diagnosis and online monitoring process of metal cutting chatter is from various monitored signals of the cutting process to early identify the "symptoms" that system is tending to be unstable. To do this, suitable characteristic quantities, need to be calculated in real-time according to the monitored signal being sampled, so as to reflect the extent of the systems tending to be unstable, and constantly to be compared with the threshold preset, so that, before the arrival of violent chatter alarm or control signals can be accurately and timely given out.

H. Shi, *Metal Cutting Theory*, Springer Series in Advanced Manufacturing,
https://doi.org/10.1007/978-3-319-73561-0_15

15.1 The Selection and Analysis of Monitoring Signals

15.1.1 The Experimental Device and the Monitory Signal Selection

In this chapter, vibration acceleration signal and cutting force signal are selected to be the chatter monitoring signals.

In order to study the changes in signals sampled from a cutting process, the experiment was carried out. The test was performed on a medium lathe. The test conditions are as follows: the work-piece: 900 mm in length, 48 mm in diameter, 45# steel in material, it is supported in three jaw chuck on one end and by a center on the other end. Cutter: cemented carbide insert tool, material is YT15; Cutting tool geometry parameters are: tool cutting edge angle $K_r = 45°$ or $90°$, tool minor cutting edge angle $K_r' = 45°$, rake angle $\gamma_o = 10°–13°$, clearance angle $\alpha_o = 8°–10°$. Cutting parameters: spindle revolutions $N = 550–800$ r/min, feed rate $f = 0.08–0.1$ mm/r, depth of cut $a_p = 0.5–1$ mm.

The block diagram of experimental set-up is shown in Fig. 15.1. The accelerometers are installed in y and z directions of the tool holder and the z direction of the tail center respectively to measure the vibration acceleration of the machine. A KISTLER three direction dynamometer is installed at the tool holder to measure the cutting force.

In the above cutting conditions, the turning process starts to be stable, and when the cutting tool moves to the middle part of the work-piece, the system suddenly loses its stability and chatter was generated.

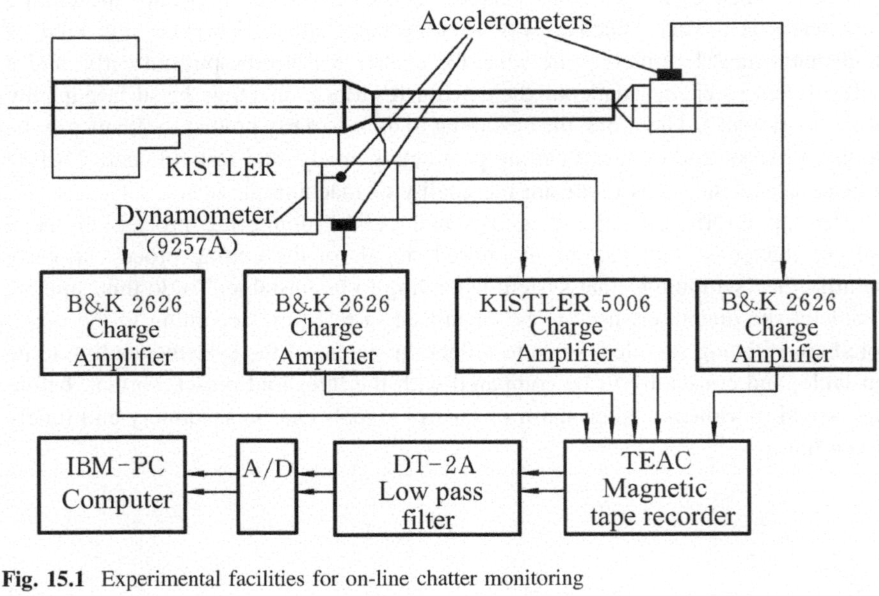

Fig. 15.1 Experimental facilities for on-line chatter monitoring

Test shows that the acceleration signal in z direction sampled from tail center is more sensitive to reflect the changes of chatter build process than the rest of the acceleration signal, therefore, in the following analysis this signal is used. Besides, the cutting force signal in z direction is also used.

15.1.2 The Characteristics Changes During Transition from Stable Condition to Chatter

Several experiments show that there are two changes in vibration signal during the transition process. First, the amplitude of vibration signals in the time domain was increasing, that is, the vibration energy is increasing; second, in the frequency domain, the main frequency band of the vibration signals moves from the high band to the low band. These two changes are the physical basis for the early diagnosis of chatter.

15.1.2.1 The Characteristic Changes of Vibration Signal in Time Domain

Figure 15.2a, b are the time histories of the two monitored signals: the vibration acceleration sampled in z direction at tail center and cutting force F_c measured by the dynamometer (in z direction).

In the figure, the whole transition process of the cutting process from stationary to the beginning of instability and to the chatter is fully established. Obviously, in

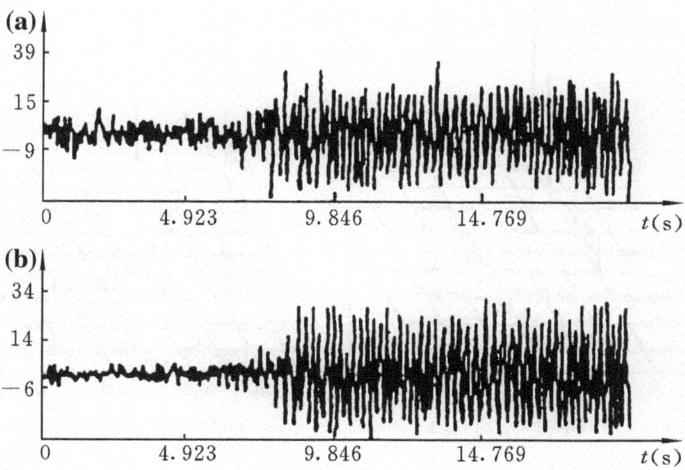

Fig. 15.2 Monitoring signals in the transition process from stable cutting to the full establishment of chatter. **a** The acceleration signal of vibration. **b** The cutting force signal

the time domain, the amplitude of vibration increases quickly, which indicates that the vibration energy increases continuously with the occurrence of chatter. This is one of the characteristics changes in time domain during the transition (the ordinates of the two figures are the charge amplifier output voltage value connected to the accelerometer and the dynamometer, respectively. The following is the same).

15.1.2.2 The Characteristic Changes of Vibration Signal in the Frequency Domain

Figure 15.3a, b are the power spectrum arrays of vibration acceleration signal a_z and the cutting force signal F_c presented in Fig. 15.2, respectively. It can be seen from the figure that in the transition process, the main frequency band of force and vibration acceleration signals moves from high frequency band to low frequency band, and the frequency domain energy is changed from dispersion to concentration. This change reflects a more profound trend in the establishment of chatter.

Its physical meaning is that for medium-sized machine (along with its work-piece and fixture) the chatter frequency (near the natural frequency of the unstable mode) is generally in the range of 200–300 Hz; And the frequency of forced vibration and background noise caused by various internal and external vibration sources is generally higher than the natural frequency of the unstable mode, and is dispersed, not concentrated at a certain frequency. When the system is in a stable state, the forced vibration and background noise are the main ones, so the main frequency band of the signal is in the high frequency band, and the spectrum energy is more dispersed. After chatter increase sharply, vibration of the frequency near to an unstable mode of the machine tool structure immediately rise, and

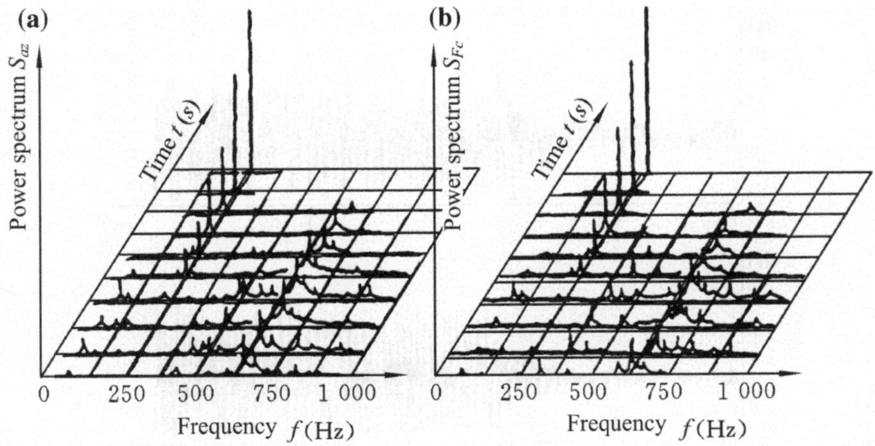

Fig. 15.3 Power spectrum arrays of monitored signals from stable cutting to the establishment of stable chatter. **a** The Power spectrum arrays of acceleration signal of vibration. **b** The Power spectrum arrays of cutting force signal

occupied the main position, so the signal frequency zone will move to low frequency domain and the energy concentrated near the chatter frequency. This is another characteristic change of the monitoring signals in the transition process.

15.1.2.3 Correlation Analysis of Monitoring Signals

The above analysis is based on the self-change of the each of the monitoring signals in the process of chatter building. But from the regeneration chatter point of view, when the chatter is fully established, between the cutting force signal and the vibration signal there must exist some correlation, and the chatter establishment process should also be establishment process of such correlation. Therefore, the correlation analysis between the cutting force and the vibration signals is an effective way to find the chatter gestation process. So, the correlation analysis between vibration acceleration signal a_z and cutting force signal F_c is conducted.

The cross-power spectrum between vibration acceleration a_z and cutting force F_c can be expressed as:

$$G_{aF}(f) = C_{aF}(f) - \mathrm{j}Q_{aF}(f) = |G_{aF}(f)|\mathrm{e}^{-\mathrm{j}\theta_{aF}(f)}, \tag{15.1}$$

where, $|G_{aF}(f)|$, $C_{aF}(f)$ and $Q_{aF}(f)$ are the amplitude, real part and imaginary part of the cross-power spectrum, respectively. $\theta_{aF}(f)$ is the phase spectrum, which shows the phase difference between a_z and F_c. $\theta_{aF}(f)$ can be calculated according to the following formulae:

$$\theta_{aF}(f) = \begin{cases} \arctan\frac{Q_{aF}(f)}{C_{aF}(f)} & (C_{aF}(f) > 0) \\ \pi + \arctan\frac{Q_{nF}(f)}{C_{nF}(f)} & (C_{aF}(f) < 0, Q_{aF}(f) > 0) \\ -\pi + \arctan\frac{Q_{aF}(f)}{C_{aF}(f)} & (C_{aF}(f) < 0, Q_{aF}(f) < 0) \end{cases} \tag{15.2}$$

$\theta_{aF}(f) < 0$ shows that signal a_z is lag behind signal F_c; $\theta_{aF}(f) > 0$ shows that signal a_z is ahead of signal F_c;

Figure 15.4a–c show the amplitude of cross-power spectrum during stable cutting, transition from stable cutting to chatter, fully established chatter, respectively. The figure shows that the cutting process from smooth to chatter, the energy of the dominant frequency of the spectrum moves from the high frequency band to low frequency band, and from distributed into centralized. Moreover, the correlation between a_z and F_c signals has risen sharply, which is reflected in the increase of the amplitude of cross-power spectrum.

In this chapter, the multiple test results show that under testing conditions, when stable cutting, at the main frequency range the phase difference $\theta_{aF}(f)$ of a_z and F_c is between 75° and 135° (a_z is ahead of F_c); When chattering, $\theta_{aF}(f)$ is in the range of −90°–−180° (a_z is lag behind signal F_c).

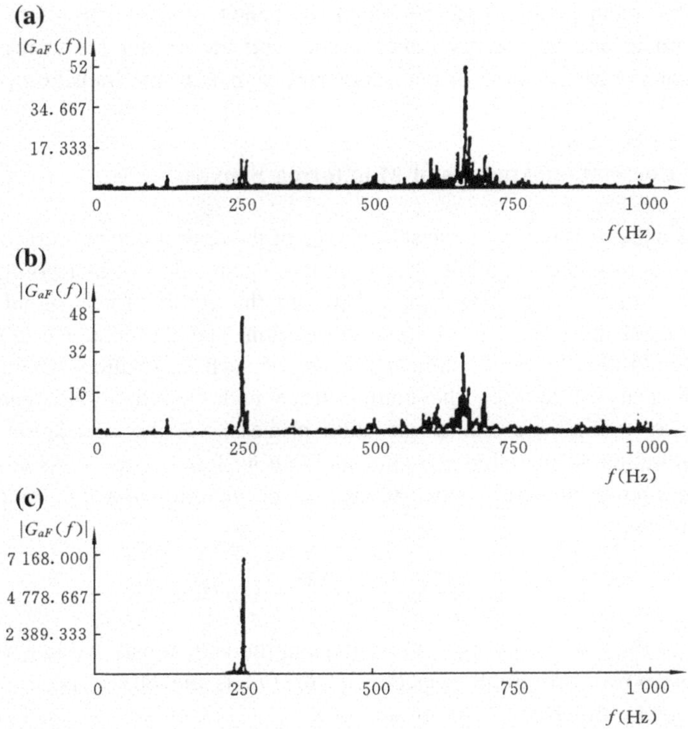

Fig. 15.4 The cross power spectrum amplitude of vibration acceleration and cutting force. **a** Stable cutting, **b** transition from stable cutting to chatter, **c** fully established chatter

To sum up, the cutting process from smooth to instability, the correlation between vibration acceleration a_z and F_c changes will happen in two ways: one is the main frequency band of cross spectrum moves from high frequency to low frequency, and the amplitude increases sharply; another is phase difference $\theta_{aF}(f)$ of a_z and F_c is moving from 75°–135° to −90°–−180°.

15.2 The Fast Algorithm of Characteristic Quantity

The characteristic changes in the time domain and frequency domain are the physical bases for the early diagnosis of chatter. However, in order to use the computer to quantitatively identify these changes in the cutting process, it is necessary to find the appropriate characteristic quantities to describe these changes of the signal. Obviously, the selected characteristic quantities not only need to be able to reflect the characteristic changes of the signal, but also the calculation must simple, so that it can be calculated in real time.

15.2.1 The Characteristic Quantities Describing the Variation of the Signals in the Time Domain

Signal's variance σ^2 is obviously an ideal characteristics quantity, which reflects the trend of signal amplitude increase in time domain, it should be calculated according to the formula

$$\sigma^2 = \frac{1}{N}\sum_{i=1}^{N}(x_i - \bar{x})^2, \tag{15.3}$$

where, x_i is sampled data, N is the number of sampling points, $\bar{x}$ is the average of x_i,

$$\bar{x} = \frac{1}{N}\sum_{i=1}^{N} x_i.$$

As shown in Fig. 15.5, with the intensification of chatter vibration, the variance σ^2 of acceleration signal a_z soared, therefore, σ^2 can reflect the characteristics of the energy change of signal in time domain.

It is worth noting that if σ^2 is taken as the only characteristics, and according to it to set of "chatter impending" criterion, it is not reliable. The reason is due to accidental disturbance, the amplitude of signal may jump up at some point, making σ^2 over threshold, and so causing misjudgment. If, on the other hand, we raise the threshold of σ^2 although can avoid misjudgment, but monitoring system thus become slowness, having no respond before the real chatter is impending, and so lost the chance of putting out the chatter in time. Therefore, other characteristics need to be sought.

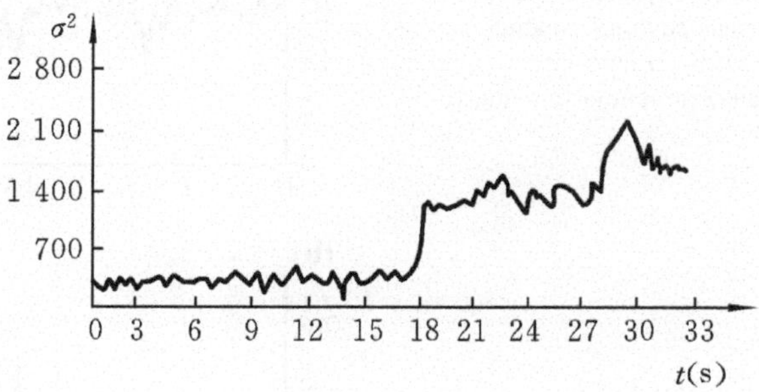

Fig. 15.5 The variance of the vibration acceleration signal increases with the development of chatter

15.2.2 The Characteristic Quantities Describing the Variation of the Signals in the Frequency Domain and Its Fast Algorithm

It is easy to understand the mean square frequency calculated according to the following formula

$$\mathrm{MSF} = \frac{\sum_{i=1}^{n} f_i^2 S(f_i)}{\sum_{i=1}^{n} S(f_i)} \tag{15.4}$$

is an ideal characteristic quantity that reflects the change of the monitoring acceleration signal in the frequency domain. The characteristic quantity is a weighted average of vibration frequency square, and the amplitude of power spectrum is the weight. When the chatter occurs, the MSF decreases sharply due to the lower shift of the signal main band, as shown in Fig. 15.6a.

Besides the MSF, another characteristics quantity, i.e., the one step autocorrelation function ρ_1, can also reflect the characteristics change of signals in the frequency domain. We are now going to analyze this characteristic quantity. Originally, ρ_1 is a characteristic quantity in the time domain, but it will be proven that, when sampling interval Δ is properly chosen, ρ_1 can also reflect the change of dominant frequency of the signal.

First, suppose that the signal consists of only a single sine wave,

$$x_i(t) = \sin(\omega_i t + \theta_i),$$

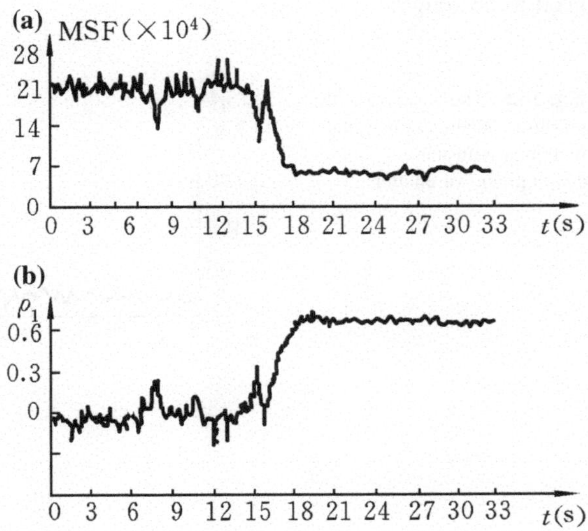

Fig. 15.6 The characteristics reflecting the changes in frequency domain. **a** The mean square frequency (MSF). **b** One step autocorrelation function (ρ_1)

Here, we regard θ_i as a random variable, and it is also assumed that the process is Ergodic, then the one step autocorrelation function of the signal is

$$\rho_{1i} = \cos 2\pi f_i \Delta \tag{15.5}$$

Up to the Nyquist frequency $f_n = 1/(2\Delta)$, the relationship between the ρ_{1i} and f_i is shown in Fig. 15.7. Obviously, ρ_{1i} is monotonically decreasing as the f_i rising.

Second, assume that the signal contains a variety of harmonics:

$$x(t) = \sum_{i=1}^{n} x_i(t) = \sum_{i=1}^{n} A_i \sin(\omega_i t + \theta_i)$$

Then, the one step autocorrelation function is

$$\rho_1 = \frac{\sum_{i=1}^{n} A_i^2 \cos 2\pi f_i \Delta}{\sum_{i=1}^{n} A_i^2} = \frac{\sum_{i=1}^{n} A_i^2 \rho_{1i}}{\sum_{i=1}^{n} A_i^2}. \tag{15.6}$$

Here ρ_1 is clearly, the weighted average of the one step autocorrelation functions of all the harmonics, but with each harmonic power A_i^2 as weight. And it is also monotonically increasing with the dominant frequency decreasing while the cutting process goes from a smooth cutting condition to a chattering condition, as shown in Fig. 15.6b.

In fact, there is a close inner link between the two parameters ρ_1 and MSF. Note that in the Eq. (15.6), A_i^2 is actually the value of the power spectrum of the signal at f_i, which is denoted as $S(f_i)$, so the (15.4) can be rewritten as

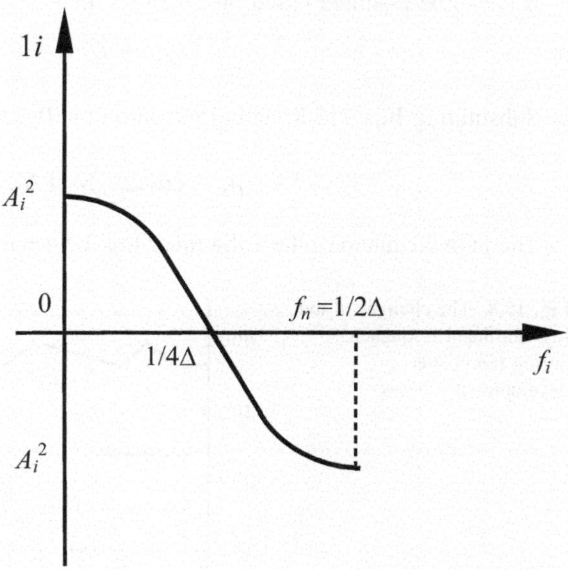

Fig. 15.7 The relationship between ρ_{1i} and the f_i within the scope of the Nyquist frequency

$$\rho_1 = \frac{\sum_{i=1}^{n} (\cos 2\pi f_i \Delta) S(f_i)}{\sum_{i=1}^{n} S(f_i)}. \quad (15.7)$$

Compared with Eq. (15.4), their structures are very similar, only in Eq. (15.7) the variable f_i^2 is replaced with $\cos 2\pi f_i \Delta$, therefore both ρ_1 and MSF can reflect the change of the main frequency band of the signal.

To reveal the relationship between ρ_1 and MSF, the concept of "dominant frequency" is introduced here. The concept can be defined differently, and one is the dominant frequency in the sense of mean square frequency:

$$f_{\mathrm{MSF}} = \sqrt{\mathrm{MSF}}; \quad (15.8)$$

The other is the dominant frequency in the sense of one step autocorrelation function:

$$f_{\rho 1} = \frac{1}{2\pi\Delta} \arccos \rho_1. \quad (15.9)$$

Figure 15.8 represents the changes of the dominant frequency in the two senses during the transition process. As can be seen from the figure, the two dominant frequencies are generally close to each other.

f_{MSF} compared with $f_{\rho 1}$ slightly smaller, this is because when according to Eq. (15.4) to calculate MSF, the power spectrum $S(f_i)$ only took the value $f_i \leq 600$ Hz, while omitting the high frequency part, then make the calculated f_{MSF} smaller than its real value. In fact f_{MSF} and $f_{\rho 1}$ are very close to each other, they all reflected a common trend: when chatter occurs, the dominant frequency of vibration obviously move to low frequency.

It is can be assumed based on Fig. 15.8, that

$$f_{\mathrm{MSF}} \approx f_{\rho 1}. \quad (15.10)$$

Substituting Eqs. (15.8) and (15.9) into (15.10) leads to

$$\rho_1 \approx \cos 2\pi \sqrt{\mathrm{MSF}} \Delta. \quad (15.11)$$

The above equation reflect the inner link between MSF and ρ_1.

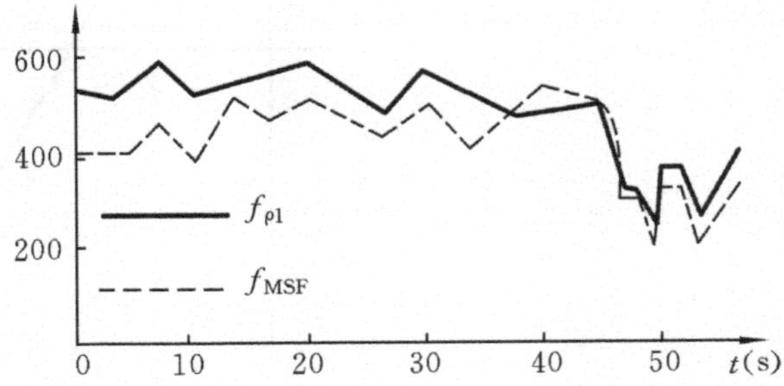

Fig. 15.8 The changes of the two dominant frequencies during the chatter development process

When calculating the ρ_1 and MSF does not need to calculate power spectrum $S(f_i)$, and they can be calculated directly from the sampling data, according to the following formula

$$\mathrm{MSF} = \frac{D}{4\pi^2 B}, \tag{15.12}$$

$$\rho_1 = \frac{C - A^2}{B - A^2}, \tag{15.13}$$

where,

$$A = \sum_{i=0}^{N} x_i,\ B = \sum_{i=0}^{N} x_i^2,\ C = \sum_{i=0}^{N} x_i x_{i-1},\ D = \sum_{i=0}^{N} \dot{x}_i^2,$$

where, x_i ($i = 0, 1, 2, \ldots, N$) is sampled data, and $\dot{x}_i = \frac{x_i - x_{i-1}}{\Delta}$, ($i = 1, 2, \ldots, N$) is the first difference, and $N + 1$ is the number of samples and Δ is the sample interval.

The quantities of A, B, C and D can be recursively calculated at the sampling interval.

Once the sample finished, the above amounts A, B, C, D have been calculated, then MSF and ρ_1 can be immediately worked out by Eqs. (15.12) and (15.13).

As for by σ^2 expressed by Eq. (15.3) can be calculated according to the following equation:

$$\sigma^2 = \frac{1}{N}\sum_{i=1}^{N} x_i^2 - \left(\frac{1}{N}\sum_{i=1}^{N} x_i\right)^2 = \frac{B}{N} - \left(\frac{A}{N}\right)^2 \tag{15.14}$$

Now back to compare the mean square frequency MSF and one step autocorrelation function ρ_1, as mentioned above, these two characteristics quantities both can describe the signal changes in the main frequency band or the movement of the dominant frequency, however, from the point of calculation workload and calculation speed, ρ_1 is obviously better than that of MSF.

Admittedly, the calculation of MSF does not necessarily require the power spectrum $S(f)$ as shown in Eq. (15.4), which can be calculated directly from the monitoring data according to Eq. (15.12). But, the calculation of MSF need to compute numerical differential (or difference), it costs a lot of time, and the calculation of numerical differential or difference also easily affected by the contingency of the local error. The calculation of ρ_1, however, does not need to calculate differential or difference, so as to save calculation time and prevent the influence of local error.

Computing speed for on-line monitoring of the process of often plays a decisive role, this is why we want to put forward one step autocorrelation function ρ_1 to replace mean square frequency as a characteristics quantity reflecting the changes in frequency domain of the monitored signal.

15.3 Early Diagnosis of Machine Tool Chatter

15.3.1 The Pattern Vector and the Pattern Plane Describing the Changes in System Stability

The method of pattern recognition is used to diagnose the instability of the cutting process. The pattern vector is composed from the characteristics quantities of σ^2 and ρ_1,

$$X = \begin{pmatrix} x_1 \\ x_2 \end{pmatrix} = \begin{pmatrix} \rho_1 \\ \sigma^2 \end{pmatrix}, \tag{15.15}$$

where, two components x_1 and x_2 of vector X are representing ρ_1 and σ^2, respectively. The pattern space corresponding to this pattern vector is a plane, and the end point of this pattern vector corresponds to a point on the plane, which is called feature point. The abscissa axis x_1 reflects the state of the signal in frequency domain, and the ordinate x_2 reflects the state of the signal in time domain. Therefore, it is possible to track the position variation of the feature point described by these two characteristics as coordinates in this plane to determine whether the cutting process is becoming unstable.

Figure 15.9 indicates that under various cutting conditions (including stable cutting and chatter occurring), the distribution of the feature points on the pattern plane during many tests. It can be seen from the figure that in stable cutting, the feature point (○) is distributed in the lower left corner of the pattern plane, while when chatter occurs, the feature points (●) are distributed in the upper right region. Therefore, the changes of the stability of the system can be monitored by observing the moving of feature points on the pattern plane.

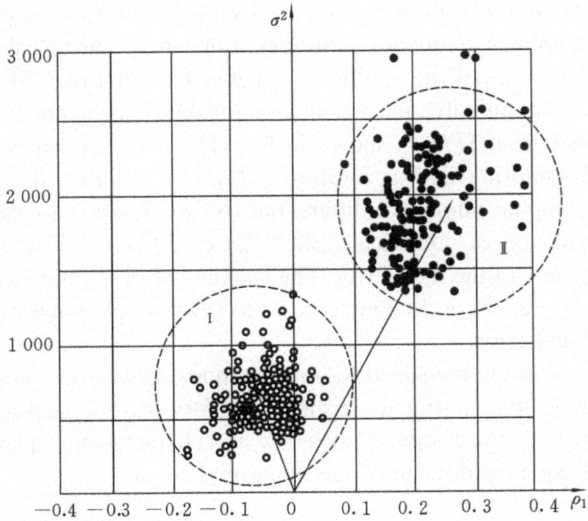

Fig. 15.9 The distribution of feature points in the pattern plane under table cutting and chatter state

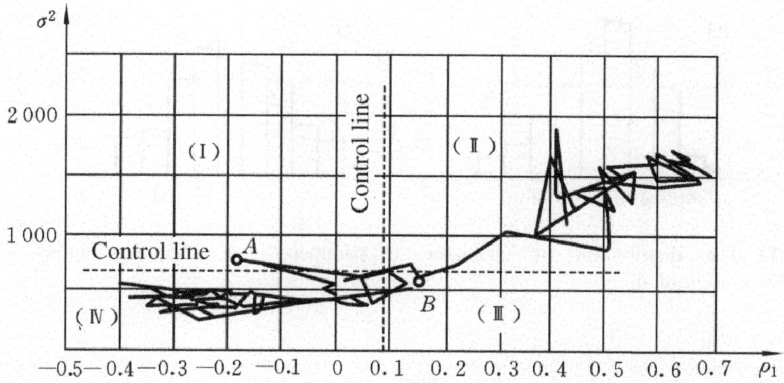

Fig. 15.10 The moving trajectory of the feature points on the pattern plane

Figure 15.10 represents the flow trajectory of the feature point from stable cutting to chattering. This trajectory intuitively reflects the change of stability of the cutting system. When the cutting system is stable, σ^2 and ρ_1 are relatively smaller, the feature point is moving randomly in the lower left corner area IV of the pattern plane; When cutting process tend to be unstable, the value of σ^2 and ρ_1 rose rapidly, and the feature point rushed out of the area IV quickly and moving towards the upper right corner II.

After the chatter was fully developed and stabilized at a certain level, the feature points were randomly moving in area II. It is important to note that in chatter occurrence and development process, feature points are normally starting from regional IV, first moved to the lower right region III, at this point only vibration frequency beginning to move down, but there is no obvious rise in amplitude, and then, with the rise of the amplitude, feature point moves from region III to region II.

Thus, comprehensively considering the two characteristics quantities of σ^2 and ρ_1 on the pattern plane, is advantageous to seizing the symptom of instability, and to the prediction of the chatter earlier.

15.3.2 The Criterion for the Early Diagnosis of Chatter in the Pattern Plane

Due to influences of many factors, the σ^2 and ρ_1 are random variables. Figure 15.11a, b are the distributions of σ^2 and ρ_1, respectively, under the condition of stable cutting. The figures show that σ^2 and ρ_1 are approximately obey the normal distribution. Let's assume that they are independent of each other, and therefore they joint probability density is

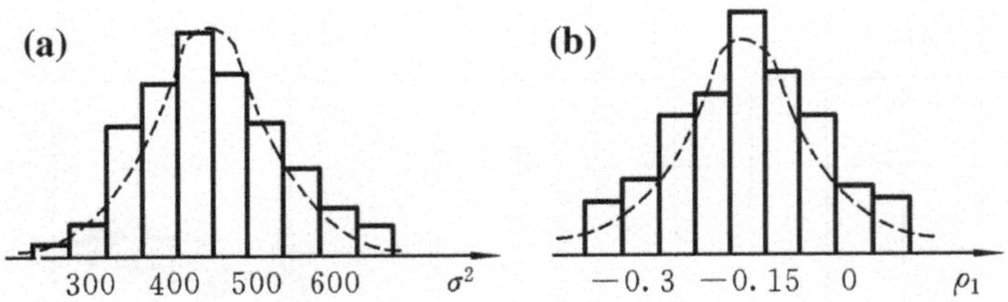

Fig. 15.11 The distribution of characteristic parameters. **a** The distribution of σ^2. **b** The distribution of ρ_1

$$f(\sigma^2, \rho_1) = \frac{1}{2\pi\sigma_1\sigma_2} \exp\left\{ -\frac{1}{2}\left[\frac{(\sigma^2 - \mu_1)^2}{\sigma_1^2} + \frac{(\rho_1 - \mu_2)^2}{\sigma_2^2} \right] \right\}, \tag{15.16}$$

where, σ_1^2, σ_2^2 are the variances, and μ_1, μ_1 are the mean values of σ^2 and ρ_1, respectively. If take a confidence level of 99.74%, then the confidence interval of σ^2 and ρ_1 on the pattern plane will be an ellipse, its minor semi-axis and major semi axis will respectively $3\sigma_1$ and $3\sigma_2$, as shown in Fig. 15.12.

In fact, using a computer to determine whether a pattern vector endpoint, i.e., the feature point belongs to an ellipse region is very troublesome, and for the sake of utility, the elliptic confidence region is replaced by the rectangular region shown in Fig. 15.1.

The key to the early diagnosis of machine chatter is to detect early the tendency of the feature point to be moved from region *IV* to region *II* in the cutting process, and the criterions are

$$\sigma^2 \geq \Sigma \cap \rho_1 \geq P, \tag{15.17}$$

where,

$$\Sigma = \mu_1 + 3\sigma_1, P = \mu_2 + 3\sigma_2. \tag{15.18}$$

Fig. 15.12 The distribution of the characteristic quantity in the pattern plane

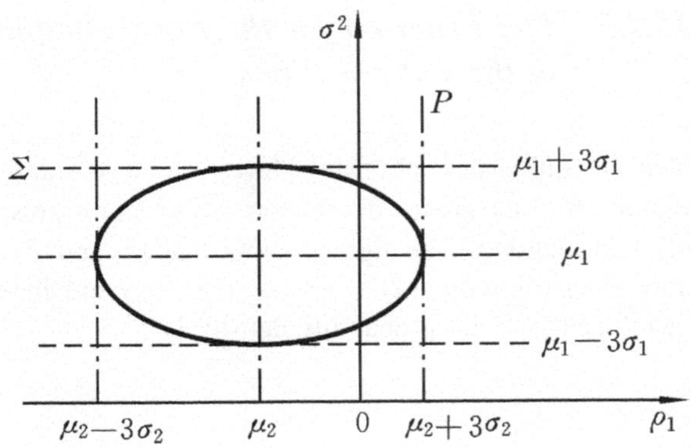

Figure (15.10) shows only σ^2 and ρ_1 exceed their threshold values Σ and P, respectively, i.e., the feature point falls into the region II can we predict the chatter is coming. The regions I and III only mean that σ^2 alone or ρ_1 alone exceed its threshold value, as points A, B in Fig. 15.10, it is not enough to judge a coming of chatter.

15.4 An Experimental System of Computer Monitoring and Controlling Machine Tool Chatter

Computer monitoring and controlling is a key technology in computer aided automatic machining system which is developing nowadays. Here, through an experimental system established to illustrate how to link early diagnosis of chatter and online chatter control together, so as to realize computerized chatter online monitoring and in process control.

15.4.1 Computerized Monitoring and Controlling Strategy for Machine Tool Chatter

As mentioned in previous chapters, the shape and location of the stability lobes in the stability charts and the present cutting parameters are closely related to the adjustment of cutting parameters for the purpose of vibration suppression. And they are affected by a variety of factors (such as changes in the machine configuration, tool wear, etc.), so it is difficult to determine in advance the reasonable adjustment direction and the adjustment amount of cutting parameters. Therefore, to adjust the cutting parameters online properly, you can only take the method of “error-and-try”. The established experimental system is by means of adjusting spindle speed and feed rate to suppress the chatter. The method is that after a trial change in the cutting parameters, computer evaluates the effect of the change:

If the computer determines that the amplitude falls below the threshold value, then the chatter has been successfully suppressed, the computer stops adjusting the cutting parameters and returns the online monitoring and diagnosis;

If the amplitude is not decline, increased instead, this indicates that the direction or the value of adjustment are not appropriate, should adjust the quantity or adjust the direction, until the amplitude dropped to below the threshold value, and the system returns to on-line monitoring and diagnosis;

If the chatter is not suppressed in the allowable adjustment of the system, an alarm is issued so that the operator can intervene.

The block diagram of online monitoring strategy is shown in Fig. 15.13.

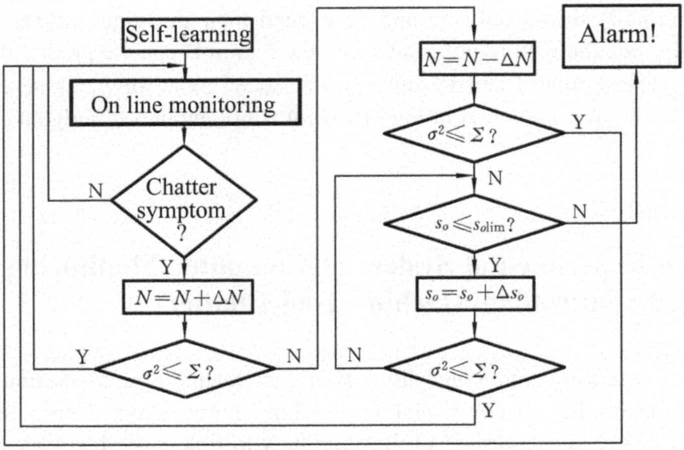

Fig. 15.13 The flow chart of on-line monitoring and in-process control system of machine tool chatter

15.4.2 *An Experimental System of Computerized Online Monitoring and Controlling of Machine Tool Chatter*

The block diagram of the experimental system is shown in Fig. 15.14 (The experiments were conducted many years ago, the equipment and the devices used are out of date. However, the principle and the ideas should still be valuable). The system has the following functions:

(1) Can in process adjust the cutting parameters;
(2) Can monitor and control the chatter of the machine tool;
(3) Can online display the vibration signal and the pattern plane, i.e., the stability state of the cutting process.

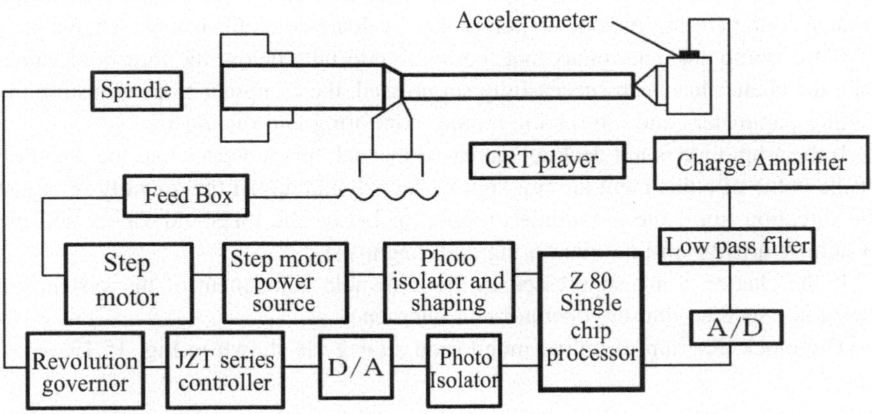

Fig. 15.14 Block diagram of on-line monitoring and control system for machine tool chatter

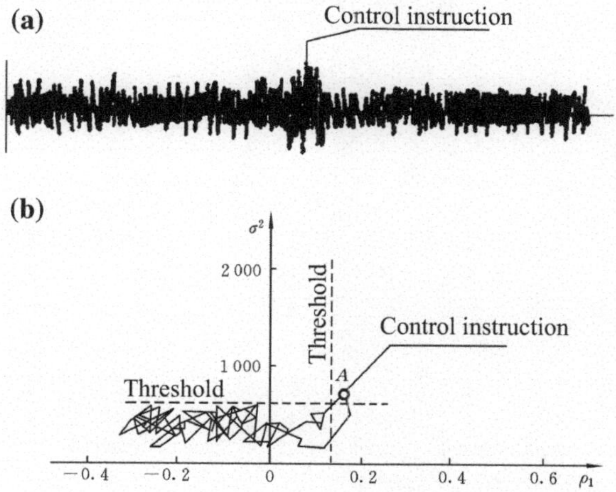

Fig. 15.15 The effect of online monitoring and control of machine tool chatter. **a** The signal in time domain. **b** The trajectory of feature point

Figure 15.15a, b respectively represent the monitoring signal and the pattern plane during chatter monitoring and controlling experiments on this system.

As shown in Fig. 15.15b, the spindle speed $N = 650$ r/min, the feed rate is $s_o = 0.08$ mm/r, and the cutting depth of cut is $a_c = 0.5$ mm, the cutting process tends to be unstable and the feature point enter the unstable region. Monitoring unit of the system detected the early signs of chatter at point A, and immediately issued a control command, the feed rate is adjusted to $s_o = 0.2$ mm/r, the trajectory of feature point was immediately drawn back to the stable region. Figure 15.15a shows that the further growth of vibration signal is stopped.

Many experiments on the system show it is possible and effective to use the pattern plane to describe the dynamic stability of a cutting process, to monitor and to early diagnose chatter in the cutting process and to suppress or prevent chatter by adjusting the cutting parameters in real time.

The experiment also shows that it is feasible to use the technology above described to ensure a machining conditions without chatter or low chatter.

Chapter 16
On-Line Chatter Prevention Technology for NC Machine Tools

The modern CNC machine tools have provided a new idea and new possibility for the on-line cutting chatter control—to suppress the chatter with the small continuous periodic disturbance of the spindle speed. This approach is easy to implement and does not require any additional hardware or devices. Moreover, theoretical analysis and cutting test have fully demonstrated the effectiveness of this method.

Based on the non-linear theoretical model of machine tool chatter, described in Chap. 13, in the Chap. 14, we introduced the methods and techniques of step adjustment of the cutting parameters (spindle revolutions and feed rate) to suppress chatter. In this chapter, the methods of small continuous periodic disturbance imposed on spindle revolutions to suppress machine tool chatter were in-depth studied, both the cutting experiment and compute digital simulation were conducted to reveal the possibility and the law of using such a method to on-line chatter prevention for NC machine tools (the digital simulation method is the same as described in Chap. 13).

16.1 The Model of Cutting System and Its Kinetic Parameters

In order to study the cutting chatter process and the vibration suppression effect of the spindle speed disturbance, a non-linear mechanical model of the cutting system must be established.

Let's tack slender shaft turning process as an example, its schematic diagram is shown in Fig. 16.1a, the dynamic model of the system is given in Fig. 16.1b, it has

Part of this chapter is from the study of Mr. Zhang Haiyan, a doctoral student from the University of Michigan. His Ph.D. work was partly under my direction when I was a visiting professor in the university (1992.9–1993.9).

H. Shi, *Metal Cutting Theory*, Springer Series in Advanced Manufacturing,
https://doi.org/10.1007/978-3-319-73561-0_16

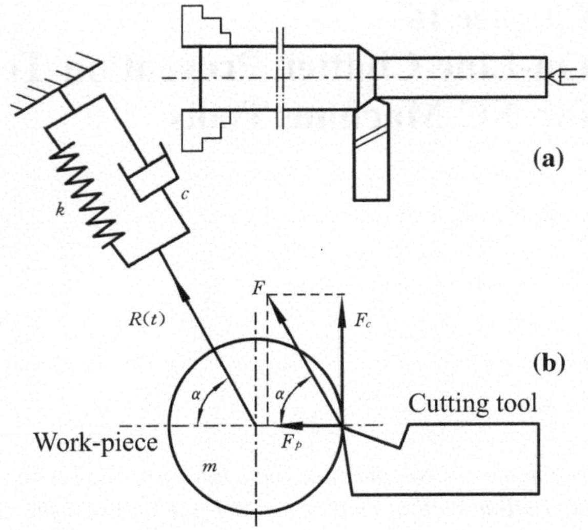

Fig. 16.1 A slender shaft turning system model. **a** Schematic diagram of slender shaft turning process. **b** The single-degree-of-freedom mechanical model of the turning system

been simplified to a single degree of freedom system. Although a practical process system always has multiple degrees of freedom and multiple modes, however, when the system is unstable, it is often that one mode plays dominant role. Therefore, the system can be simplified into a single degree-of-freedom model when modeling.

In order to compare the results of computer simulation with actual cutting test results, the dynamic parameters of the physical model must be measured in advance. In this slender shaft turning system, the flexibility of the work-piece is much greater than that of the machine tool structure. Therefore, the flexibility of the machine can be omitted, and the flexibility of the work-piece was measured only. The natural frequency ω_o and the damping rate ξ of the slender shaft with the clamping method shown in Fig. 16.1a were measured by means of hammering method.

A PCB vibration hammer (type SP205) was used to knock the cutting place of the shaft (i.e., the middle part of the shaft), the vibration signal of the work-piece is detected by the acceleration sensor, through charge amplifiers, the measured signal are recorded by a Teac tape recorder, and then was analyzed by a computer [92].

16.1.1 Determination of Damping Ratio ξ

The first wave peak height $A_o = 42.75$ m/s^2 is measured from the recorded vibration history, and after 16 vibration cycles, the amplitude of the vibration is reduced to $A_n = 0.9$ m/s^2, from which the logarithmic attenuation rate can calculated as [79],

$$\delta = \frac{1}{n}\ln(\frac{A_o}{A_n}), \tag{16.1}$$

And from this, the damping ratio can be calculated as

$$\xi = \frac{\delta}{\sqrt{4\pi^2 + \delta^2}} = 0.0384. \tag{16.2}$$

16.1.2 The Determination of Natural Frequency ω_o

The spectrum analysis was performed for the measured vibration signals and the peak frequencies were found to be four: 85, 101, 133 and 207 Hz. The frequency of the cutting chatter will be close to one of these frequencies. In simulation, ω_o was assigned these four values, respectively.

16.1.3 The Determination of Static Stiffness k

The installation of the slender work-piece on the lathe is shown in Fig. 16.1a, with a chuck holder at one end, which is roughly equivalent to the fixed support; the other end is supported by the tailstock center, which is between the solid support and the simply support. It is known that the stiffness at the center of the "solid-simply" supported beam is $k_1 = 192EI/L^3$, and the stiffness at the center of the "solid-solid" supported beam is $k_2 = 768EI/7L^3$, where, $E = 2.156 \times 9.8 \times 10^{10}$ is the elastic modulus of the work-piece material, $I = \pi d^4/64$ is the cross sectional moment of inertia of the work-piece, the diameter of work-piece is $d = 50$ mm, and its length is $L = 600$ mm. Substituting the above data in, it can be worked out that $k_1 = 33,598$ N/mm and $k_2 = 58,796$ N/mm. According to the consistency between some simulation data and test data, fit a value between the above two $k = 45,000$ N/mm as the static stiffness. In fact, the value is close to the median 46,197 of k_1 and k_2.

16.1.4 The Determination of Cutting Force Model Parameters

The cutting force components that affect the radial deformation of the work-piece should be cutting force F_c and back force F_p as shown in Fig. 16.1b (feed force F_f has very small influence on the radial deformation of the work-piece, it can be ignored).

Table 16.1 Experimental data for fitting the empirical formula of cutting force

f (mm/r)	0.05	0.10	0.15	0.20	0.25	0.30	0.35	0.40
F_c (N)	86.99	157.32	212.80	277.23	286.67	374.28	390.93	441.89
F_p (N)	74.39	134.56	181.58	237.10	247.73	320.08	339.43	377.88
F (N)	114.46	207.02	280.00	364.77	381.15	492.48	522.27	581.43

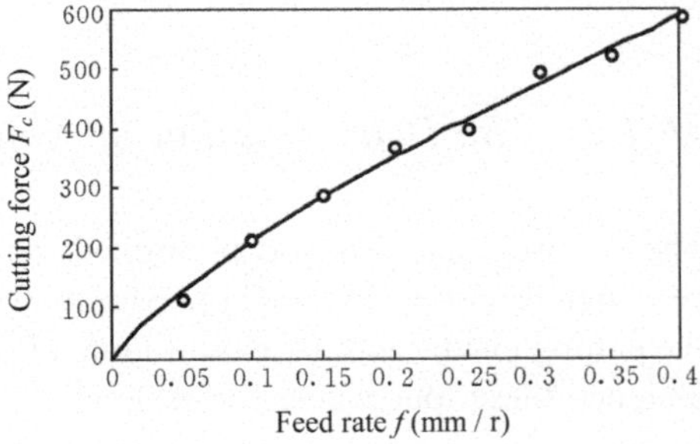

Fig. 16.2 Experimental data of cutting force and its fitting curve. Solid lines—Fitted curve, open circle—Experimental data

The resultant force of F_c and F_p is $F = \sqrt{F_c^2 + F_p^2}$. The relation of F with depth of cut a_p and feed rate f is

$$F = Ca_p f^y. \tag{16.3}$$

where, parameters C and y are fitted from measured data in turning test. In turning test, the spindle revolutions $N = 470$ r/min, cutting speed $V = 60$ m/min, depth of cut $a_p = 1$, and $f = 0.05$–0.40 mm/r. The experimentally measured data are given in the Table 16.1.

The parameters in the Formula (16.3) are fitted as $C = 1198.71$ (N), $y = 0.77$ from the data in the table. Thus, the formula of cutting force becomes:

$$F = 1198.71a_p f^{0.77}. \tag{16.4}$$

The fitted curves and measured data are shown in Fig. 16.2.

16.1.5 The Evaluation of Penetration Rate Factor

The definition of penetration rate factor c was described in Sect. 14.1.2.2 and [80, 90, 91]. Since the measurement of value c is very difficult, we refer to the value range of c in the relevant literature, and compare the results of prediction and actual measurement evaluated $c = 0.004$.

16.2 Computer Digital Simulation: The Vibration Suppression Effect of Spindle Speed Disturbance

This section is based on the above theoretical model by means of computer digital simulation to study the effect of spindle speed disturbance on the vibration suppression effect.

16.2.1 The Amplitude of Spindle Speed Disturbance

If the amplitude of the disturbance is too small it cannot produce a vibration suppression effect. As shown in Fig. 16.3, in the same conditions, the amplitude of the spindle speed disturbance is 44 r/min, which cannot suppress the chatter, but only has obvious modulation effect. When the amplitude of the spindle speed disturbance increases to 64 r/min, the chatter is completely suppressed.

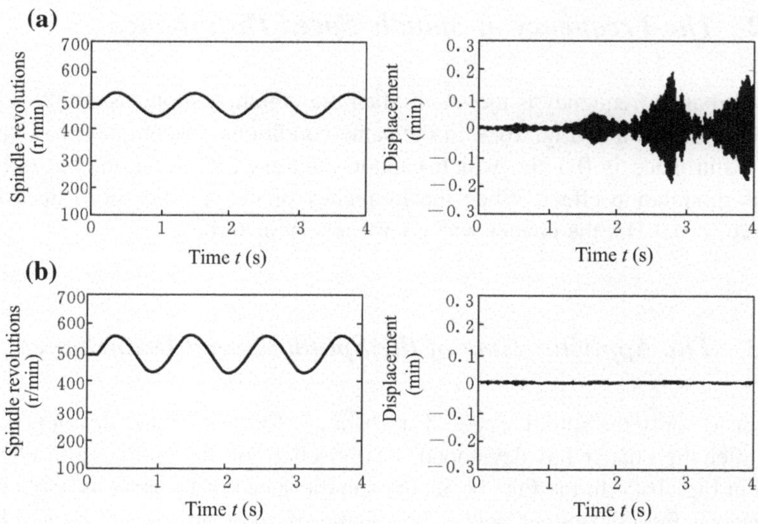

Fig. 16.3 The influence of the amplitude of spindle revolutions disturbance on the vibration suppression effect (simulated result). **a** $N = 490 + 44\sin(2\pi \times 0.9t)$, $f = 0.1$ mm/r, $a_p = 4$ mm. **b** $N = 490 + 64\sin(2\pi \times 0.9t)$, $f = 0.1$ mm/r, $a_p = 4$ mm

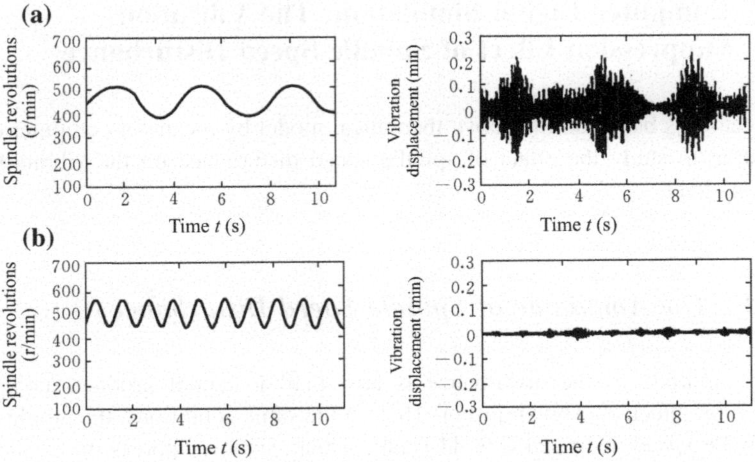

Fig. 16.4 The influence of the frequency of spindle revolutions disturbance on the vibration suppression effect (simulated result). **a** $N = 490 + 60\sin(2\pi \times 0.3t)$, $f = 0.1$ mm/r, $a_p = 4$ mm. **b** $N = 490 + 60\sin(2\pi \times 0.9t)$, $f = 0.1$ mm/r, $a_p = 4$ mm

16.2.2 The Frequency of Spindle Speed Disturbance

The disturbance frequency is too small, then the vibration suppression effect is not possible. As shown in Fig. 16.4, in the same conditions, the frequency of spindle speed disturbance is 0.3 Hz, which cannot suppress the vibration, but only has obvious modulation effect. When the frequency of the spindle speed disturbance increased to 0.9 Hz, the chatter was completely undisturbed.

16.2.3 The Applying Time of the Spindle Speed Disturbance

Whether to apply the spindle speed disturbance before the chatter developed, or to apply after the chatter has developed, has no effect on the suppression effect, as shown in Fig. 16.5. In the Fig. 16.5a, the spindle speed disturbance is applied after the chatter is fully developed, and *b* the spindle speed disturbance is applied before the chatter is fully developed, so the effects of both are the same.

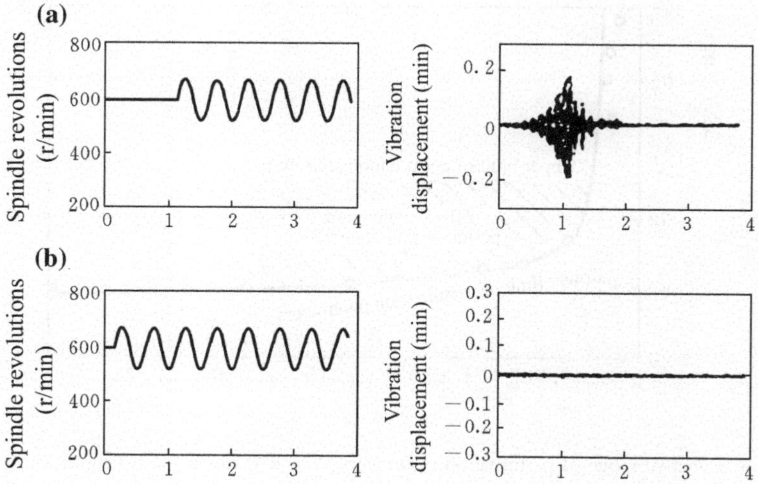

Fig. 16.5 The influence of the applying time of the spindle speed perturbation on the effect of vibration suppression (simulated result). **a** After the chatter is developed, the spindle speed disturbance was applied, $N = 590 \rightarrow N = 590 + 40\sin(2\pi \times 2t)$, $f = 0.1$ mm/r, $a_p = 4$ mm. **b** Before the chatter developed, the spindle speed perturbation was applied, $N = 590 + 40\sin(2\pi \times 2t)$, $f = 0.1$ mm/r, $a_p = 4$ mm

16.3 The Effective Range of Spindle Speed Disturbance Parameters

Generally speaking, the spindle speed disturbance can be expressed as

$$N(t) = N_o + \Delta N \sin(2\pi f t), \tag{16.5}$$

where, N_o for nominal revolutions, i.e., the spindle revolutions before undisturbed; ΔN for the amplitude (r/min) of disturbance sinusoidal signal, namely the spindle speed fluctuation; f is the frequency of disturbance (Hz).

16.3.1 The Lower Limit of the Disturbance Parameters

It is good to have a smaller amplitude ΔN of spindle speed disturbance and a smaller frequency f, of course, however, such as the digital simulation experiments proved, if ΔN or f is too small, they will failed to reach the effect of vibration suppression. So for ΔN and f there is a minimum limitation, below the minimum value, chatter cannot be suppressed. Simulation experiments also show that the minimum value of ΔN and the minimum value of f are related to each other, and

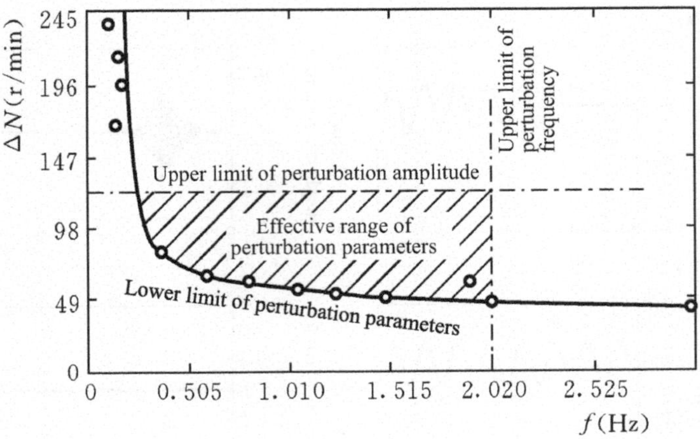

Fig. 16.6 Effective range of perturbation parameters, $N = 490$ r/min, $f = 0.1$ mm/r, $a_p = 4$ mm

their relationship is shown in Fig. 16.6, the relationship between the smallest ΔN and the smallest f can be fitted into a hyperbola.

16.3.2 The Upper Limit of the Disturbance Parameters

Upper limit of the ΔN are mainly restricted by the processing quality, the spindle revolution fluctuations should not have an adverse impact on machining quality of parts. In addition, the fluctuation of rotational disturbance should not obviously shorten the tool's life. As recommended by literatures [93] and According to the test results, when the $\Delta N/N \leq 25\%$ or less, the disturbance will not produce obvious adverse effect on machining quality and tool life. Here, therefore, take the large amplitude of the disturbance of spindle speed for $\Delta N_{max} = 0.25\ N = 122$ r/min.

The upper limit of spindle speed disturbance frequency is mainly limited by the response speed of the spindle drive system of the machine tool.

The machine tool used in the test is a Profiturn 15 CNC lathe. Spindle speed disturbance signals is input from the additional input terminal of the numerical control system (override), which is superimposed on the spindle speed signal of normal, they jointly drive the spindle DC motor, and through a belt drive and a sliding gear drive to drive the spindle assembly and to drive the work-piece to rotate, and to achieve rotational speed disturbance. Its system and structure are shown in Fig. 16.7.

The test system of the frequency response function between the input of the spindle speed disturbance signal and the actual instantaneous speed (output) of the work-piece is shown in Fig. 16.8. The input sinusoidal disturbance signal in the test and the corresponding spindle speed response are shown in Fig. 16.9 (some examples).

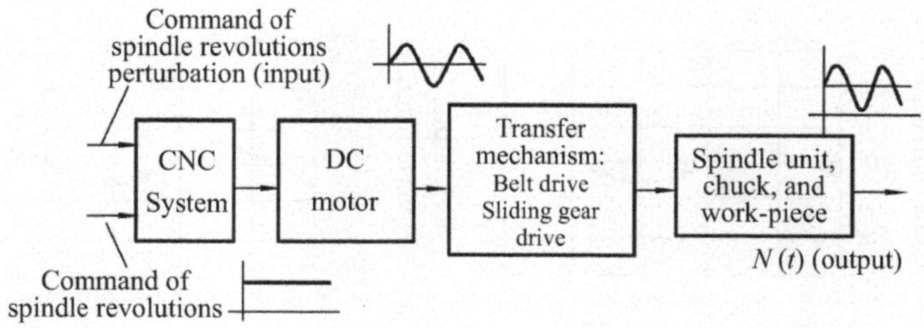

Fig. 16.7 Spindle revolutions perturbation system and its input and output

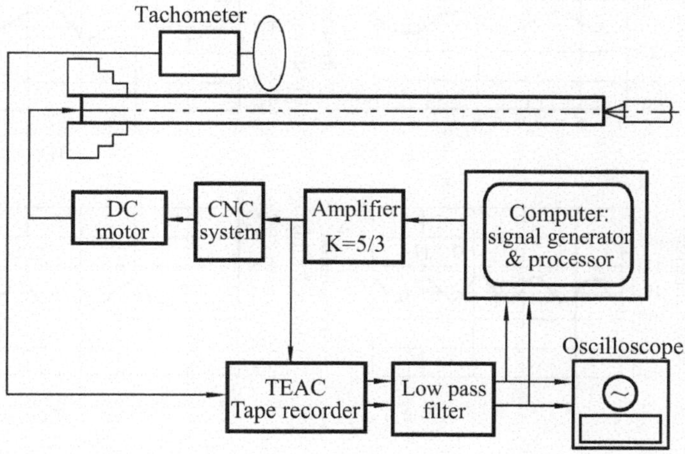

Fig. 16.8 The system testing the response characteristic of the spindle system to sinusoidal disturbance signal on its spindle revolutions

The frequency response characteristics measured are shown in Fig. 16.10, where, figure *a* is the amplitude characteristic and figure *b* is the phase characteristic. As can be seen from the diagram, the system is basically a second-order system. After fitting, its natural frequency ω_o is about 3.5 rad/s (f_o = 0.557 Hz), damping rate $\xi = 0.5$. This is a system with very large inertial, and it is not possible for it to respond to the higher frequency drive signals. In fact, when the frequency of the excitation is greater than 2 Hz, the amplitude of the response decreases sharply. Therefore, the maximum speed disturbance frequency of spindle is set to f_{max} = 2 Hz, as shown in Fig. 16.6.

It needs to be explained that the numerical control machine used in the experiment was an older type of CNC machine tools. The response frequency band of the newly designed machine spindle system may be wider. However, the spindle system of NC machine tool is designed for constant speed cutting, and its response ability to harmonic disturbance is always limited. Therefore, it is impossible to use the higher frequency speed disturbance. The answer to fundamentally solving this

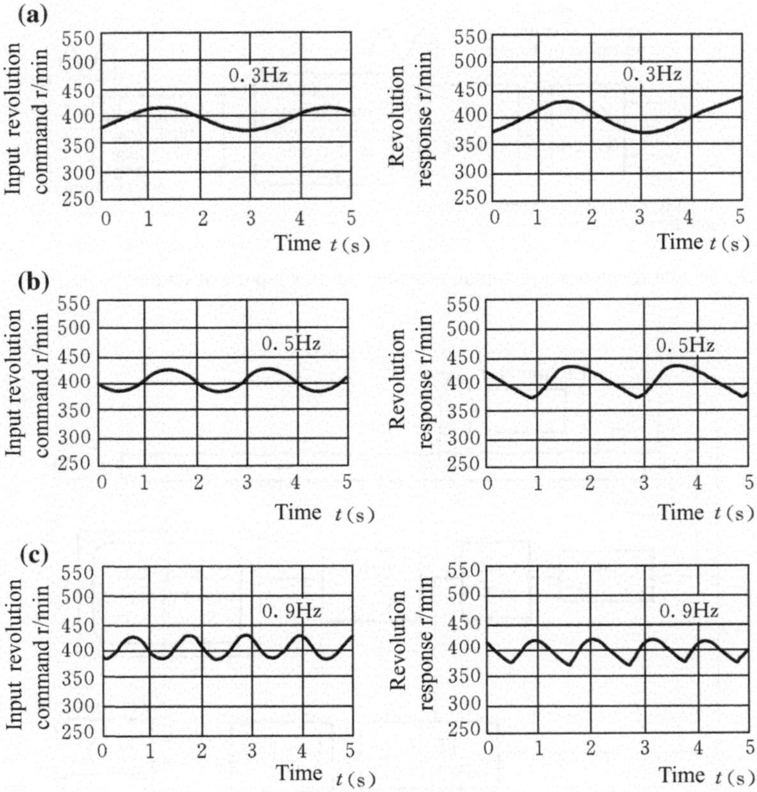

Fig. 16.9 The sinusoidal disturbance signal on the spindle system and its corresponding response (examples, tested results)

problem is only a special vibration tool holder designed and to replace the spindle rotation speed disturbance of the spindle system with the vibration of the tool itself. This is beyond the scope of this book, and here we will not go further at this issue.

16.4 The Optimal Disturbance Parameters

Nondimensionalizing the disturbance amplitude ΔN and the disturbance frequency f, i.e., let $\mu = \Delta N/\Delta N_{max}$ and $\lambda = f/f_{max}$, we can change Figs. 16.6, 16.7, 16.8, 16.9, 16.10 and 16.11. The plane of the figure is the parameter plane, the ordinate and the abscissa represent the parameters μ and λ, respectively. A point (λ, μ) on the plane

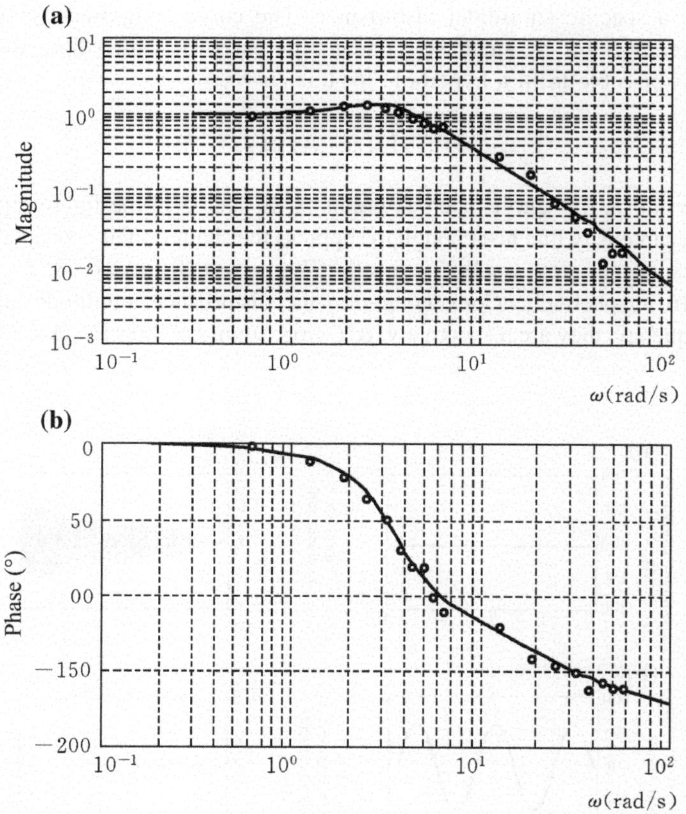

Fig. 16.10 Tasted frequency response characteristic of the spindle. **a** Magnitude. **b** Phase

Fig. 16.11 The disturbance parameter plane and the limit disturbance curve

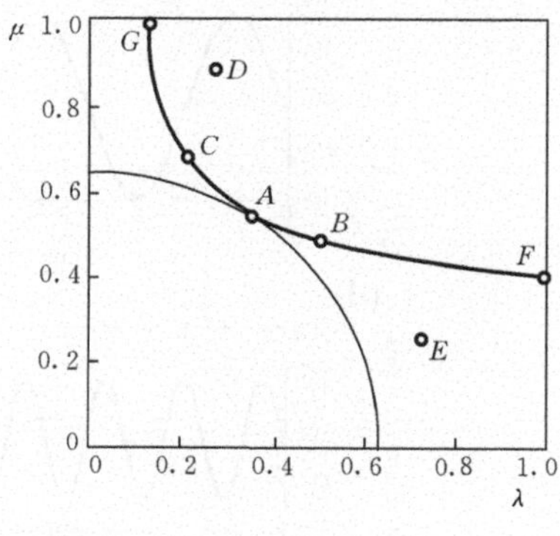

represents a specific sinusoidal disturbance. The curve composed by smallest μ under different λ (or by smallest λ under different μ) is called limit disturbance curve, it is possible to fit a hyperbola for this curve,

$$\mu = 0.3917(\lambda - 0.101)^{-0.24}. \tag{16.6}$$

Each point on this curve, such as points A, B, and C in the figure represents a disturbance intensity that just enough to suppress the chatter: their coordinates on the parameters plane are: $\lambda = 0.33$, $\mu = 0.5579$; $\lambda = 0.20$, $\mu = 0.6857$; $\lambda = 0.50$, $\mu = 0.4816$, respectively. Conversing into the disturbance amplitude and disturbance frequency, they are respectively: $\Delta N = 68$ r/min, $f = 0.66$ Hz; $\Delta N = 84$ r/min,

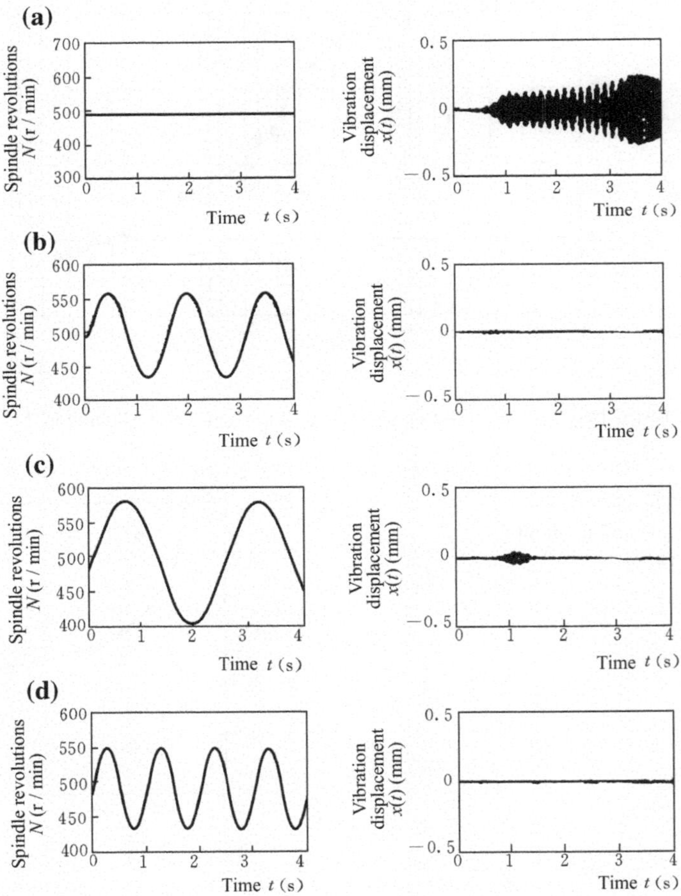

Fig. 16.12 The spindle revolution disturbance in the limit case and its chatter suppression effect (simulated results). **a** Constant revolutions (without perturbation). **b** The case corresponding to point *A*. **c** The case corresponding to point *B*. **d** The case corresponding to point *C*

$f = 0.40$ Hz; $\Delta N = 59$ r/min, $f = 1.00$ Hz. The revolution disturbance and its vibration suppression effect corresponding to these three points are shown in Fig. 16.12 (the constant speed cutting, i.e., no disturbance, is also included in the figure, for a comparison).

Since these points are at the limit disturbance curve, if the disturbance amplitude or the disturbance frequency are slightly reduced, the suppression is invalidated.

The points in the upper right of the limit curve represent the effective disturbances, such as point *D*, it can effectively suppress chatter; in the lower left corner of the curve, such as *E*, is an invalid disturbance, not sufficient to inhibit the chatter. Compared to the points on the limit curve, points *D* represent greater disturbance amplitude or higher disturbance frequency, which is not necessary. Therefore, from the point of view of the suppression of chatter, the upper right area of the limit curve is not practical; On the other hand, the area at the lower left of the limit curve cannot suppress chatter, so only the limit curve itself represents the correct parameters for suppressing chatter. Points on the limit curve, however, are not equally good: the points on both ends of the curve (i.e., the points F and G in Fig. 16.11), or the points close to the both ends, represent effective disturbance parameters though, they need larger disturbance amplitude, or high disturbance frequency, it is difficult to implement.

As mentioned earlier, technically we hope that neither amplitude μ nor frequency λ of the disturbance are too high, so, only those point in the middle part of the curve represents reasonable disturbance parameters. Hereby we can regard the point which satisfies the condition that $\sqrt{\lambda^2 + \mu^2}$ is minimum is the optimum point. To find this point, in Fig. 16.11 we make a circle with the origin as its center and the tangent point *A* of the circle with the limit curve must be optimum point.

Figure 16.11 is the case with the depth of cut $a_p = 4$ mm, When the a_p takes different values, a series of limit curve families are obtained, as shown in

Fig. 16.13 The limit disturbance curve family and their most advantage point series for different depth of cut

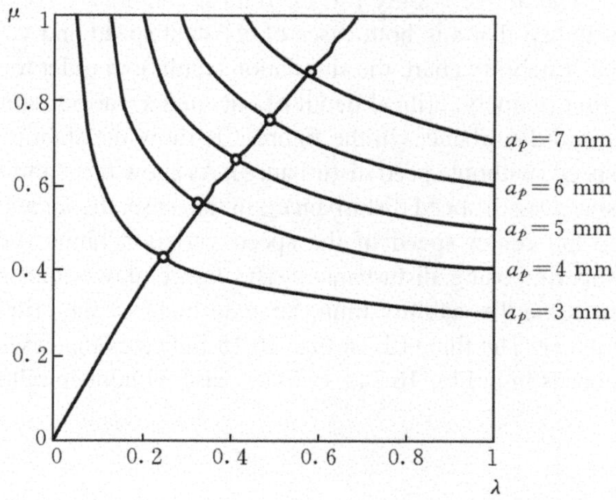

Fig. 16.13. In the figure, an optimum point is found for each limit curve by the above principle. These optimum points form an optimal line, the parameter equation fitted of which is

$$\begin{cases} \mu = 0.1785a_p^{0.8091} \\ \lambda = 0.0822a_p^{0.9880} \end{cases}, \tag{16.7}$$

Or, the equation shows the relationship of disturbance parameters μ and λ

$$\mu = 1.3568\lambda^{0.8189}. \tag{16.8}$$

In terms of the optimal perturbation parameters, the [95] has a more in-depth analysis from the view point of the energy and spindle system rotational acceleration, readers who are interested in the issue, read the literature pp. 101–110, and we will not go in detail here.

16.5 The Stability of the Cutting System Is Enhanced by the Spindle Speed Disturbance

This section studies the influence of spindle speed disturbance on the stability of cutting system. As shown in Fig. 16.14, the simulation results show that under the three constant revolutions, N = 430, 490 and 550 r/min, the cutting system is not stable, however, the cutting process with the disturbance of variable cutting speed $N = 490 + 60\sin(2\pi \times 0.9t)$ is stable, though, its speed is changing in the unstable speed range 430–550 r/min. This indicates that spindle speed disturbance can really enhance the stability of the cutting system.

The above is only for a certain revolutions (N = 490 r/min). In Fig. 16.15, the stability charts in both cases of constant speed and variable speed cutting are given (*WN* stability chart, the simulation results), in order to verify the increase in stability limit (namely, critical depth of cut) under various spindle speed due to the spindle speed disturbance. In the figure, Os show the stability limit under constant cutting speed (without speed disturbance); Δs show the stability limit under variable cutting speed (with speed disturbance), in this case, the location of symbols Δs is according to the center speed of the speed variation range (i.e., the constant speed before superimposing disturbance). The figure shows that speed perturbation can really increase the stability limit, i.e., can increase the critical depth of cut of starting to chatter. The three □s in Fig. 16.15 indicates the cases of the three constant cutting speeds in in Fig. 16.14a–c. In the case without spindle speed disturbances (constant

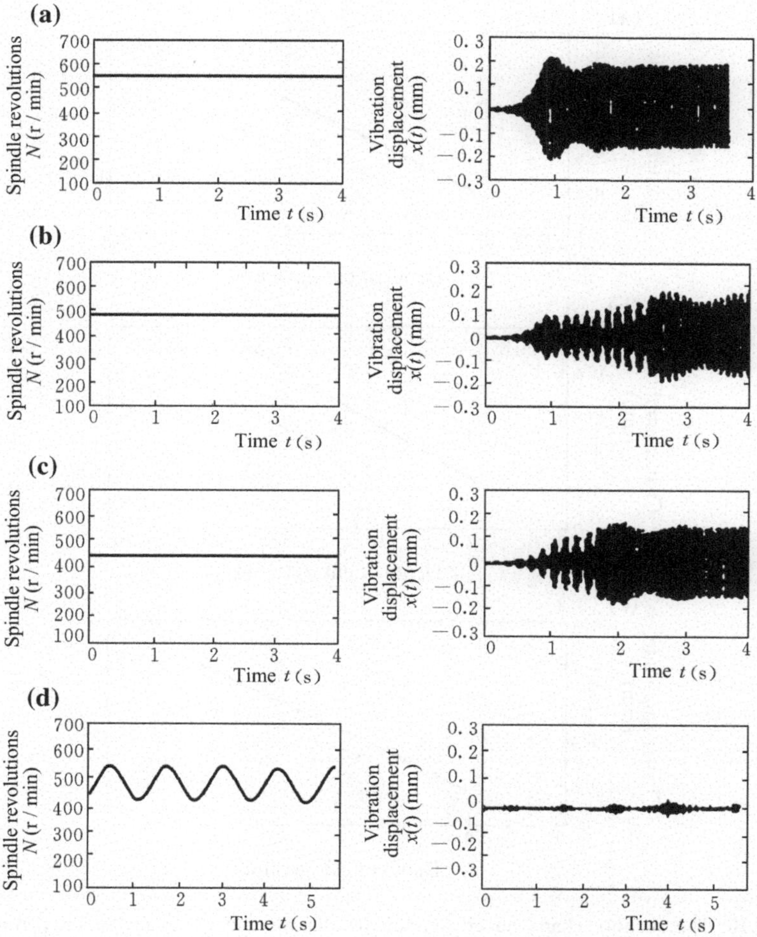

Fig. 16.14 The spindle revolution disturbance can enhance the stability of the cutting system (simulation results)

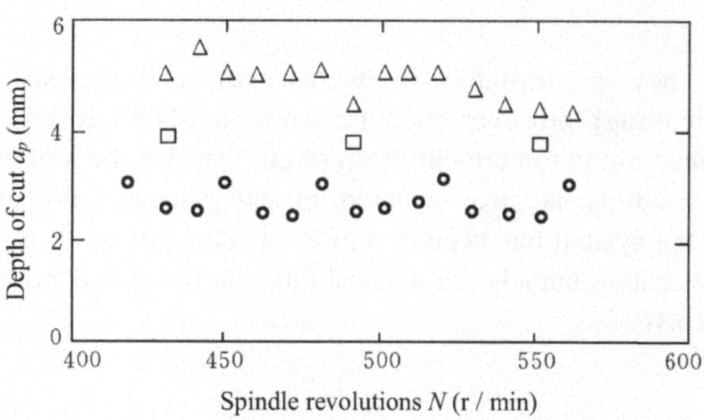

Fig. 16.15 Stability chart under constant spindle revolutions or with spindle revolution perturbation (*WN* stability chart, simulation results). Open circle—Constant cutting speed (without speed disturbance), open triangle—Variable cutting speed (with speed disturbance), open square—The three constant cutting speeds in Fig. 16.14

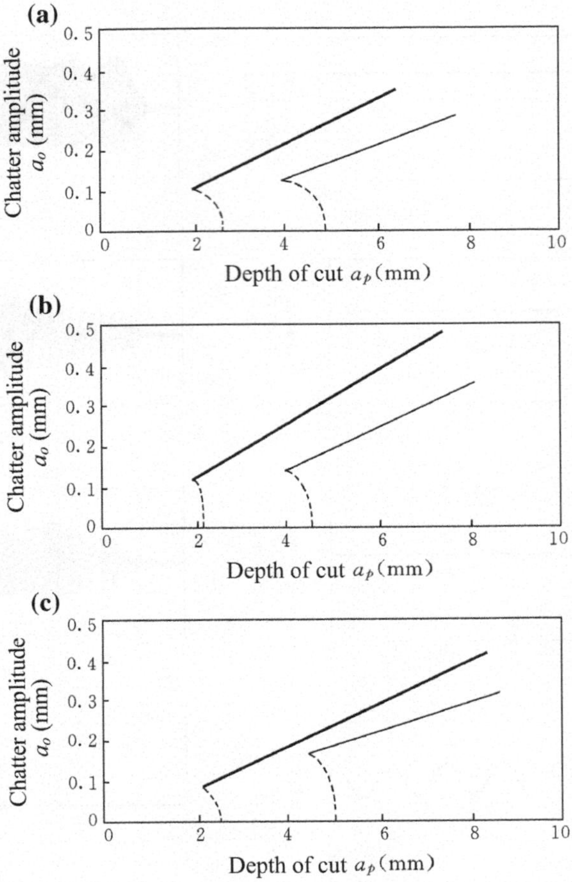

Fig. 16.16 The stability charts under constant revolutions and under revolution perturbation (*AW* stability charts, simulation results). **a** $N = 430$, $N = 430 + 60\sin(2\pi \times 0.9t)$, $a_p = 4$ mm, $f = 0.1$ mm/r. **b** $N = 490$, $N = 490 + 60\sin(2\pi \times 0.9t)$, $a_p = 4$ mm, $f = 0.1$ mm/r. **c** $N = 550$, $N = 550 + 60\sin(2\pi \times 0.9t)$, $a_p = 4$ mm, $f = 0.1$ mm/r, Thick lines—Constant revolutions, without perturbation, Thin lines—Variable revolutions, with perturbation

speed cutting), they are unstable, in the case of rotational speed disturbance (variable speed cutting), however, they are stable, as shown in Fig. 16.14d.

Besides the increase in the critical depth of cutting, the effect of enhancement of cutting system stability is also reflected in the decrease of stabilized chatter amplitude after the system has been destabilized. This point can be obtained from another stability chatter, namely the *AW* stability chart, resulted from simulation, as shown in Fig. 16.16.

Bibliography

1. 师汉民. 分析和计算刀具角度的一种新方法. 华中工学院学报. 1978;6(3):1–17.
2. 师汉民. 金属切削刀具几何角度的一种图解法. 华中工学院学报. 1980;8(4):1–10.
3. 师汉民. 分析和计算刀具角度的一种新方法. 工具技术. 1980;(2):49–55.
4. Shi H. A new method for analyzing and calculating angles on cutting tools. Int J Mach Tool Des Res. 1982;22(3):177–96.
5. Shi H. Graphic determination of geometric angles on metal-cutting tools. Int J Mach Tool Des Res. 1986;26(2):99–112.
6. Shi HM, Zhang HS. The fundamentals and the method of optimizing the cutting edge shape on standard twist drills. In: Pei Z, editor. Proceeding of the fourth international conference on metal cutting, non-conventional machining and their automation. Beijing: Beijing Institute of Technology Press; 1989. p. T13–18.
7. 熊良山, 师汉民, 张华书. 一种切削力小、能耗低的麻花钻. 华中理工大学学报. 1991;19(增刊):89–94.
8. 师汉民,熊良山, 张华书. 计算机辅助麻花钻刃形优化与前角控制. 见: 中国高校金属切削研究会学术委员会主编. 中国高校金属切削研究会第四届学术年会论文集. 北京: 机械工业出版社; 1991. p. 295–9.
9. Shi H, Zhang H, Xiong L. A Study on Curved Edge Drills. J Eng Ind, Trans ASME. 1994;116(2):267–73.
10. 师汉民. 切屑的流出、卷曲与折断: [研究生学位论文]. 武汉: 华中工学院. 1965.
11. 师汉民. 切屑的流出、卷曲与折断 .华中工学院学报. 1965;5(7):66–106.
12. Wang J, Sha J, Ni J, et al. Development of new drilling force models. In: Devor RE, Dornfield DA, Ehmann K, editors. Proceeding of international congress and exposition & the first S. M. Wu symposium on manufacturing science. Dearborn: Society of Manufacturing Engineers; 1994. p. 112–5.
13. Wang J, Shi H, Sha J, et al. The interference of chip ejection under non-free cutting. In: Stephenson DA, Stevenson R, editors. Proceeding of international congress & exposition & the winter annual meeting of ASME. Warren dale: The Material, Metal & Materials Society & ASME; 1994. p. 139–56.
14. Wang J, Sha J, Ni J, Shi H. Development of new drilling force models. In: Zhou Z, editor. Proceeding of the international symposium on manufacturing science and technology for the 21st Century & the first S. M. Wu memorial symposium. Beijing: International Academic Press; 1994. p. 171–8.
15. 师汉民, 陆涛, 陈永洁 .非自由切削中的排屑干涉. 见: 袁哲俊 主编. 金属切削理论与技术的新进展.武汉: 华中理工大学出版社; 1995. p. 17–20.
16. Shi H, Wang J. A model for non-free-cutting. Int J Mach Tools Manuf. 1995;35(11):1507–22.

H. Shi, *Metal Cutting Theory*, Springer Series in Advanced Manufacturing,
https://doi.org/10.1007/978-3-319-73561-0

17. Shi H, Lu T, Wang X. On the nature of minimum energy of metal cutting process. In: Editorial Committee, editors. Proceeding of the international workshop on automobile manufacturing science and technology & the second S. M. Wu symposium on manufacturing science .上海: 上海 科学技术文献出版社; 1996. p. 87–92.
18. 陆涛, 王西彬, 师汉民. 切削过程的非自由度系数的研究. 华中理工大学学报. 1996;24 (6):38–40.
19. 陆涛, 王西彬, 师汉民. 金属切削过程的分叉与突变现象. 华中理工大学学报. 1996;24 (9):1–4.
20. 王西彬, 师汉民, 陆涛. 非自由切削的流屑角及其分叉. 中国机械工程. 1996;7(4):74–6.
21. 王西彬, 师汉民, 陆涛. 圆弧刃切削的非自由系数及其改善. 工具技术. 1996;30(4):13–6.
22. 王西彬, 师汉民, 陆涛. 切削过程的分叉与突变. 机械工程学报. 1997;33(6):20–5.
23. 师汉民. 自由切削法与自由切削刀具. 中国机械工程. 1998;9(7):47–50.
24. Shi H, Wang X, Lu T. Bifurcation and catastrophes in metal-cutting processes. J Manuf Sci Eng, Trans ASME. 1998;120(4):817–20.
25. 闫兴, 师汉民 .非自由切削的基础理论和自由切削法 .中国学 术期刊文摘(科技快报). 1998;4(9):1147.
26. 师汉民. 现代刀具切削过程中的排屑干涉. 中国科学(E辑). 1999;28(2):186–192.
27. Shi H. Chip-ejection interference in cutting processes of modern cutting tools. Sci China E. 1999;42(3):275–81.
28. 师汉民 .金属切削过程中的分叉与突变现象— ——兼论切削过 程的可控性问题 .应用力学学报. 1999;16(1):15–20.
29. Shi H, Wang X, Lu T. Non-free cutting and its degree of freedom confinement. J Manuf Sci Eng, Trans ASME. 1999;121(1):150–53.
30. 闫兴, 师汉民. 自由切削法及其应用. 见: 袁哲俊主编. 中国高 校切削与先进制造技术研究会第六届年会论文集. 北京: 机械工业出版社; 1999. p. 26–29.
31. 闫兴, 师汉民, 陈永洁, 等. 自由切削刀具的设计原理及实例. 硬质合金. 1999;16(3):162–5.
32. Shi HM, Tobias SA. Theory of finite amplitude machine tool instability. Int J Mach Tool Des Res. 1984;24(1):45–69.
33. 师汉民. 关于机床自激振动的一个非线性理论模型 (第一部分). 应用力学学报. 1984;1 (1):1–14.
34. 师汉民. 关于机床自激振动的一个非线性理论模型 (第二部分). 应用力学学报. 1984;1 (2):75–88.
35. 师汉民. 影响机床颤振的几个非线性因素及其数学模型. 华中工学院学报. 1984;12 (6):102–12.
36. Shi H. Effects of some nonlinear factors on machine tool chattering and their mathematical models. J Huazhong Univ Sci Technol. 1986;8(1):9–22.
37. Shi HM, Xie YY, Lu TK. Some new discoveries by the nonlinear theory of machine tool dynamics. In: Wen B, editor. Proceeding of international conference on mechanical dynamics. Shengyang: Press of Northest University of Technology; 1987. p. 474–79.
38. 师汉民, 谢月云, 卢天贶. 切削颤振过程的微机数字仿真. 华中工学院学报. 1987;15 (2):49–56.
39. 卢天贶, 师汉民. 机床颤振中 "双频颤振" 的研究 .国防科技大学学报. 1988;10(4):67–73.
40. 梅志坚, 杨叔子, 师汉民, 等. 机床非线性颤振的描述函数分析. 应用力学学报. 1991;8 (3):69–78.
41. 梅志坚, 杨叔子, 师汉民. 描述切削过程动态稳定性的一个非线性理论模型, 华中理工大学学报. 1991;19(增刊II):57–62.
42. 吴雅, 师汉民, 梅志坚, 等. 金属切削机床颤振理论与控制的新进展 .中国科学基金. 1993;(2):99–105.
43. 师汉民. 机床自激振动的非线性理论 (编号: TG0535). 见: 中国科学文库编委会主编 .中国科学文库(金属学、金属工艺卷). 北京: 科学技术文献出版社; 1997. p. 978–80.
44. 梅志坚, 师汉民, 刘经燕, 等. 金属切削过程颤振早期诊断的一个有效判据. 华中工学院学报. 1985;13(5):87–94.

45. 杨叔子，刘经燕，师汉民，等．金属切削过程颤振预兆的特性分析．华中工学院学报．1985;13(5):79–86.
46. 刘经燕，梅志坚，郁国平，等．金属切削过程振动信号的分析与监视．振动与冲击．1987;(1):59–72.
47. 梅志坚，刘经燕，师汉民，等．金属切削过程颤振的计算机模式识别及在线监控．华中工学院学报．1987;15(2):41–8.
48. 梅志坚，杨叔子，师汉民，等．一种无颤振机床技术的研究．机床．1987;(9):9–12.
49. 昌松，梅志坚，师汉民，等．机床颤振预兆早期诊断的频率矩心判别法．华中理工大学学报．1988;16(3):91–4.
50. 梅志坚，杨叔子，师汉民，等．机床颤振的计算机控制技术的研究．工业控制计算机．1988;(5):17–20.
51. 梅志坚，师汉民，昌松，等．信号功率谱特征变化的时域快速诊断．华中理工大学学报．1988;16(3):85–90.
52. 梅志坚，杨叔子,师汉民 .机床颤振的早期诊断与在线监控 .振动工程学报．1988;1(3):8–17.
53. 卢天贶，师汉民．车床动态特性的实验研究与颤振控制．动态分析与检测技术．1989;(1):33–8.
54. 梅志坚，杨叔子，师汉民，等．机床非线性颤振的描述函数分析．应用力学学报．1991;8(3):69–78.
55. 梅志坚，杨叔子，师汉民．描述切削过程动态稳定性的一个非线性理论模型．华中理工大学学报．1991;19(增刊II):57–62.
56. 吴雅，师汉民,梅志坚，等 .金属切削机床颤振理论与控制的新 进展 .中国科学基金．1993;(2):99–105.
57. Zhang H, Ni J, Shi H. Machining chatter suppression by means of spindle speed variation (PartI: the numerical simulation). In: Devor RE, Dornfield DA, Ehmann K, editors. Proceeding of international congress & exposition & the first S. M. Wu symposium on manufacturing science. Dearborn: Society of Manufacturing Engineers; 1994. p. 161–7.
58. Zhang H, Ni J, Shi H. Machining chatter suppression by means of spindle speed variation (Part II: the experimental investigation). In: Devor RE, Dornfield DA, Ehmann K, editors. Proceeding of international congress & ex position & the first S. M. Wu symposium on manufacturing science . Dearborn: Society of Manufacturing Engineers; 1994. p. 169–75.
59. Shi H, Wang Y. Stability of metal chips and their breaking. In: Shi H, editor. Proceeding of the second international metal cutting conference. Wuhan: Huazhong University of Science and Technology Press; 1985. p. 140–51.
60. Efimov NV. Higher geometry. Moscow: Mir Publisher; 1980.
61. 钟集．高等几何．北京：高等教育出版社; 1983.
62. 成都工具研究所．GB/ T 12204—90 金属切削基本术语．北京：中国标准出版社; 1991.
63. 中国机械工程学会 生产工程分会 编译,主编:彭晋龄,机械制造术语辞典 第四分册:金属切削的基本术语．北京:中国标准出版社; 1994.
64. Boothroyd G. Fundamentals of metal machining and machine tools. London: McGraw-Hill; 1975. p. 167–84.
65. Wang J. Development of new drilling force model for improving drill point design. Dissertation. Ann Arbor: The University of Michigan; 1994.
66. Merchant ME. Mechanics of the metal cutting process, I. Orthogonal cutting and a type 2 chip. J Appl Phys. 1945;16(2):267–317.
67. Prigogine I, Stengers I. 从混沌到有序．曾庆宏，沈小峰 译．上海：上海译文出版社; 1987.
68. Nicolis G, Prigogine I. 非平衡系统的自组织 .徐锡申 译．北京：科学出版社; 1986.
69. Shaw MC. Metal cutting principles. Oxford: Clarendon Press; 1984.
70. Stephenson DA, Agapin IS. Calculation of main cutting edge forces and torque for drills with arbitrary point geometries. Int J Mach Tools Manuf. 1992;32(4):521–38.
71. Watson AR. Drilling model for cutting lip and chisel edge and comparison of experimental and predicted results—I–IV. Int J Mach Tools Manuf. 1985;25(4):347–404.

72. Wiriyacosol S, Armarego EJA, Lorenz G. Trust and torque prediction in drilling from a cutting mechanics approach. Ann CIRP. 1979;28(1):87–91.
73. 闫兴. 自由切削刀具设计理论与方法的研究及其应用: [学位论文]. 武汉:华中科技大学机械学院; 2000.
74. 钱伟长. 非线性力学的新发展 —— 稳定性、分叉、突变、混沌. 武汉: 华中理工大学出版社; 1988.
75. 陆涛. 现代刀具设计基础理论及精密加工表面微观形貌的研究: [学位论文]. 武汉: 华中理工大学机械学院; 1997.
76. Grzesik W. The mechanics of continuous chip formation in oblique cutting with single-edged tool—part I theory. Int J Mach Tools Manuf. 1990;30(3):359–71.
77. Grzesik W. The mechanics of continuous chip formation in oblique cutting with single-edged tool—part II experimental verification of the theory. Int J Mach Tools Manuf. 1990;30(3):373–88.
78. Lin GCI. Prediction of cutting forces and chip geometry in oblique machining from flow stress properties and cutting condition. Int J Mach Tools Des Res. 1978;18:117–30.
79. 师汉民. 机械振动系统—— 分析·建模·测试·对策 (第三版, 第十一章, 第十三章). 武汉: 华中科技大学出版社; 2013.
80. Tobias SA. Machine tool vibration. New York: Wiley; 1965.
81. Tobias SA, Fishwick W. The chatter of lathe tools under orthogonal cutting conditions. Trans ASME. 1958;80(5):1079–88.
82. Kegg RL. Cutting dynamics in machine tool chatter. J Eng Ind, Trans ASME. 1965;87(4):464–70.
83. Lemon JR, Ackermann RC. Application of self-excited machine-tool chatter theory. J Eng Ind, Trans ASME. 1965;87(4):471–9.
84. (日)星鐵太郎. 机械加工びびり现象. 东京: 工业调查会; 1979.
85. (日)星鐵太郎. 机械加工の振动解析. 东京: 工业调查会; 1990.
86. Hahn RS. On the theory of regenerative chatter in precision grinding operations. Trans ASME. 1954;76(4):593–7.
87. Hooke CJ, Tobias SA. Finite amplitude instability—a new type of chatter. In: Tobias SA, editors. Proceeding of the 4th international machine tool design and research conference. Manchester: Pergamon Press; 1964. p. 97–109.
88. Saravanja-Fabris N, D'Souza AF. Nonlinear stability analysis of chatter in metal cutting. J Eng Ind, Trans ASME. 1974;96(2):670–5.
89. Tlusty J, Ismail F. Basic nonlinearity in machining chatter. Ann CIRP. 1981;30(1):299–304.
90. Hanna NA. Prediction of chatter during face milling. Dissertation. Birmingham: University of Birmingham; 1968.
91. Hanna NA, Tobias SA. A theory of nonlinear regenerative chatter. J Eng Ind, Trans ASME. 1974;96(1):247–55.
92. Zhang H. Chatter modeling analysis and control for CNC machine systems. Dissertation. Ann Arbor: The University of Michigan; 1996.
93. Hoshi T, Sakisaka N, Moriyama I, et al. Study for practical application of fluctuating speed cutting for regenerative chatter control. Ann CIRP. 1977;25(1):175–9.
94. Hooke CJ. Finite amplitude instability of Lathe tools. Ph.D. dissertation. University of Birmingham; 1963.
95. 师汉民. 金属切削理论及其应用新探 (Metal cutting theory and practice—a new perspective). 华中科技大学出版社; 2003 (ISBN: 2902-8/TG. 53).

Index

H. Shi, *Metal Cutting Theory*, Springer Series in Advanced Manufacturing,
https://doi.org/10.1007/978-3-319-73561-0

The manufacturer's authorised representative in the EU is Springer Nature Customer Service Centre GmbH, Europaplatz 3, 69115 Heidelberg, Germany. If you have any concerns regarding our products, please contact ProductSafety@springernature.com

Printed and bound by CPI Group (UK) Ltd, Croydon, CR0 4YY
05/10/2026
02228144-0013